MW01617149

# HUSSERL AND THE PROMISE OF TIME

This book is the first extensive treatment of Husserl's phenomenology of time-consciousness. Nicolas de Warren's detailed analysis of texts by Husserl, some only recently published in German, examines Husserl's treatment of time-consciousness and its significance for his conception of subjectivity. He traces the development of Husserl's thinking on the problem of time from Franz Brentano's descriptive psychology, and situates it in the framework of his transcendental project as a whole. Particular discussions include the significance of time-consciousness for other phenomenological themes: perceptual experience, the imagination, remembrance, self-consciousness, embodiment, and the consciousness of Others. The result is an illuminating exploration of how and why Husserl considered the question of time-consciousness to be the most difficult, yet also the most central, of all the challenges facing his unique philosophical enterprise.

NICOLAS DE WARREN is Associate Professor of Philosophy at Wellesley College.

# MODERN EUROPEAN PHILOSOPHY

# HUSSERL AND THE PROMISE OF TIME: SUBJECTIVITY IN TRANSCENDENTAL PHENOMENOLOGY

NICOLAS DE WARREN

*Wellesley College*

CAMBRIDGE
UNIVERSITY PRESS

CAMBRIDGE UNIVERSITY PRESS
Cambridge, New York, Melbourne, Madrid, Cape Town,
Singapore, São Paulo, Delhi, Mexico City

Cambridge University Press
The Edinburgh Building, Cambridge CB2 8RU, UK

Published in the United States of America by Cambridge University Press, New York

www.cambridge.org
Information on this title: www.cambridge.org/9781107405134

First published 2009
First paperback edition 2011

*A catalogue record for this publication is available from the British Library*

*Library of Congress Cataloguing in Publication Data*
Warren, Nicolas de, 1969–
Husserl and the promise of time : subjectivity in transcendental
phenomenology / Nicolas de Warren.
p.   cm. – (Modern European philosophy)
ISBN 978-0-521-87679-7 (hardback)
1. Husserl, Edmund, 1859–1938. 2. Time. 3. Phenomenology. I. Title. II. Series.
B3279.H94W37 2009
115.092–dc22

2009030224

ISBN 978-0-521-87679-7 Hardback
ISBN 978-1-107-40513-4 Paperback

*Pour Clara*

# CONTENTS

# ACKNOWLEDGMENTS

Numerous friends and colleagues have supported my writing of this book in various ways.

For my years at Boston University and beyond, I am especially grateful to Daniel Dahlstrom for guidance and encouragement; Alfredo Ferrarin for his erudition and companionship; Erazim Kohak and Krystof Michalski for first showing me the way of philosophy.

I am grateful to Dieter Lohmar for time at the Husserl Archives in Cologne; Michael Kelly for conversation and corrections; Andrea Staiti, Lilian Alweiss, and Naveh Frummer for insightful comments on earlier drafts of this book; and Nolen Gertz for his sharp editorial eye.

Whether in print or in person, I have learned much about Husserl from colleagues: Rudolf Bernet, John Brough, Richard Cobb-Stevens, John Drummond, Burt Hopkins, Claudio Majolino, Dermot Moran, Tom Nennon and Robert Sokolowski.

I should also like to thank my departmental colleagues – Ann Congleton, Ifeanyi Menkiti, Catherine Wearing and Ken Winkler – for their breadth of mind and sound judgment as well as others who have prevailed in making Wellesley College a home for intellectual curiosity: Sarah Bishop, Thomas Cushman, Richard French, Tom Hodge, Jonathan Imber, Lidwien Kapteijns, Jim Noggle, Tim Peltason, Jim Petterson, Larry Rosenwald and Margery Sabin.

I am especially indebted to Ken Haynes for the generosity of his intelligence; Lisa Rodensky for teaching me in so many ways; Brendan Reay for the ancients and the moderns; Andy Kingston and David Polan for imparting the little that I know – and hear – of music; Luigi Caranti for our many Kantian disputations, especially those under the Italian sun; and Hilary Gaskin for her saintly patience.

I count myself fortunate in having in James Dodd both a friend and a colleague; there is very little here that did not mature in the company of his friendship during the past fifteen years.

Lastly, thank you to my parents, for being curious about and supportive of my intellectual endeavors; and to Clara, who saved me, on so many occasions, from myself.

# ABBREVIATIONS

Volume numbers (if any) and page numbers appear after the abbreviation below. Page numbers in square brackets refer to the English translation, which I have sometimes silently amended. When no page number appears in square brackets, the translation is my own.

CM    Husserl, E. *Cartesianische Meditationen*

DP    Brentano, F. *Deskriptive Psychologie*

EB    *The Encyclopedia Britannica Article*, in Husserliana IX. English translation in: *Psychological and Transcendental Phenomenology and the Confrontation with Heidegger (1927–1931)*, trans. T. Sheehan and R. Palmer (Kluwer: Dordrecht, 1997)

FTL    Husserl, E. *Formale und transzendentale Logik*

Hua    Husserl, E. *Husserliana*

LU    Husserl, E. *Logische Untersüchungen*

PE    Brentano, F. *Psychologie vom empirischen Standpunkt*

RZK    Brentano, F. *Philosophische Untersuchungen zu Raum, Zeit und Kontinuum*

SP    Derrida, J. *La voix et le phénomène*

# Introduction

## THE PROMISE OF TIME: SUBJECTIVITY IN HUSSERL'S TRANSCENDENTAL PHENOMENOLOGY

There are times when I entertain the thought that Edmund Husserl's life possessed a secret affinity with Raymond Queneau's fictional character Valentin Brû. In *Le Dimanche de la vie*, Private Brû spends his afternoons tracking the hands of a clock, working hard to keep his mind "clear of the pictures that everyday life deposits in it" in an effort to think, with his eyes open, of nothing. Yet, Monsieur Brû's effort to catch sight of time repeatedly fails, as he is hopelessly distracted by events taking shape in time: "clusters of commonplace words go crackling through a wasteland of automatic movements or of colorless objects."[1] Brû's gaze continually falls short of time: "*for the moment*, he doesn't notice anything at all. He stares at a branch, or a pebble, but he loses sight of time. Time has pushed the hand on ten minutes farther and Valentin hasn't caught it at it. And since the branch and the pebble, *nothing* has happened."[2] Time stares him in the face, yet occurs behind his back, catching him by surprise; he is too late for time's passage, always failing to see time on time.

I imagine Husserl also spending his afternoons in the pursuit of time, waiting to catch time in the act. But Husserl did not watch time in the Sunday of Life. In the fall of 1917 and spring of 1918, Husserl, with one son dead and the second grievously wounded in the war to end all wars, vacationed twice in Bernau (Germany) to pursue his phenomenological analysis of the constitution of time, "the most difficult of all problems confronting the human mind" (Hua XXIV, 255). In some of Husserl's most challenging writings, commonly known as the "Bernau Manuscripts," the inexpressibly familiar consciousness

---

1 *The Sunday of Life*, trans. Barbara Wright (London: Calder, 1976), 119.
2 *Ibid.*

of time's passage is meticulously described as centered on the axis of the unceasing renewal of the now. The emergence of each now takes the form of a primordial hope, waiting for another now yet to come in the wake of a now already past. And yet, the novelty of each now is never entirely expected or identical with the now just past; each now is never entirely caught by the passage of time. We repeatedly expect another now, but its arrival always surpasses our expectation, catching us, so to speak, from behind. In such a description, Husserl comes to see consciousness repeatedly arising in a renewed awakening of time itself. This seeing of time brings together the consciousness of time with the time of consciousness under the heading of "inner or original time-consciousness," which Husserl discovers as the constitutive foundation for the possibility of an intelligible, and thereby meaningful, world of human experience.

This interplay of renewal and reflection permeates Husserl's entire phenomenological enterprise. A few years after the First World War, in 1923 and 1924, Husserl published, in the Japanese journal *Kaizo*, three essays on the theme of renewal.[3] In these essays, Husserl calls for the renewal of European culture through the advent of philosophy as a rigorous science, envisioned as the project of transcendental phenomenology. The themes of these essays point forward to the *Crisis of the European Sciences and Transcendental Phenomenology*, in which this call to awakening, this coming to one's senses, or reason (Husserl's preferred term for philosophical reflection from these later years – *Besinnung* – contains these various shades of meaning), becomes once again repeated, this time with greater urgency, conviction and breadth of philosophical comprehension.[4] But these essays also point back to the *Logical Investigations*, and in particular to the *Prologemena*, where this call for the renewal of thinking already signals, in a less pronounced tone than either in the *Kaizo* essays or in the *Crisis*, Husserl's entrance into philosophy. "The philosopher is the man who wakes up and speaks," as Merleau-Ponty, undoubtedly thinking of Husserl, among others, once eloquently stated.[5]

---

3 Hua XXVII.
4 See James Dodd, *Crisis and Reflection: An Essay on Husserl's* Crisis of the European Sciences (Dordrecht: Kluwer, 2004).
5 Maurice Merleau-Ponty, *In Praise of Philosophy and Other Essays* (Evanston, IL: Northwestern University Press, 1970), 63.

The apparent distance between Husserl's macroscopic vision of his transcendental enterprise and Husserl's microscopic vision of time-consciousness, patiently pursued in research manuscripts that remained largely unpublished during his lifetime, could not be greater – how does the question of time-consciousness become, through the enterprise of Husserlian phenomenology, the question of philosophy itself? How does the awakening of phenomenology to time-consciousness encapsulate the promise of phenomenological philosophy? In *The Promise of Time: Subjectivity in Husserl's Transcendental Phenomenology* I propose an answer to these questions in trying to understand how and why Husserl considered the question of time-consciousness to be the most difficult, yet also the most central, of all the challenges facing his unique philosophical enterprise.

Despite the unquestionable importance of Husserl's analysis of time-consciousness for transcendental phenomenology and its far-reaching significance for the development of philosophy, Husserl's historically unparalleled undertaking to fathom the enigma of time-consciousness still lacks the critical and detailed treatment that it deserves. More than one hundred years after Husserl's 1905 lectures on the phenomenology of inner time-consciousness, our understanding of Husserl's investigation resembles the paradox it was meant to clarify: we understand it as long as we are not asked to say exactly what we understand of it.

As Husserl often remarked, Augustine first uncovered the profound difficulties posed by the question "What is time?" in discovering its entanglement with the question "Who am I?" As Augustine famously asked, "What, then, is time? I know well enough what it is, provided that nobody asks me; but if I am asked what it is and try to explain, I am perplexed."[6] Augustine's dilemma suggests that in failing to articulate what I take for granted in my experience of time, there is a failure to understand in both a specific and fundamental sense. Specifically, I am unable to explain how I perceive the passage of time, even though I implicitly grasp time's passage as self-evident, and without question, in the course of my ordinary existence. Yet I also fail to understand what is fundamental about *myself* in so far as I am subject to the passage of time. My own presence in time becomes

6 Augustine, *Confessions*, trans. R. S. Pine-Coffin (London: Penguin, 1961), XI, 14.

implicated – open to question in an unexpected manner – in failing to illuminate the presence of time; time eludes me, much as I elude myself. As Augustine remarks elsewhere in the *Confessions*, I have become a question for myself, unknown to myself in light of the questionability of time. Augustine's formulation of the question "What is time" constitutes in fact a double question: how is the consciousness of time's passage – the puzzling perception of a time that is, that is no longer, that is yet to come – possible? What am I to understand of myself as a being for whom time is at all in question?

Despite important contributions towards our understanding of Husserl's phenomenology of time-consciousness, a principal reason for our less than adequate comprehension of its argument and significance stems from a failure to recognize, and thus explore, the dual significance, comparable to Augustine's, of the problem of time-consciousness for transcendental phenomenology. As I argue in this book, understanding what is specific about the importance of time-consciousness for Husserl's phenomenology reveals what is fundamental about subjectivity for Husserl's philosophical project; understanding what is fundamental for transcendental subjectivity – its temporality – reveals what is specific to Husserl's transcendental phenomenology.

A phenomenological investigation of transcendental subjectivity is meant to usher a general reformation of modern philosophy. As Husserl once expressed his ambition, the goal of transcendental phenomenology was to fulfill "the secret longing of Western philosophy." This historically unprecedented demonstration of how (transcendental) subjectivity represents *the* fundamental problem of philosophy characterizes the unique *promise* of Husserl's phenomenological enterprise. As Husserl stated, "my calling is the study of pure subjectivity." Indeed, Husserlian phenomenology is arguably the last concerted attempt of modern philosophy to lay systematic claim to the centrality of subjectivity for philosophical reflection. Whether such a claim represents a decisive ending or a renewed beginning is, as this book hopes to suggest, yet to be definitively resolved, despite vigorous pronouncements of the "death of subjectivity" during the later half of the twentieth century. The question of subjectivity remains unavoidable as long as our puzzlement with time remains inescapable; indeed, the evasion of subjectivity is the evasion of time itself.

Husserl never wrote a complete presentation of his phenomenology of time-consciousness. Even a cursory assessment of his writings on this subject matter reveals a complicated patchwork of different strata of analysis, composed at different stages in the development of his phenomenological philosophy and for different purposes, but mainly for the purpose of private reflection. Aside from occasional side-glances thrown in the direction of time-consciousness in his published writings and university lectures, the bulk of Husserl's writings on time-consciousness are in the form of research manuscripts. Consisting mostly of notes, condensed reflections and fragmentary lines of analyses, none of which were immediately intended for publication, these writings present a distinct set of problems for any reconstruction and critical engagement with Husserl's phenomenology of time-consciousness. The experimental, fluid character of Husserl's infinite task of phenomenological clarification is nowhere more pronounced than in his repeated efforts at providing an analysis of time-consciousness. Indeed, it is significant that Husserl never considered himself to be in a position to undertake its definitive elaboration; it is as if the extant writings on time-consciousness functioned as placeholders, intricate promissory notes, for a more complete engagement to come. Seen in this light, it is therefore not inappropriate to compare these writings, and the type of thinking they contain, to an artist's compositional sketchbooks, albeit of a peculiar kind of artist engaging in peculiar form of intellectual creation. In these writings, a basic conceptual framework becomes delineated and solutions to individual problems begin to take shape; multiple drafts of specific insights, detailed studies of particular phenomenon, and an array of conceptual devices are developed in view of a final composition, painstakingly brought into greater relief, that never fully came to light. To adopt Paul Klee's formula about painting, in Husserl's writings on time-consciousness, the aim is not "to render the visible, but to render visible."

An interpretation of Husserl's analysis of time-consciousness must remain mindful of this fact that Husserl never produced a complete presentation of his phenomenology of time-consciousness nor did he ever sanction any stage in the development of his analysis as definitive. Given the scope of Husserl's thinking and the corresponding vastness of his literary production, this book offers an interpretation of Husserl's analysis and its philosophical significance rather than an encyclopedic presentation or a survey of the historical development of Husserl's analysis of time-consciousness, even if the parameters of its chronological unfolding must nonetheless contextualize our discussion and debate. For reasons explained separately, this book does not provide a comprehensive account of all three *Husserliana* collections of writings on time-consciousness.[7]

Any study specific to Husserlian phenomenology written in English must take into consideration the limited availability of Husserl's thinking in English translations, the continued expansion of the *Husserliana* (currently at 37 volumes), supplementary volumes of manuscript material, and the wealth of still unpublished manuscripts in the Husserl Archives, all of which create a textual and interpretative labyrinth of daunting complexity. An interpretation cannot restrict itself, as Paul Ricoeur once argued, to those texts prepared by Husserl for publication, since these writings, despite enjoying authorial approval, often understate the originality and breadth of Husserl's thinking. Equally, the temptation should be avoided to over-compensate the patent disparity between Husserl's published and unpublished writings by relying too heavily on the embarrassment of riches to be found in the unpublished and, more significantly, largely unedited materials in the Husserl Archives. Evidence for any number of readings can be found in Husserl's literary remains; like other philosophers of enduring significance, Husserl was a thinker of many lives and after-lives. While attempting to strike a balance between relying on published and unpublished texts, this book takes its cardinal points of reference from either principal publications or lecture courses that exist in English translation. An exception is the Bernau Manuscripts that are discussed in chapter 5, and which have yet to appear in English. In choosing to engage Husserl's thinking in this way, my aim is to offer an interpretation in light of which readers can *return* to Husserl's writings. I have always considered these writings as providing a space and a vocabulary in which, and with which, to think; hence, Husserl's strange openness, despite a technical jargon – his infamous "phenomenologicalese" – that causes much frustration for novices and adepts alike.

*The Promise of Time* is divided into three sections. In the first section (chapters 1 and 2), I frame Husserl's phenomenology of time-consciousness as both a specific problem about the consciousness of temporal passage and as a fundamental problem for a phenomenological understanding of transcendental subjectivity. In chapter 1, "The ritual of clarification," Husserl's framework of transcendental

---

7 Cf. Appendix "Note on textual sources."

phenomenology is introduced, with an emphasis on the centrality of intentionality for Husserl's understanding of subjectivity in its "world-constituting" function. This chapter paints a general portrait of Husserl's transcendental project, as it took shape in its literary expressions in *Ideen I* and the *Cartesian Meditations,* while also present-ing, in quick strokes, the details of a phenomenological conception of transcendental subjectivity in light of which the theme of time-consciousness ascends to its transcendental significance. Chapter 2, "A rehearsal of difficulties," examines what Husserl understood as the problem of time-consciousness, prior to its transcendental inflection, by investigating Brentano's theory of "original association," which served as the point of departure for Husserl's phenomenological analysis. Chapter 2 reconstructs Brentano's theory of original associ-ation on its own terms, primarily on the basis of his lecture-courses on descriptive psychology from the 1880s, and examines its relationship to other key concepts in Brentano's descriptive psychology (e.g., intentional relation and inner consciousness). In addition to under-standing Brentano's struggle with the perception of time, chapter 2 stresses the role of Stern's concept of "mental presence-time" and his critique of "momentary consciousness" for the formation of Husserl's thinking, especially in shaping the *sense* in which time-consciousness becomes a problem for phenomenological reflection. Whereas chapter 1 moves from the problem of phenomenology to the problem of time-consciousness, chapter 2 moves from the problem of time-consciousness to the problem of phenomenology.

In section two, I explore in detail what Husserl identified as "the great problem" of phenomenological or inner time-consciousness (the temporality of the stream of consciousness) and the self-constitution of absolute time-consciousness. Section two (chapters 3, 4, and 5) examines Husserl's phenomenological analysis of time-consciousness in *On the Phenomenology of Inner Time-Consciousness (1893–1917)* (chapters 3 and 4) and *Die Bernauer Manuskripte über das Zeitbewußtsein (1917/18)* (chapter 5). I propose to read these two volumes together, not as separate sets of documents but as forming a constellation of texts exhibiting the multi-faceted development of Husserl's analysis. Chapter 3, "The ghosts of Brentano" enters into the 1905 lectures *On the Phenomenology of Inner Time-Consciousness* by taking up the two issues first rehearsed in Husserl's critique of Brentano: the perception of time and the time of perception. In addition to tracking the sophisticated translation of the problem of the perception of time into a framework of phenomenological reflection, I discuss in detail Husserl's meticulous descriptions of the consciousness of temporal passage, as primarily focused on the "running-off" or "flowing-away" of the now. Husserl's earliest efforts to provide a phenomenological clarification of time-consciousness is pursued to its aporetic conclusion in which the ghosts of Brentano's previous failures in the face of time resurface within Husserl's own thinking. Chapter 4, "The retention of time past," continues along the tracks established in chapter 4, yet expands the scope of inquiry. In this chapter, I explore Husserl's phenomenological analyses of the imagination and remembrance, and argue for their significance in relation to the problem of time-consciousness. Indeed, I suggest that Husserl's discovery of the "double-intentionality" of time-consciousness, as first discovered in its retentional form, is motivated by his earlier discovery of the double-constitution of the imagination and remembrance. Through his reflections on the imagination and remembrance, Husserl comes to recognize the myriad senses in which different forms of consciousness, along with their intentional correlates, are constituted in absolute time-consciousness. With this insight, Husserl is able to progress beyond the impasse of his earlier analysis of time-consciousness to the mature framework of the three-fold declension (original presentation, retention, protention) of absolute time-consciousness. Chapter 5, "The impossible puzzle," turns to the "Bernau Manuscripts" and continues to develop Husserl's phenomenology of time-consciousness. In the Bernau writings, the *transcendental* landscape of time-consciousness is fully *mapped*, yet not explored. Chapter 5 focuses on the formulation of the distinction between "near" and "far" retentions, Husserl's increased attentiveness to protention and the reshaping of his basic conception of original time-consciousness, as exemplified in a startling reconfiguration of his time-diagram. Chapter 5 ends with a reflection on the impossible puzzle of absolute self-constituting time-consciousness.

In section three (chapters 6 and 7), I expand the reach of my interpretations of time-consciousness developed in section two. Chapter 6, "The lives of Others," considers Husserl's refutation of transcendental solipsism as the formulation of his refutation of idealism as well as the constitution of the Other as an alter-ego, as developed in the *Cartesian Meditations*. I argue for a significant correction in Husserl's phenomenology of the Other that brings to bear insights gained from the reflections on time-consciousness developed in the Bernau Manuscripts. Chapter 7, "The life of consciousness," serves as both a continuation of developing the ramifications of

Husserl's analysis of time-consciousness for his genetic conception of subjectivity and as a concluding discussion of the specific interpretation developed in *The Promise of Time*. The tone and style of discussion in this chapter is suggestive, not conclusive; the pace of discussion quicker, the arc of reflection more open. From this vantage-point at the end, looking back to the course of the preceding chapters while also looking beyond the scope of this book, the *point* of this book crystallizes Husserl's remark that, "an authentic analysis of consciousness is, so to speak, a hermeneutics of the life of consciousness." And in understanding this remark as encapsulating the argument of this book, my hope is to shed much needed light on Husserl's claim that, with transcendental phenomenology, "the Delphic expression γνῶθι σεαυτὸν has acquired new meaning."

*The Promise of Time* presents an interpretation of Husserlian phenomenology; as such, I allow myself to be led as much as I allow myself to lead. As with any interpretation, I am deliberately selective and critically self-serving; selective since my aim is not to deliver a comprehensive treatment, but to plot a course through Husserl's thinking; self-serving since my trajectory reflects the choices of paths taken, or not-taken. Throughout the writing of this book, I have been mindful of my conviction that a philosopher should neither over-simplify the complexity of a problem nor over-complicate the economy of its solution. The relentless and restless character of Husserl's thinking exacerbates this challenge of striking a balance between attending to the richness of a problem and heeding a sought-after clarity of what constitutes an understanding. Any serious engagement with Husserl's transcendental phenomenology cannot avoid this tension; it is the very life of his thinking. My hope is to have struck a balance between both poles without compromising or mistaking one for the other.

1

# THE RITUAL OF CLARIFICATION

*Under what obligation do I lie of making such an abuse of time?*
— David Hume

## An untimely provocation

Husserl was a philosopher who believed that the course of Western history could never forsake the idea of philosophy first ushered into the world with Greek thinking, even as he recognized that he came to philosophy in an historical epoch in which that unique window of human possibility was rapidly, and perhaps entirely, coming to a close. From the seminal *Logical Investigations* to the unprecedented *Crisis of the European Sciences*, Husserlian phenomenology struggled to define itself against the current of an age that unthinkingly abdicated a responsibility towards the highest perfection of reason in the name of a reason paradoxically called modern. Husserlian phenomenology is therefore best approached as an untimely provocation, the thrust of which is encapsulated in the charge that "the spirit of radicalism has been lost under the title of philosophy" (Hua VIII, 10). This call to radicalism takes on different forms. Its most visible banner is the motto "back to the things themselves" that commonly informs our view of Husserl's enterprise, to the extent that the generalized term "phenomenology" is often taken as synonymous with *any* invocation of lived experience and the "first person point of view" in contemporary philosophical discourse. The primary concern, however, that animates Husserlian phenomenology is neither "lived experience" nor "consciousness" *as such*, but a problem first broached in the *Logical Investigations* that continually defined, with increased sophistication and breadth, the center of Husserl's gravity, and to which he untiringly

10

returned once again in the unfinished *Crisis of the European Sciences*: how is knowledge possible?

That the problem of knowledge is central to Husserl's enterprise should not surprise us, since any significant claim to philosophy cannot avoid the quest for what could also be called the "thinkability of the world," the way in which the world, as the ordered totality of all possible experience, becomes rediscovered through the origin of its intelligibility. This primary concern with the problem of knowledge must be taken in its transcendental significance as the issue of how knowledge is at all possible or, in the formulation just proposed, of how experience is at all "thinkable," open *as* the possibility of being known. Yet Husserl does not accept the meaning of this transcendental directive without question, despite its familiarity in its Kantian expression or in the derived neo-Kantian form that shaped Husserl's academic environment. Instead, Husserl invokes a broader imperative to refashion the possibility of knowledge, from the beginning and as a beginning, for the first time in the history of philosophy. Though this ambition might strike us as paradoxical, or even absurd, Husserl's enterprise is predicated on the claim that genuine philosophical thinking, as the knowledge of how knowledge is at all possible, has yet to begin. As Husserl remarks, "I do not say that philosophy is an incomplete science. I say more directly that philosophy is not at all a science, that it has not at all begun to become a science."[1] On this point, Heidegger is entirely correct: phenomenology is neither a school nor a movement that may or may not be historically surpassed. Phenomenology is nothing less than the permanent possibility of thinking itself.

Husserl in this fashion announces the enterprise of phenomenology as a protest against the false entitlements (and enticements) of thinking we know how to answer questions we in fact do not yet know what it means to ask. If philosophy has genuinely yet to begin, it is because we have yet to learn how to articulate the traditional questions of philosophy in a rigorous fashion. In Husserl's view, "the genuine sense of philosophical problems has not once been brought to scientific clarity."[2] Under the mantle of the transcendental reduction, Husserl's methodological response to the unscientific condition of philosophy can be characterized as patiently recasting the basic

1 "Philosophie als strenge Wissenschaft," 8 [167].      2 *Ibid.*, 8 [167].

questions of philosophy, taken in the traditional ontological form of "what is such-and-such," into transcendental questions of givenness, of how, and under what conditions, objects with a determinate sense of being so-and-so are at all given to consciousness (this perceptual object as being so-and-so, this imaginary object as being so-and-so, etc.).

In a manner evocative of Kant, Husserl's transcendental phenomenology transposes the problem of knowledge into the problem of givenness by exploiting the transcendental solidarity between the objectivity of knowledge and the subjectivity of experience, whereby the enabling conditions of knowledge are exposed as just those conditions under which objects of experience are at all given to consciousness. We can only meaningfully experience something that we can come to know as meaning something for us. In a manner evocative of Leibniz – the least appreciated of Husserl's historical sources – Husserl's transcendental critique delimits "an intellectual space" (*geistige Raum*), or space of reason, that becomes mapped through a systematic and eidetic clarification of ontological regions, formal and material, on the basis of the intuitive apprehension of apriori forms of experience, as correlated to constitutive acts of transcendental subjectivity. Husserl in this fashion combines the transcendental criticism of Kant with the Leibnizian visions of a *mathesis universalis* and a *charateristica universalis*, in which the equivocations and vagueness underpinning the factionalism of philosophy are translated into a unified and universal discourse of scientific philosophy, a "theory of all theories."[3] Husserl's phenomenology of reason moves, however, beyond the historical alternatives of a discursive ordering of semantic representations (or a system of signs), a speculative chain of metaphysical reasons (or intellectual intuitions), or a systematic construction of experience based on principles of reason. In a manner evocative of Brentano, phenomenological reflection reconfigures the transcendental space of reason into what Husserl calls the "field of transcendental experience," by way of reflective (in the strict phenomenological meaning discussed below) descriptions of the constitutive performance of subjectivity in view of its intentional objects of experience. The method of transcendental reduction – the inalienable

---

3 Cf. CM, 47 [5]: "To be sure, we still have philosophical congresses. The philosophers meet but, unfortunately, not the philosophies. The philosophers lack the unity of an intellectual space in which they might exist for and act on one another."

centerpiece of Husserl's philosophical reformation – uncovers an array of indispensable and sedimented conditions of givenness that are not only constitutive of experience, but that, in terms of their apriori and constitutive significance, are themselves recovered *as* experience, as the concrete life of transcendental subjectivity in its function of world-constitution. This discovery of transcendental subjectivity as the origin of the world takes the form of spelling out its transcendental "ABC" or "alphabet," the grammar of its own becoming (the conditions of possibility for subjectivity), as well as an embracing apriori correlation between possible modes of consciousness and possible objects of experience (conditions of possibility of objectivity), in and through which the openness of the world attains its sense of being as an intelligible world of ordered experience.

Husserl's effort to renew the promise of philosophy in the form of transcendental phenomenology occurs at a historical juncture when this promise appears either displaced by the triumph of cultural world-views over conceptual reflection or deformed through the accomplishments of the natural sciences taken out of proper philosophical context. Against these dual tendencies of enfeeblement and falsification, no other philosopher of the twentieth century so conscientiously responded to the modern crisis of rationality. From a Husserlian perspective, a loss of radicalism defines the philosophical thoughtlessness of our age despite the success of the natural sciences and the clamor of critical reasoning and deconstructive defiance that pervades academic philosophy today; indeed, radicalism is rarely expected of philosophy or philosophers, unless philosophy fashions itself into a participatory discourse of ethical or political critique, or philosophers directly engage themselves in ethical or political issues of pressing urgency. Husserl's call for the radicalism of transcendental phenomenology *should* ring strange to our ears – in what sense does transcendental phenomenology express a genuine form of radicalism?

Husserl's radicalism is theoretical as well as ethical, and gives new life to the Cartesian realization "that once in my life I had to raze everything to the ground and begin again from the original foundations."[4] Like Descartes, Husserl's quest for renewal demands "a spiritual preparedness" (*eine seelische Bereitschaft*) as the *conditio sine qua non*

---

4 René Descartes, *Meditations on First Philosophy*, trans. R. Ariew and D. Cress (Indianapolis: Hackett, 2006), 9.

for seizing the idea of philosophical reflection as the highest perfection of human existence. Responding to the call of philosophy requires "resolution" (*Entschluss*) in the form of "absolute self-responsibility" towards the project of reason that defines the historical prerogative of Western civilization. Of course, such a demand for a new beginning will invariably enter into conflict with entrenched habits of mind. Not unlike other philosophers of substance, Husserl challenges the distinctly *philosophical* prejudice against conceiving anew what it could fundamentally mean to think philosophically. This conceit is legitimated under the name of academic philosophy, much of which, for Husserl, represents the institutionalization of thoughtlessness. As Husserl remarks, "Instead of a unitary living philosophy, we have a philosophical literature growing beyond all bounds and almost without coherence. Instead of a serious discussion among conflicting theories ... we have a pseudo-reporting and a pseudo-criticizing, a mere semblance of philosophizing seriously with and for one another" (CM, 46 [5]).

In Husserl's unsparing critique, the original ambition of philosophy has been deformed by the bureaucratization of thinking or falsified into a "scientistic" notion of philosophy through the illicit appropriation of methods from the natural sciences. Husserl emphatically insists on the rigor of philosophy, yet remains suspicious of any philosophical pretension to the methods and models of the exact sciences – "one cannot define things in philosophy in the same manner as one does in mathematics" (Hua III, 6 [xxii]). Yet Husserl is not the equivalent of a philosophical Luddite. His philosophical thinking draws from his underlying mathematical intelligence, ever keen to pursue conceptual clarification and argumentative discipline. Nor does he reject the professionalism of the modern philosopher – Husserl envisions generations of phenomenologists hard at work in the systematic realization of transcendental phenomenology and, to that effect, fashions a technical vocabulary that gives new meaning to the praise of German engineering. Husserl's ambition, however, is to inaugurate a new experience of thinking, a new kind of scientific intelligibility (*eine neuartige Wissenschaftlichkeit*), and a new form of philosophical life. This project of renewal is centered on the foundation of seeing that Husserl proposes as the singular principle – "the principle of all principles" – by which the meaning of knowledge is restored to the experience of truth: "*that every originary presentive intuition is a legitimizing source of cognition*, that *everything originarily* (so to speak in its 'personal' actuality [*leibhaften Wirklichkeit*]) *offered* to us *in 'intuition' is to be accepted simply as what it is presented as being*, but also *only within the limits in which it is presented there*" (Hua III, 43–44 [44]). In this robust form, the radicalism of seeing is the radicalism of reason: "the primary fundamental form of rational consciousness of originarily presentive 'seeing'" (Hua III, 282 [326]). Philosophical reflection is the ritual of clarification; when this ritual becomes a form of life, clarification becomes the vocation of the philosopher, a call to the responsibility of seeing, and seeing again.

Even though Husserl's ritual of clarification takes its cue from a traditional impulse to fashion philosophy into a rigorous science, one should not fail to be surprised by the "absolute foreignness" of Husserlian phenomenology. As Husserl understands, "The transcendental reduction is a kind of transformation of one's whole way of life, one that completely transcends all life experience heretofore and that, due to its absolute foreignness, is hard to understand both in its possibility and actuality. The same holds correspondingly for a transcendental science" (EB, 133). This foreignness is evident from the envisioned shape of transcendental philosophy as a science of transcendental subjectivity. When placed against the broader canvas of the history of philosophy, the proposal of transcendental subjectivity as the "original field of all reason and forms of reason" can be seen as a post-Kantian transformation of an original Socratic ideal of philosophy as self-knowledge (Hua VII, 28). As Husserl writes, "*self-knowledge, but only radically pure or transcendental self-knowledge, is the source of all the knowledge that in the ultimate and highest sense is genuinely and satisfyingly scientific*" (Hua VIII, 167). In the narrower confines of twentieth-century philosophy, this bold claim, despite its suggestive Socratic allure, that "the all embracing science of transcendental subjectivity" is the one and only genuine philosophy, might strike our ears as *anachronistic* – as it should, in the specific sense of confronting us with a philosophical claim that appears remote from the current of the times and "all habits of thinking up to now" (*die gesamte bisherigen Denkgewohnheiten*) that, indeed, the inauguration of phenomenological thinking seeks to throw out of joint (Hua III, 5 [xxi]).

## The banality of the world

Aristotle's famous pronouncement notwithstanding, there is no obvious sense in which the world becomes philosophically questionable or interesting. Wonder at the meaningful order of the world or disappointment at our human failings are not sufficient to motivate an attitude towards the world in which nothing is taken for granted, an attitude that Husserl stipulates as the *conditio sine qua non* for the launching of his philosophical enterprise. But if this demand for estrangement is meant to usher philosophy, in the form of Husserlian phenomenology, into the world, how can phenomenological philosophy make good on its promise to rediscover the world in the form of a fundamental question, of how the world is at all open to knowledge? A philosophical beginning entirely removed from an experience of the world would amount to nothing but a gesture, the beginning of an "empty generality" (Hua VIII, 22). Yet how can the world occasion an experience in which nothing of the world is taken for granted? Under what conditions can the world become fundamentally incomprehensible? This elusiveness of the world for a fundamental – philosophical – form of questioning reflects the obviousness of the world itself, the sense in which we unquestionably take for granted the givenness of the world as the basis for any experience of the world, including its own questionability.

Under the heading of the natural attitude, Husserl designates the unshakable *familiarity* of being at home in the world, as that unquestionable acceptance of the givenness of the world as the ground for any possible experience. This unspoken acceptance of the world as such manifests itself as an underlying "attitude" (*Einstellung*) that comes to us "naturally," not, however, in the sense of an instinct or disposition expressive of our human nature, but as the "obviousness-ness" that the world manifests in so far as it is experienced. What is more obvious than the "fact" that the world reveals itself first and foremost in experience? Husserl, however, contends that this openness of the world in experience (or the openness of the world called experience) presupposes an inarticulate, and yet intelligible attitude that we necessarily take for granted, and which is more intimate than any theoretical or philosophical reflection. This "positing" (*Setzung*) of the natural attitude is not an explicit act of decision nor a "world-view" that we deliberatively project: it is the ground upon which we stand. The natural attitude is *lived* in the mundane course of our perceptual experience of individual objects in space and time. Perceptual experience reveals to me the presence of the world, and it is this positing of the world as presently given in experience – that all experience occurs in the form of the present – that constantly serves as the ground for any experience at all. As Husserl writes, "The natural world, the world in the habitual meaning of this word [*die Welt im gewöhnlichen Wort-sinn*], is and was continually present for me [*ist und war immerfort für mich da*], as long as I lived in the world unquestionably [*solange ich natürlich dahinlebte*]" (Hua III, 61 [54]).

The concept of world implicit in the notion of a natural attitude admits of a plurality of worlds, each of which is anchored in the perceptual experience of individual objects – this soccer ball, these cleats, this tree, etc. – in space and time. And yet, of the plurality of possible worlds that I may experience, certain worlds are neither directly anchored in perceptual experience, nor are the objects of these worlds comparable to the kinds of perceptual objects on the basis of which cultural objects are grounded. The world of pure number theory, for example, does intersect or overlap with other worlds of my mundane experience. In Husserl's manner of descrip-tion, although the world of numbers is not of the same ontological kind as the world of soccer (for example), such a world and its distinctive kind of ideal objectivities, as opposed to the real objectivity of perceptual experience, is nonetheless a possible world of *experience* for me, if we construe the term experience broadly to cover any form of *givenness to consciousness*. Of all the possible worlds and the form of their objectivities (real, ideal and imaginary), worlds need not collide or intersect; an imaginary world may never enter into direct connec-tion with the world of number theory much as the world of number theory remains remote from the world of soccer. And yet, in so far as each world of possible experience possesses the form of a world in which objects of meaning can be given to me in correlating acts of consciousness – whether these acts of consciousness are perceptual experiences, acts of the imagination, or acts of calculation – all possible worlds, as worlds of *experience*, are situated within an all-encompassing horizon of the world at large, or what Husserl progres-sively calls, over the course of his phenomenological development, "the natural world," "the natural attitude," the "general thesis of the natural attitude," or the "life-world." Although, to be sure, Husserl's (later) conception of the life-world cannot be simply identified as equivalent to his (earlier) conception of the natural attitude, the arc of Husserl's thinking, from the natural attitude to the life-world, from the *Ideen* to the *Crisis*, strives to articulate the "question of the being world," which represents the distinctive and fundamental way in which Husserl frames the transcendental problem of how knowledge is at all possible, or, in other words, "the thinkability" of the world. The world can be said in many ways, yet each of these ways of being a world is situated within the horizon of one and the same world, which functions as a constant pre-supposition, or "pre-understanding," of experiencing as such, silently accompanying any possible experience of the world, whether imaginary, real or ideal, whether the world of unicorns, soccer or mathematics.

The openness of the world in the natural attitude is primarily manifest as a world *to be discovered* in perceptual experience. In our most immediate experience of the world, perceptual objects are encountered in sensible intuitions, or perceptions. These objects are given as themselves (it is the table that I see, not an image or mental surrogate) to me as *meaning something for me*, in other words, as bestowed with the sense of being so-and-so. As I scan my surroundings and take a rapid perceptual inventory of my room, I perceive this object here as a black table, this object there as a pen, etc. I implicitly take the objects of my perceptual environment as there, or present, regardless of whether I am in fact looking at them explicitly. I take the objects of possible perceptual experience as existing independently of me; these are objects that I can discover in my world, objects that I do not imagine or invent, but objects that I encounter as other than me. Yet, even though I unquestioningly take the objects of my experience as other than me, this does not signify that objects of experience cannot become, over the course of experience, questionable for me. As I write these lines, I take for granted that my house still stands in the place where I last saw it and that my cat is still sleeping on the couch. Of course, upon my return home, I could discover my house destroyed and my cat dead, or finally realize that my house is truly a trailer and my cat truly a lemur. But my surprise and sadness is precisely that these objects are no longer there for me in the manner in which I had experienced them to be; in the former instance, these objects no longer exist; in the latter instance, these objects in their manner of being so-and-so no longer confirm my expectation of their being so-and-so. Husserl contends that a naïve realism pervades the natural attitude. What I take for granted is not only the givenness of objects as other than me, as objects that I encounter in the strong sense of "standing-opposed to me" (*Gegen-stand*), but also the world-at-large as "opposed to me," as independent of my consciousness. The natural attitude defines an unquestioned assumption that the sense of what there is, of "beings," is divided into the being of things independent from me and the region of my own conscious-being that, on the one hand, finds itself in the world among things, but which, on the other hand, is also a perspective onto the world. On this view, the natural attitude gives rise to two kinds of philosophical theories: subjective idealism or naïve realism. Each of these positions is tacitly framed by an unexamined naturalism and its dualism of the "space of reasons" and the "myth of the given," to evoke a contemporary idiom.

Within the natural attitude, I encounter not only the perceptual world of things, but also a plurality of worlds: the world of values, the world of practical activities, and objects within these worlds are encountered as being so-and-so, as beautiful, practical, etc. The natural attitude is also an attitude that I adopt towards the lives of others. I know that I am not alone in the world, and I implicitly understand that those around me have feelings, thoughts – the same modes of consciousness as I have. Even if I cannot know what the Other is truly thinking or see the world from her perspective, I accept as a given that we both have perspectives on the same world, and that objects there for me – this table, this pen, etc. – are also there for the Other, even if seen from different perspectives than my own.

The natural attitude is not only an attitude that I take towards the world of experience, it is also an attitude that I take towards myself in so far as I experience myself as perceiving, thinking, etc. My consciousness is directed towards objects of experience, to which I relate in a variety of ways: theoretical acts of explanation and observation, perceptual acts of looking, hearing and touching; acts of imagining, etc. Within the natural attitude, Husserl thus already identifies that consciousness is the consciousness of an object as being so-and-so, as meant; by the same token, Husserl acknowledges that such acts of consciousness are lived, or experienced, in a pre-reflexive manner. When looking at this tree in my backyard, my consciousness is directed towards the tree, and not towards my own act of perception. I am, however, aware of myself as perceiving this tree, yet this self-awareness (or self-consciousness) is not itself thematic, though I can, through a further act of reflection, make my perceptual act into the theme, or object, of my consciousness, in which case I am no longer immersed in my directedness towards the tree, but redirected towards myself as perceiving the tree. Whether I am actively reflecting on my consciousness or living through my acts of perception directed towards their objects, my own consciousness is given to me in the form of my actual self-givenness; for not only is the world given to me in the form of presence, but I am myself present in the world, given to myself through the very consciousness in and through which I am directed, or open, to the world of experience. As Husserl writes: "As what confronts me, I continually find the one spatio-temporal actuality to which I belong like all other human beings who are to be found in it and who are related to it as I am" (Hua III, 63 [57]).

In sum, the world's sense of being (*Seinsinn der Welt*) "is the constantly pre-given world that is in advance valid in its being, but not from any particular point of view, interest or other kind of universal purpose. Every purpose presupposes the world, including the universal purpose to understand it in scientific truth" (Hua VI, 461). If the world as taken for granted – the world-at-large – is not itself a particular world defined by an interest of the kind that defines imaginary worlds, the world of sport, and the world of mathematics, how can the world as such become a theme of philosophical inquiry? As Husserl states, "there is normally no occasion for us in which to make the world into an explicit theme as a universal world (Hua VI, 495). The world-at-large is unquestionable; there is no possibility of escape; there is no interest in terms of which the world could admit to a fundamental form of questioning. In this sense, the world does not have the form of a question to be discovered, but rather, the form of a foundation that is unquestionable; it defines the limits of meaningful questioning.

And yet, the world-at-large has arguably always been the object of fascination. Is wonder at the ordered givenness of the world the first awakening of a theoretical attitude that questions the obviousness of the world-at-large? Moreover, could we not argue that the project of modern science takes as its explicit object the "total concept of objects of possible experience and knowledge of experience" (Hua III, 1 [xvii])? Indeed, the human interest to know leaves no region of being unexamined. Every region of being – the human, the biological, etc. – is the object of theoretical investigation, and within this universal expanse of being, different regions correspond to determinate scientific enterprises. Nature is the subject matter of physics; life is the subject matter of biology; consciousness is the subject matter of psychology. When stated in this manner, the aggregate of all possible sciences – the material sciences such as physics and biology as well as the formal sciences such as mathematics – encompass the entire range of possible senses of being. In what is undoubtedly his most strident claim, Husserl contends, however, that the natural sciences, and, indeed, the idea of science as such, remain naïve in their tacit and unquestioned acceptance of the world-at-large, of how knowledge is at all possible within an encompassing horizon of being. The natural sciences presuppose the *possibility of experience* much as my dealings with the mundane course of ordinary experience. This shared naïveté feeds from the obscurity of the *givenness* of objects within the horizon of the world-at-large. This obviousness of the natural attitude can be formulated as the obviousness that all knowledge is the accomplishment of subjectivity in the sense that any object of knowledge draws its sense as an object from the sense of its possible givenness to a possible consciousness. As Husserl writes, "that all knowledge is the accomplishment of a knowing subjectivity is purely obvious [*eine pure Selbstverständlichkeit*] and yet it is the origin of all sorts of confusions and all types of absurd metaphysics" (Hua VIII, 38). Whether formulated in the discourse of the natural attitude or of the life-world, from the *Ideen* to the *Crisis of the European Sciences*, the question of transcendental phenomenology remained, in various degrees of expression and lucidity, the question of the sense of being of the world (*Seinsinn der Welt*), namely, "enigma of all enigmas," regarding "the world-problem of the profound essential connection of reason and beings" (Hua IV, 12).

## The thinkability of the world

This unquestioned presupposition of the world in the natural attitude would seem to preclude the possibility of a phenomenological inquiry into the "world-problem" of the essential correlation between reason and beings. Husserl's unwavering commitment to a radical philosophical beginning requires a circumstance in which the world becomes incomprehensible in a fundamental manner, that is, in which nothing of the world is taken for granted – including established forms of knowing, such as the natural sciences and historical expressions of philosophical reflection, which, on Husserl's view, remain ensnared in the naïveté of the natural attitude. And yet, despite such a radical stricture on his own philosophical enterprise, Husserl must nonetheless retain the *idea of knowledge as such*, without which the possibility of any questioning, including his own, would become meaningless. It is only in the light of the idea of intelligibility as such that the world can become incomprehensible as such, that is, questionable in a fundamental manner (CM, 50 [9]). Thus, although Husserl suspends any pre-existing historical conception of knowledge, he nonetheless turns, and needs to turn, to a clarification of the idea of knowledge as such, in its essential meaning, or sense, and possibility. As the essential sense of knowledge, Husserl distills the structure of intentionality and the centrality of evidence; intentionality is the movement of knowledge towards truth. Knowledge targets truth, or, in other words, the intentional object as meant. As Husserl argues, *"The concept of intentionality whatever* – any life-process of consciousness-of something or other – and *the concept of evidence, the intentionality that is the giving of something-itself, are essentially correlative"* (FTL, 160). Husserl establishes a connection between knowledge, intentionality and consciousness: knowledge has the structure of intentionality in so far as knowledge targets truth; intentionality, as an accomplishment of sense, is constituted *as consciousness.* With this connection in hand, an analysis of knowledge folds into an analysis of consciousness, taken, however, in its essential form as intentionality. In addition to an internal connection between intentionality and evidence, what defines the uniqueness of Husserl's insight into the intentionality of consciousness is the claim that consciousness intends its objects in terms of their sense or meaning (*Sinn*). An intentional object is intended as something, that is, as being so-and-so. This insight into the solidarity between objectivity and sense allows Husserl to transform the traditional problem of knowledge into the transcendence of the objectivity of sense. This solidarity of objectivity and sense underpins the robust sense in which objects transcend consciousness. The recognition that consciousness intends an object as meant, that is, as being so-and-so, further stipulates that the intended object may in fact not be given to consciousness in the manner in which it is intended, or "meant." Yet, given this intrinsic distinction within the structure of intentionality between empty and fulfilled intentions, consciousness *always* possesses a relation to its intentional object, as intended in an empty manner, regardless of whether the object is in fact given or not. The difference between an intentional object as meant and an intentional object as itself given is characterized in terms of the distinction between an empty intention and fulfilled intention. When I turn to look outside my window and perceive a tree, the tree itself is given to me in a manner that fulfills the empty intention of having been told that a tree stands outside my window without having yet seen it.

As the essential structure of knowledge, intentionality restores the sense of truth, as evidence, to the sense of what it means to know. Genuine knowledge must be based on evidence, or, the self-givenness of the intentional object as being so-and-so. Husserl, however, does not consider that there is one sense to the concept of evidence; on contrary, different kinds of objectivities admit different senses of what counts as evidence (e.g., the self-givenness of a perceptual object is of a different order than the self-givenness of a mathematical object), yet evidence as such, whether of a perceptual object or a mathematical object, must be grounded in the intuitive self-presence of the object itself. As Husserl writes, "*Category of objectivity and category of evidence are perfect correlates. To every fundamental species of objectivities* – as intentional unities maintainable throughout an intentional synthesis and, ultimately, as unities belonging to a possible 'experience' – *a fundamental species of 'experience,' of evidence corresponds* ..." Phenomenological reflection takes up "the task of exploring all these modes of the evidence in which the objectivity intended to *shows itself* ..." (FTL, 161). In this manner, Husserl's ontologically capacious conception of evidence reflects the manifold ways in which objects can be given as meant. In Husserl's definition: "Evidence is ... an experience of being and of being-so-and-so [*eine Erfahrung von Seiendem und So-Seiendem*], precisely in the form of encountering face-to-face something itself in an intellectual seeing [*Es-selbst-geistig-zu-Gesicht-Bekommen*]" (CM, 52 [12]). Evidence is the showing of being itself: evidence is not a "subjective feeling" – it is an experience in which an object is *revealed or shown* as being so-and-so. This emphasis on evidence as the showing of something as being so-and-so opens evidence to the full range of different senses of being; different regions of being are defined by different senses of evidence. In turn, the concept of evidence is internally differentiation into "adequate" and "apodictic" forms of self-givenness. Adequate evidence is evidence of something itself that excludes every doubt that it is so-and-so, yet it still remains open to the possibility of imagining that state of affairs as not-being (CM, 55 [15]). "Apodicticity," by contrast, is evidence of "higher dignity" in which the field of possibility is closed in a consciousness of impossibility. As Husserl writes: "Evidence, which in fact includes all experiencing in the usual and narrower sense, can be more or less perfect" (CM, 52 [12]). In this sense, since knowledge is always a claim to evidence, or, in other words, a movement towards evidence, every intention of knowing "strives" for fulfillment in the intended object itself.

Knowledge, in striving for truth, strives for "finality" (*Endgültigkeit*). As Husserl notes, "*Thus evidence is a universal mode of intentionality, related to the whole life of consciousness.* Thanks to evidence, the life of consciousness has an *all-pervasive teleological structure*, a pointedness towards 'reason' and even a pervasive tendency towards it ..." (FTL, 160).

Husserl's philosophically refreshing conception of evidence is inseparable from an equally novel conception of the apriori. As Husserl remarks, "The truly fundamental cognition in this connection – a cognition foreign to all previous psychology and all previous transcendental philosophy – is that *any straightforwardly constituted objectivity ... points back, according to its essential sort ...* to a correlative *essential form* of manifold, actual and possible, *intentionality ... which is constitutive for that objectivity.*" In other words, "*the whole life of consciousness is governed by a universal constitutional Apriori, embracing all intententionalities...*" (FTL, 246). It would not be an over-statement to claim that the trajectory of Husserl's thinking, from the *Logical Investigations* to *The Crisis of the European Sciences*, can be summarized as the attempt to demonstrate the transcendental truth of the proposition, just quoted, that "*the whole life of consciousness is governed by a universal constitutional Apriori, embracing all intententionalities.*" In light of this significant claim, it is surprising that Husserl's phenomenological conception of *thinking*, in its rigorous sense as the reflective grasp of essential apriori structures, received relatively little theoretical exposition, in deference, perhaps, to its tacit exposition in the praxis of phenomenological research. In *Logical Investigations*, Husserl argued that categorial and ideal objects are given to consciousness in an intuitive and fulfilled manner in categorial intuitions. Husserl distinguishes between two kinds of categorial intuitions: synthetic and ideative (or "general" or "universal"). Whereas categorial intuitions of the first kind (e.g., the state of affairs "the cat is on the mat") require a foundation in sensible presentations – in the perception of a cat sitting on a mat – categorial intuitions of the second kind, in which universal and ideal objects (e.g., mathematical objects) are grasped, are independent of any kind of sensible presentation. Such ideal objects of meaning are nonetheless apprehended intuitively; such a form of categorial intuition is, however, not to be conflated with an abstraction or a generalization, nor with an intellectual intuition. In writings subsequent to the *Logical Investigations*, Husserl revises this conception of categorial thinking, and develops in its wake the method of eidetic variation and "intuition of essence" (*Wesenschau*). In the process of eidetic variation, thinking progressively apprehends an invariant form in a play of variation, as facilitated through the imagination in which the structure of the possible becomes described and explored, and ultimately yielding to an intellectual grasp of an essential structural form. This performance of eidetic variation is exemplified in Husserl's time-diagrams. These diagrams should not primarily be seen as visual representations; more significantly, these diagrams *model* the dynamic of time-constitution. The actual construction of the diagrams repeats in thinking the stages of the constitutive accomplishment of time-consciousness under description.

## Transcendental reductions

The trajectory of a philosopher's thinking is a movement of questioning that unfolds without ever becoming entirely transparent to itself. Husserl's fashioning of his phenomenological enterprise through the methodological development of the transcendental reduction represents such a trajectory of self-discovery. Over the course of his thinking, Husserl remained perpetually dissatisfied with his method of reduction, which he continually recalibrated and diversified in pace with the expanding reach of phenomenological reflection. The reduction is not an instrument that we bring to the world, but a mode of reflection that uncovers a way of questioning *into* the world that requires a fundamental shift in attitude, a breaking of our naïveté, through which the obviousness of experience turns into a philosophical problem. As Merleau-Ponty remarks: "The problems of the reduction are not for him [Husserl] a prior step or preface to phenomenology; they are the beginning of inquiry. In a sense, they are inquiry, since inquiry is, as he said, a continuous beginning."[5] Moreover, the method of reduction does not exist in the singular. The reduction exists in the plural, and each kind of reduction admits different degrees of achievement.

   Once the idea of knowledge has been clarified in the manner sketched above, we can return to the unquestioned givenness of the world and confront its apparent self-evidence with a demand for "thinkability" or "intelligibility" in light of which the givenness of the

5 Maurice Merleau-Ponty, "The Philosopher and his Shadow," in, *Signs* trans. R. C. McCleary (Evanston, IL: Northwestern University Press, 1964), 161.

world becomes "incomprehensible," that is, open to question. In light of this recognition of experience as implicitly the claim of evidence, a phenomenological critique of the experience begins in the recognition of the obviousness of the world as implicitly a claim to evidence. As Husserl repeatedly insists: "More than anything else the being of the world is obvious [*selbstverständlich*]" (CM, 55 [17]). And yet, this obviousness of the world is not at all self-evident when placed in the light of the idea of demand of evidence. As Tran Duc Thao insightfully notes: "The principle of evidence is not a dogmatic theory but a simple starting point that outlines the plan to follow in achieving a philosophical critique."[6] In this questioning of the obviousness of the world, Husserl turns to perceptual experience – for it is in perceptual experience that the actuality of the world is given to me in the most apparently self-evident manner. As Husserl writes, "The being of the world, by reason of the evidence of natural experience, must no longer be for us an obvious matter of fact; it too must be for us, henceforth, only an acceptance-phenomenon [*Geltungsphänomen*]" (CM, 58 [18]). What is more self-evident than the fact that I am here sitting in bed, next to the television and with a computer keyboard in hand?

The suspension, or *epoché*, of the natural attitude, is not equivalent to *doubting* the givenness of the world; it is not a negation of the world, but rather a form of questioning that invokes a fundamental shift in attitude: we inhibit our naïve acceptance of the world, and shift our attention from "what the world is" to "how it is at all given to me." This shift to what Husserl calls the "phenomenological attitude" involves a self-induced modification of consciousness to the extent that consciousness withholds its own (implicit) acceptance of the world as given in order to *see*, and thus to question, *itself* as the ground of acceptance of the world. The reduction is as much a questioning of the world as a questioning of consciousness itself in relation to the world. The suspension of the natural attitude therefore does not exclude or destroy the world – despite Husserl's easily misunderstood characterization of the "destruction of the world" in the *Ideen*. As Husserl writes, "Meanwhile the world experienced in this reflectively grasped life goes on being for me (in a certain manner) 'experienced'

6 Trân Duc Tháo, *Phenomenology and Dialectical Materialism*, trans. D. Herman and D. Morano (Dordrecht: Riedel, 1951), 90.

as before, and with just the content it has at any particular time" (CM, 59 [19]). Instead, even in this extreme formulation of a destruction of the world (meant to expose the contingency of experience on foundational acts of transcendental subjectivity), the reduction transforms the traditional problem of knowledge into the problem of transcendence by way of a proper understanding of intentionality in and through which objects of experience are constituted. The defining insight behind the method of reduction is that the discovery of transcendental subjectivity functions as the counterpart to the discovery of the intentionality of consciousness. Transcendence belongs intrinsically to the sense of the world, yet this transcendence only acquires its sense as transcendence from my experiencing, as transcendence for consciousness. The connection between "transcendence" and "transcendental" is here clearly circumscribed: the basic problem of transcendental phenomenology is the problem of transcendence, and the ego, or consciousness, "who bears within him the world" is transcendental in this sense, as intentionality, as the ground of the world. As Husserl states, "The objective world, the world that exists for me, that always has and always will exist for me ... derives its whole sense and its existential status, which it has for me, from me myself, *from me as the transcendental ego*" (CM, 65 [26]).

Husserl can in this regard legitimately and fruitfully speak of transcendental experience since the suspension does not break our relatedness to the world, but rather transforms the sense of that relatedness. As Husserl writes, "the perception of this table still is, as it was before, precisely a perception of this table" (CM, 71 [33–32]). In the natural attitude, the intentionality of consciousness remained obscurely evident or misinterpreted in terms of a naïve realism. The suspension is thus required in order to formulate a proper conception of intentionality of consciousness that does not falsify the meaning of transcendence as the genuine transcendence of sense for consciousness. In other words, "Each *cogito*, each conscious process, we may also say, *'means' something or other* and bears in itself, in this manner peculiar to the *meant*, its particular *cogitatum*" (CM, 71 [33]). The reduction recovers the intentionality of consciousness, and in this sense, translates the problem of knowledge into the (transcendental) problem of transcendence. As Husserl writes, "in all of modern psychology there has never been an intentional analysis which was fully carried through. Obstacle: naturalizing of consciousness. Naturalism over-powered intentionality" (Hua IX, 219–220). The transcendental reduction thus carries two purposes: it discloses the field of transcendental subjectivity in showing how transcendental subjectivity is not worldly; it must show how the being of the world is constituted in consciousness. Both aspects are connected: showing the accomplishment of the transcendence of consciousness also tells us how consciousness is given in a manner that is different from objects of the world.

In altering the sense of our relatedness to the givenness of the world, the suspension of the natural attitude encompasses *"the whole world, including ourselves [uns Menschen]"* (Hua III, 69 [63]). But what does Husserl mean when he insists that the suspension of the natural attitude is not simply the questioning of how the world is at all given to me, but that, in this suspension, "human beings" are also suspended, that is, no longer taken for granted? We can immediately dispatch with any absurd idea that Husserl "denies" or "excludes" human beings, and thus leaves us with some "spiritual" consciousness; Husserl wants to question what it is to be at all given as a human being, by which he means, as a psycho-physical entity. The attitude towards the world is also a taking for granted of my own self and consciousness, and my body, as a part or natural occurrence in the world. Indeed, what is more obvious than the consciousness of myself in the world? I take for granted my own self-givenness in the world, and thus not only the world as "always there for me" (*immerfort für mich da*) but myself as "always there for me" (*immerfort für mich da*) (Hua III, 61 [54]). In suspending ourselves as humans, we no longer want to take the sense of our own subjectivity as consciousness for granted. The accomplishment of the *epoché* is not only world-directed, it is also self-directed, or, in other words, it is necessarily self-directed in so far as I am world-directed, and find myself in the world.

Over the course of his writings, Husserl developed three different ways of formulating the reduction, and he stressed each of these ways in different manners throughout the development of his thinking. Each of these ways opens converging angles on transcendental subjectivity as the field for Husserl's searching phenomenological descriptions and their eidetic shaping into the science of transcendental phenomenology. The three paths to the reduction represent three dimensions of transcendental subjectivity. The Cartesian path opens subjectivity as the field of transcendental experience, or immanence, and a *"new idea of the grounding of knowledge"* (CM, 66 [27]). Subjectivity is here conceived as a *foundation*. The Kantian path opens subjectivity as the transcendental prerogative of constitution, as centered on the guiding question of how experience is at all possible for consciousness in the form of its possible intelligibility. Subjectivity is here conceived as *world-constituting*, but also, as we shall discover in Husserl's unique brand of transcendental thinking, as *self-constituting*. The Brentanian path (i.e., through intentional psychology) opens subjectivity as a field of experience or givenness. Subjectivity is here conceived as the *concreteness of experience*, or "lived experience." All three paths to the reduction, and their corresponding dimensions of subjectivity, reflect distinct tendencies within modern philosophy (Descartes; Kant; Locke, Hume and Brentano); each tendency achieves transcendental clarification in Husserlian phenomenology. As indicated, all three tendencies stress a particular conception of subjectivity: subjectivity as foundation or origin; subjectivity as constituting; subjectivity as descriptive field of experience. In Husserl's thinking, all inflections or dimensions of subjectivity are present, yet each develops at a different pace in his thinking and with varying emphasis; each of these dimensions ultimately comes into greater resolution through the problem of time-consciousness in its definition of the "new being" of transcendental subjectivity.

Husserl's characterization of transcendental subjectivity as a "new region of being" is, however, misleading since transcendental subjectivity is, in fact, not a "piece" or particular region of being, but the ground upon which beings as such, including subjectivity itself as a constituted psycho-physical entity, can be disclosed. As Husserl indicates: "the absolute region of being of absolute or 'transcendental' subjectivity is not a partial region in the total region of reality" (Hua III, 72). "Absolute" designates the sense in which transcendental subjectivity is the "foundation" of *constitution*, or, in other words, absolute in the sense of constitution, as itself the activity, or performance, of constitution. In this sense, absolute subjectivity, as the "movement," so to speak, of constitution is "separate," or "distinct," from the world as constituted, yet it is not separate in the sense of exteriority or beyond. For the world as a totality, there is no exteriority: "the world is in itself a totality that, in its very meaning, does admit of an extension." And yet, as we shall discover in our discussion of absolute time-consciousness, "absolute or transcendental subjectivity carries in a particular and unique way the real totality of the world (of all possible worlds) in itself" in terms of its "intentional constitution" (Hua III, 72). Transcendental subjectivity is thus neither outside nor inside the world; it carries, or better, is the world in its constitutional unfolding.

Indeed, subjectivity is both inside and outside the world in the sense soon to be explored as both constituting and constituted in time-consciousness.

The transcendental reduction discovers the "infinite realm of new being" by which Husserl does not mean a being different than my subjectivity or consciousness in the natural attitude, but rather, a new sense of my own being, as transcendental subjectivity, that is obscured in the natural attitude. The contrast between the "new sense of being" of transcendental subjectivity and its mundane manifestation – transcendental subjectivity in its self-forgetting – is drawn explicitly in terms of a difference between temporality and spatiality. As Husserl remarks, "it belongs to the essence of conscious life to shelter within itself neither the spatial outside-of-one-another, inside-of-one-another, or throughout-one-another, nor the spatial totality, but rather an intentional implication and motivation, an intentionally mutual self-enclosure of intentional objects, such that according to form and principle no analogue whatsoever obtains in the physical domain" (Hua IX, 36). In light of this characterization of the "sense of being" of transcendental subjectivity as temporality – a claim that will occupy our thinking throughout this study – one can argue that the most significant of the three ways to the reduction is the third path through intentional psychology, in other words, through Brentano. The third path to the reduction does not replace the Cartesian and Kantian paths, yet both of these paths depend on the success of the third path in a manner that the third path does not depend on the first two. Indeed, this significance of the third path to secure a description of transcendental *experience* accounts for why Husserl's phenomenology of time-consciousness, and precisely in its transcendental significance, takes its bearings from Brentano and an initially psychological statement of the problem of time, rather than, as for example in Heidegger, with Kant or even Descartes. The overcoming of Brentanian descriptive psychology through the problem of time-consciousness leads to the articulation of transcendental subjectivity in its self-constituting temporality. In this way, Husserl is able to critique his own Cartesian path as well as secure the Kantian path to the reduction with a proper conception of transcendental subjectivity that had eluded Kant. Moreover, as we shall explore in chapter 6, Husserl's phenomenological refutation of idealism depends on a unique combination of his analysis of time-consciousness and the constitution of transcendental inter-subjectivity. As Husserl remarks, "the deep source of all our

errors is equating immanent temporality with objective, concrete
temporality – an equation that seems to press itself on us as self-
evident. The image of the stream plays a trick on us. Intentional
analysis of immanent temporality actually destroys this image and at
the same time places its legitimate sense before us" (Hua IX, 315). The
transcendental reduction is the reduction to a new sense of temporality
or time-consciousness; it is as much the reduction of time to consciousness as of consciousness to time, both of which depend on overcoming, or seeing-through, the master metaphor of time as a stream.
The entire effort of Husserl's phenomenology of time-consciousness
could, in this regard, be framed as the attempt to think through a
single metaphor in revealing its hidden, transcendental significance,
and thus redeem its phenomenological truth.

## The world of consciousness

Husserl often characterizes the opening of transcendental subjectivity
via the different yet converging paths of the transcendental reduction
as the discovery of an unexplored territory. In Husserl's evocative
metaphor, "our procedure is that of an explorer journeying through
an unknown part of the world, and carefully describing what is presented along his unbeaten paths, which will not always be the shortest"
(Hua III, 241 [235]). Although the comparison of transcendental
subjectivity to "an unknown part of the world" is misleading, even if
taken strictly as a metaphor – transcendental subjectivity is emphatically not a "part" of the world – the image of a research voyage into an
unknown territory nonetheless remains appropriately suggestive of
the project of phenomenological reflection, once the suspension of
the natural attitude and the method of reduction has transformed the
banality of the world into an uncharted field of transcendental experience, which has eluded philosophical inquiry until the advent of
its phenomenological awakening. Within this infinite expanse of
experience – infinite in the sense that transcendental experience
reflects the continual course of mundane experience from within the
transcendental attitude – Husserl fashions a method of analysis that is
both reflective and descriptive: reflective in the sense that the reduction
uncovers the structures of intentionality in and through which objects of
experience are constituted; descriptive in the sense of detailing the
structural elements, eidetic forms, and laws that underpin the constitution of objects of experience – phenomena – as they are given in

corresponding intentional acts of consciousness. In contrast to Husserl's earliest conception of phenomenological reflection as constrained to the (noetic) acts of consciousness and their underlying immanent content, within Husserl's mature conception of phenomenological reflection, the scope of immanent givenness includes the full structure of intentionality, the intentional acts of consciousness as well as their intentional objects; whereas intentional acts (along with their immanent content) are immanent in a "reell" manner, intentional objects, as exposed in phenomenological reflection, are immanent in an intentional manner, that is, they fall within the arc of intentionality as transcendence. Phenomenological reflection thus opens two correlative lines of description: the structure of objects as intended along with the complex of intending acts and their underlying sensual content. In contrast to the naïve experience of objects within the natural attitude, immanent objects of transcendental scrutiny are fundamentally reflective in character and based on evidence, or self-showing, of the phenomena itself.

Husserl's proposed transcendental method of reflection thus avoids both the formalism of Kant's transcendental critique as well as the empiricism of psychologism. In contrast to Kant's method of analysis, guided by the question of juridical legitimacy and its project of detailing how transcendental cognition *constructs* its objects of possible experience, phenomenological description does not prescribe the rules according to which objects of appearance *should* appear, but rather the strata according to which objects are *constituted* as they appear for consciousness. Consciousness does not legislate experience; instead, experience is the performance, or accomplishment, of consciousness. Husserl's reservation against the "mythic constructions" of Kantian thought does not, in the other direction, commit him to revert to a psychologism of experience that is equally an anathema to Husserl as it was to Kant. Husserl's art of description requires attentiveness towards objects under scrutiny and, thus, in this sense, reiterates the intentionality of consciousness in its world-directedness at the level of philosophical reflection, but it also attends to the *eidetic* and apriori structures of experience, as revealed in the intuitive apprehension of reflection. Such an art of description thus avoids the pitfalls of empirical arbitrariness as well as introspective impressions given its primary function of mapping the eidetic structures of consciousness and their intentional correlates in different regions of being as well as the transcendental articulation of subjectivity *as* the apriori correlation of consciousness and the world (and as distinct from a purely eidetic description). In its many facets of description and mapping, phenomenological reflection is a method that must be continually exercised and disciplined into an art of seeing – a ritual of clarification in which phenomena are seen, and seen again, in the progressive unfolding of their transcendental intelligibility.

Husserl continually practiced his art of phenomenological description for the various themes of phenomenological research in research manuscripts that famously run into thousands of pages, written in a special form of shorthand. At any given moment in the development of his thinking, Husserl's research was often further advanced in unpublished manuscripts than in the published writings, as exemplified with the problem of time-consciousness. In this regard, his published analyses are comparable to condensed summaries of extended reflections and descriptions practiced daily. This habit of practising daily the art of phenomenological description, essential to Husserl's unique manner of argumentation, exemplifies the sense in which philosophical reflection is the ritual of clarification. This repeated exercise of clarification is comparable to a disciplinary training of seeing, training the ability to see, as with a master painter or other artist whose powers of creation are equal to his powers of discernment, or seeing repeatedly. In this sense, the power of seeing becomes disciplined and shaped into a habit of seeing. This style of phenomenological research structures the progression of the *Ideen* with its internal structure of a recursive function. The progression from the natural attitude through the reduction to eidetic descriptions to transcendental articulation is not linear; there is a "zig-zag" movement, or back and forth; but what is important here is not to follow this complex movement, but rather to adopt one of Husserl's devices, namely, the recursive use of *examples* as the focal point for his reflections; with each example, different layers or aspects of general structures of conciousness come into clearer resolution and focus. Aspects of consciousness that were, in one example, tacitly presupposed or passed over in silence, or only dimly recognized, are taken up and clarified and made thematic in subsequent examples. This is exemplary of the phenomenological reflection as the ritual of clarification.

Husserl's exploration of transcendental experience is centered on the transcendence of the perceptual object. As Husserl writes, "the genuine concept of the transcendence of something physical which is the measure of the rationality of any statements about transcendence, can itself be derived only from the proper essential contents of perception" (Hua III, 111 [106]). As noted above, the suspension of the natural attitude places the experience of the world under the index of transcendental questionability; the apparently self-evident, or obvious, manner in which we experience things in the world is rendered unfamiliar, open to question. If we look back to our characterization of the natural attitude, we recognize that what seems obvious is that consciousness is a consciousness of things in the world. Yet, as Husserl remarks, "'consciousness of something' is something obvious and yet at the same time it is something incomprehensible [*Also 'Bewußtsein von etwas' ist ein sehr Selbstverständliches und doch zugleich höchst Unverständliches*]" (Hua III, 79 [73]). Consciousness of something is "incomprehensible" in the sense that consciousness is open to something other than itself: consciousness is the consciousness of what it itself is not, which is not given "in" consciousness. The consciousness of something is the enigma of transcendence (*das Rätsel der Transzendenz*): "how can knowledge reach out beyond itself, how can it make contact with a being that is not to be found within the confines of consciousness?" (Hua II, 5 [61]).

A piece of white paper lies in front of me on the table in a barely lit room. On Husserl's account, it is the paper itself that I perceive, not an image or other mental surrogate or representation that would mediate the givenness of the perceived object as such. Yet, the fact that it is the paper itself that I perceive merely states the problem that Husserl's analysis of intentionality is meant to clarify. For although I perceive the paper itself – a paper that I can touch, smell, etc. – due to the dimness of the light, I perceive this piece of paper with some difficulty, and must adjust my eyes in order to perceive the paper distinctly from the mass of other objects cluttered on my desk – books, pens, scrap paper, torn drafts of this chapter. Not only does this piece of paper appear to me relative to the particular and changeable circumstances of perception (different grades of light, etc.), but I do not actually perceive all of the paper at once, for I only perceive certain profiles of the paper – the front side with writing – whereas other profiles, or sides, of the paper remain hidden from my view. These various perspectives intend the paper as such. I can take different perspectives of the same object; moreover, this piece of paper occupies a definite spatial position (on the desk) relative to my own bodily presence (on the desk in front of me). When I move around the table, the spatial orientation of this piece of paper necessarily changes as a function of my changing bodily position. As I reach to take this piece of paper, I begin to perceive this piece of paper more clearly as a white piece of paper: it appears to me more clearly now that I am closer to it, I distinguish more erased markings, etc. These various characteristics define a recognizable "style" to the perceptual experience of objects, or what Husserl equally calls "external percep-tion" – external in the sense of the perception of a transcendent object.

As Husserl first developed in the *Logical Investigations*, all objectify-ing acts of consciousness – acts in which objects are given to me – are intentional. Consciousness is the consciousness of an object in the sense that the intentional object of consciousness (*cogitatum*) is intended, or "meant," as being so-and-so by a corresponding inten-tional act of consciousness (*cogito*); objects are given to me under a definite description of their sense: this paper as white, as a square, etc. The Husserlian thesis of intentionality hinges on the recognition that consciousness intends, or is directed towards, objects as objects of sense, and that such objects of sense are transcendent, and not con-tained in an immanent manner in consciousness, on account of their objectivity as a unity of sense. This emphasis on the question of sense is crucial for Husserl's phenomenological analysis and is the direct result of the suspension of the natural attitude and any naïve accept-ance of what it is to be or exist. The adoption of the transcendental attitude recognizes that "being" is always determined by sense; the basis for the givenness of an object is the question of its sense of being so-and-so. As unities of sense that are given to me relative to the conditions of (continuing) experience, intentional objects appear to consciousness; yet, by virtue of appearing to consciousness, such tran-scendent objects are not experienced (*erlebt*) in the specific sense that the intentional object is thus *not* contained in my consciousness, that is, not immanent to my consciousness. This distinction between the appearing intentional object and the experiencing consciousness establishes the crucial phenomenological distinction between "tran-scendence" and "immanence" – the conceptual distinction around which Husserl's answer to the problem of knowledge, as the problem of transcendence, crystallizes.

With this basic scheme in hand, "two sides" of corresponding phe-nomenological analysis are delineated, as reflecting the intentionality of *ego-cogito-cogitatum (qua cogitatum)*: a reflection on the "objective" side that attends to the form of the intentional object and a reflection on the "subjective" side that attends to the form of the subjective acts. Note that both "sides" necessarily belong together: intentionality of consciousness encompasses the entire field of transcendental experience due to the "universal apriori-correlation" (*Korrelationsapriori*) that Husserl identifies as the central discovery of transcendental phenomenology, and which, in his own self-interpretation, he had already discovered in the *Logical Investigations* (Hua VI, 169 [166]).

As I look at this piece of white paper on my desk, at any given moment, as already noted, I perceive only certain sides or profiles of the paper: I cannot perceive the back side while also perceiving the front side. And yet, even though there are numerous aspects of this piece of paper that are not actually perceived by me at this time, and from this determinate spatial position, the perception of what is actually, or authentically, perceived – this front side in the now, etc. – is nonetheless surrounded by horizons; I "know" that the piece of paper has a back side, which I could either confirm, or discover (or fail to discover, much to my surprise), by turning over the paper. I am implicitly conscious of there being a reverse side which I could perceive, and this consciousness of possibility is "pre-delineated" in the sense that it is motivated and based on what is actually given to me. Intentionality not only structures the actuality of the object, but also its possible experience, and this sense of possible experience structures, or situates, any actual perceptual experience of the object (along with corresponding or correlate horizons of possible acts of consciousness). In this regard, the perceptual experience of an object is structured by "inner horizons" that prescribe, or delineate, *possible* manners of its appearance; these horizons are empty in the sense that the back side of the paper is not given to me, and in this sense, not actually given; yet these empty horizons, in so far as they have the character of intentionality, intend the object, and thus "give" the object in its determinate absence, and, so to speak, delineate the horizons, and ways, of its possible and continued discovery. As Husserl argues, "*every actuality involves its potentialities*" that are not empty of content, haphazardly opening beyond the actually given, but that are "possibilities intentionally predelineated in respect of content" (CM, 81–82 [44]).[7]

---

7 This is the sense in which the relation between actuality and potentiality – more fully explored in chapter 7 – has the form of *motivation*: what is actually perceptually given to me *motivates* other possible acts of consciousness. Significantly, Husserl stipulates that a potential consciousness is always a modification of an actual consciousness.

A perceptual object is also situated in an environment in which I perceive other objects – this table, these pens, this computer, etc. In addition to the inner horizons of its internal articulation, perceptual experience is also structured by outer horizons that lead my perception to other objects and their surroundings. While I am currently fixed on closely observing this piece of white paper, a sudden noise from the corner of the room may lead my attention to the consequences of my cat's curiosity – a pile of books has fallen. In both cases of inner and outer horizons, perceptual experience has the fundamental character of an unfolding or process of continuous discovery and "explication" through which, or according to which, the objects of meaning are rediscovered, expectations fulfilled or disappointed. Though I necessarily experience the world in the form of the actual, experience cannot remain in the actual, as it is constantly reaching beyond, or transcending, towards horizons of possibilities. As Husserl writes, "This *intending-beyond-itself,* which is implicit in any conscious-ness, must be considered an essential moment of it" (CM, 84 [46]). Transcendence is not only the dimension of actuality, of how some-thing is given to me as present, it is also the dimension of potentiality that is structurally inseparable from actuality, such that we can speak of the transcendence of potentiality.

What becomes clear from this conception of intentionality as the structure of transcendence is a defining *tension* within Husserl's con-ception of constitution and its significance for the argument of tran-scendental phenomenology. Whereas Husserl identifies the natural attitude as the unquestioned acceptance of the givenness of the world as a world existing in itself, Husserl argues that, through the reduc-tion, one comes to understand that the world and its objects are constituted in acts of consciousness; these acts of consciousness are, in so far as they are objectifying acts, structured by intentionality, that is, transcendence. Husserl's notion of constitution has two facets: on the one hand, objects are constituted in consciousness, yet on the other hand, objects present themselves, or announce themselves (*sich bekunden*), in consciousness. This tension between the object as constituted in consciousness and as announcing (disclosing) itself in consciousness is central to Husserl's conception of intentionality, and is apparent in Husserl's description of perceptual experience. The perceptual experience of things – spatial objects – is surrounded by horizons of intentions that reach beyond, and in this sense, transcend, the actually given perspectives of the perceptual object. These horizons of continued possible experiences establish a halo of expectations that may or may not be fulfilled over the course of experience. The basic point for Husserl is that the perceptual object is always more than what we can at any given moment and under any particular circumstance perceive of it; in this regard, the "being" (*esse*) of the perceptual thing does not coincide with its being perceived (*percipi*). The object, or thing, is always more than what is perceptually, that is, intuitively, given. But one should not consider this transcendence of the perceptual thing (and by the same token, the pretension of external perception) as suggesting that the object exists "in itself" behind the profiles of its appearance; the recognition that the being of the object does not coincide with its perception does not mean that the being of the object as such is independent of possible perception. Husserl resists, or rejects, this turn to the distinction between appearance and thing in itself by arguing that the object is the *unity of a manifold of appearance*, and this manifold of appearance is both spatial (different sides, etc.) and temporal, and given in sensual *hylé* or sensations on the basis of which an object is apprehended by "animating" and "objectifying" acts of consciousness.

This continuous interplay of actuality and possibility (or potentiality, Husserl uses both terms equivalently), of fulfillment and empty intentions, can furthermore be characterized as the *movement* of perceptual experience – where "movement" is here understood in a quasi-Aristotelian manner as the realization or actualization of potentiality, as the transition from potentiality to actuality. Moreover, since intentionality is the way in which objects are intended as meant, this movement of perceptual experience is the realization of sense, or, in other words, a movement of explication in which aspects of the object's sense are "disclosed" against a horizon of implied and hidden aspects of its sense. The movement of perceptual experience is a continuous process of "explication" or "understanding," not in the fully objectified form of judgments, but in the perceptual form of a structured "explication" or "uncovering" of different aspects of an object's sense. This process of perceptual explication – as when I come to see with greater clarity the writing on a piece of paper or when I suddenly recognize that those shadows in the woods are nothing but the rustling of leaves, and not a stalker – unfolds on both sides, so to speak, of intentionality: "an uncovering that brings about, on the noematic side, an 'explication' or 'unfolding' ... and correlatively, an explication of the potential intentional processes

themselves" (CM, 83–84 [46]). In so far as the discovering or realization of aspects of an object's sense correlates to intentional acts of
consciousness, we can only also speak of the *self-explication*, or *self-
realization*, of consciousness itself, in its passage or transition from
potential consciousness to actual consciousness; in other words, consciousness discovers *itself* in and through the myriad ways in which it
discovers, and rediscovers, the world. I come to discover myself in
discovering the world: as more of the world unfolds for me, those
implicit acts of consciousness or horizons are in turn, and in a correlative fashion, further realized or actualized, and, in this sense,
understood.

Within this continuous interplay of actuality and possibility, consciousness is fundamentally "alert" (*Wach*). As Husserl notes, "it is of
the essence of a waking ego's stream of lived experiences that the
continuously unbroken chain of cogitations is continually surrounded
by a medium of non-actuality which is always ready to change into the
mode of actuality, just as, conversely actuality is always ready to change
into non-actuality" (Hua III, 79 [72–73]). The ego is "awake" – in the
sense that within the stream of its experience, it can take an orientation towards its experience, and thus towards actuality and potentiality
(non-actual). It is awake or open to possibilities; yet this openness
of consciousness is wider, so to speak, than the narrowly focused
attentiveness that the ego directs to particular objects of experience.
The "wakefulness" of consciousness must not be conflated with an
explicitly thematic apprehension of myself as conscious. Consciousness is directed towards its intentional object – it is the piece of paper
I perceive – and yet aware of itself in its perception of the paper in a
non-thematic and pre-reflexive manner. As Husserl writes, "when
living in the cogito, we are not conscious of the cogitatio itself as an
intentional object; but at any time it can become an object of consciousness; its essence involves the essential possibility of a *reflective
turning of regard*" (Hua III, 84 [78]). Inseparable from the understanding of consciousness as transcendence, as the consciousness of something being so-and-so, is that consciousness is also, throughout its
wakefulness, conscious of itself *as being conscious*. However, such self-
consciousness is pre-reflexive and pre-thematic. I am not aware of
myself as perceiving this piece of paper in the same manner in which
I perceive this piece of paper, even though my perception of this
paper is based on the consciousness of myself as perceiving; otherwise,
I could not claim that I am perceiving this paper. Moreover, Husserl

contends that it is only because I am already aware of myself in a pre-thematic manner that I can at all become an object of reflection for myself, for example, when, instead of "living in" my perceptual experience of the tree, I reflect on my perceptual act.

In this fashion, Husserl formulates a distinction between immanent perception and transcendent perception. In cases of immanent perception, as when I reflect on my perceptual act, the object of perception (the act of perception *along* with its intentional object) necessarily belongs to the same temporal stream of consciousness as the act of reflection (or immanent perception) itself; for indeed, I reflect on my perceptual act and thus objectify myself – both the act of reflection as well as its object belong to my consciousness. Husserl argues that this temporal solidarity is always in place in those cases in which an act of consciousness is directed towards another act belonging to the same consciousness (I reflect on why I said such and such) or in which an act of reflection is directed towards the "sinnliches Gefühlsdatum" – the non-intentional, lived content (to which we shall turn) – of the same consciousness. In such instances, "consciousness and its object form an individual unity of pure lived experience" (Hua III, 85 [79]). Immanent perception is, thus, defined by the coincidence of perception and being. Another way to state this insight is that immanent perception does not admit of any distance, or discontinuity, between the "object" of consciousness and the act of consciousness. Moreover, "*The unity of the stream of lived experience is the only unity determined purely by the essences proper of lived experiences themselves;* or, this being the same thing, a lived experience can be combined only with streams of lived experience to make up a whole the total essence of which embraces and is founded on the essence proper of these lived experiences" (Hua III, 86 [80]). The being of my consciousness, as defined by immanence and as a unity of a stream of experiences, is implicitly recognized as temporal through and through. By contrast, the relation between acts of consciousness and objects in external, or transcendent, perception does not possess the same form. In the case of external perception, we do not find an "authentic and essential unity" (*eigenwesentlich Einheit*) between acts of consciousness and its intentional object in the same sense in which consciousness is itself unified. With such an insistence on the dual significance of intentionality, as both "other-directed" and "self-directed," as transcendent and immanent, does the intentionality of consciousness imply a fundamental heterogeneity between objects of experience and consciousness?

In what sense can something other than or foreign – "other-being" (*das Andersein*) – to consciousness, as transcendence, nonetheless be constituted *in* consciousness that is, in its being as consciousness, a being for itself? How is this original difference between "being for itself" and "other-being" itself constituted? Stated in this fashion, the central phenomenological question of transcendence can be seen as the problem of "otherness" or "alterity," of how something other can be given to me. As Husserl notes, "we must now acquire a deeper insight into *how the transcendent stands with respect to the consciousness which is a consciousness of it*, into how this mutual relationship, which has its paradoxes, should be understood" (Hua III, 67 [86]).

## Intentionality and time-consciousness

Although Husserl argues, as just noted, that the intentional unity of the perceptual act of consciousness and its transcendent object is not of the same kind as the immanent unity of the perceptual act of consciousness itself, an essential unity must nonetheless obtain between the transcendence of the intentional object for consciousness and the immanence of consciousness for itself in order to provide a phenomenological foundation to the constitution of intentionality in its dual significance as "other-directedness" and "self-directedness," as transcendence *in* immanence. This question of transcendence in immanence represents Husserl's phenomenological formulation of the "problem of the world" regarding the essential correlation between "being and reason." Our discussion here cannot attempt to address exhaustively this central nerve of Husserlian phenomenology. Instead, the question of transcendence in immanence will continually preoccupy us in connection with the problem of time-consciousness. For the purpose of this chapter's restricted aim of situating the problem of time-consciousness within transcendental phenomenology, let us return to perceptual experience – the guiding clue of Husserl's exploration of intentionality – in light of our clearer formulation of the "enigma of transcendence."

The perceptual experience of this table, as already discussed, is constituted in a continuous interplay of actual and possible modes of intentional consciousness. Different perspectives are available to me within any given slice, so to speak, of a continuously unfolding perceptual experience. The side of the table that is actually given to me – the front of the table – is situated in a context of horizons that delineate, and thus motivate, other possible perspectives. These horizons of possible experiences are given to me in the form of empty intentions in which the hidden sides of the table are delineated in, or as, their absence. These empty horizons are inherently "fulfillable," so to speak, were I, for example, to walk around the table. This structure of partial, or inadequate, givenness – that I cannot perceive the object in its totality at once – reflects the spatiality of the perceptual objects. As Husserl remarks, "where there is no spatial being it is senseless to speak of a seeing from different standpoints with a changing orientation in accordance with different adumbrations and appearances" (Hua III, 97 [91]). This seems to suggest that the inadequate form of a perceptual object's givenness is solely a function of its spatial constitution. Moreover, it also seems to suggest that perspectives can only be spatial in nature; but as we shall discover when we turn to the issue of time-consciousness, Husserl will deliberately characterize the structure of time-consciousness in terms of variable perspectives within a landscape of temporal orientation. The inadequate character of perceptual experience, it should also be stressed, does not represent a deficiency of perception; instead, it further emphasizes the inherent temporality of perceptual experience as an open process, and as connected to the spatiality of objects. The perceptual experience of objects is surrounded by horizons of intentions that reach beyond, and in this sense, transcend, its actually given perspectives. These horizons delineate possible future lines of experience – the course of experiences to come. These horizons of possible experiences take the form of expectations that may or may not be fulfilled over the continued course of experience. Husserl's contention that "it is senseless to speak of a seeing from different standpoints with a changing orientation" in the absence of spatiality must thus, in light of Husserl's own thinking, become qualified. The inadequacy of perceptual experience is an "eidetic necessity" of its determinate form of objectivity, which, as we shall have opportunity to explore in the following chapters, is constituted *as* a form of temporality.

Throughout the unfolding continuity of variable perspectives in which the table is given to me, the table itself, as a unity of meaning or sense, is nonetheless intended across the manifold of appearances. Whether I am looking at the front of the table or walking around to take a look at its other sides, it is the same table that is the object of my perceptual experience. As a unity of sense, the perceptual object is always more than what I perceive of it. One should not consider this transcendence of the perceptual thing as suggesting that the object exists "in itself" behind its variable perspectives; the claim that the being of a perceptual object does not coincide with its being perceived does not mean that the being of the object is independent of perceptual experience, but rather, that a perceptual object must be considered as an open system of possible perceptions, which, for Husserl, are in principle inexhaustible. The perceptual object "itself" is the unity of the manifold of all of its *possible* appearances. The total system of the object as a manifold of possible appearances transcends my consciousness of these horizons; ultimately, as we shall discover in chapter 6, the horizonal structure of a perceptual object implies the life of a consciousness *other than* mine. The horizon of the object's possible manners of givenness extends beyond my consciousness not only in the sense of encompassing other possible subjects (others may also view the same objects in my view); the transcendence of an object's horizons also frames the different senses in which any actual perceptual experience of an object can either confirm or disappoint my consciousness of the object. The object is thus always more than, and thus not reducible to, the history of my consciousness of intending it. The unity of the object as such is determined by a "type" and corresponding "structure of rules" that prescribes, in the manner of a Kantian regulative idea, the harmonious synthesis of the manifold of its appearances; this "object type" and its "rule" structures the dimension of an object's possible and variable appearances, as confirmed or disavowed over the course of experience.

I look at this table in front of me, close my eyes, and open them once more. As I open my eyes, I discover the same table once more, as the same unity of sense. Yet, although the table remains identical in its intentional sense, the perceptual acts are temporally distinct: the act of perceiving the table a moment ago is no longer. Different perceptual acts, as immanent unities of consciousness, contribute to the continuous perceptual apprehension of one and same intentional object. As Husserl notes, "the perception itself, however, is what it is in the continuous flux of consciousness and is itself a continuous flux: continually the perceptual now changes into the enduring consciousness of the just-past and simultaneously a new now lights up, etc." (Hua III, 92 [87]). In closing my eyes, I interrupted my perceptual act without interrupting the identity of the table itself: I looked *twice* at *one* table. I could, of course, discover that some mischievous person has removed the table from my view while my eyes were closed, in which case I would reopen my eyes to the disappointment of my continued expectation of perceiving this table. Whereas the table is given to me in various perspectives, my act of perception, that is, the lived experience of perceiving, is neither given in perspectives nor is my lived experience an intentional object of my consciousness in the manner in which the table appears to me as an object. In the *Ideen*, Husserl characterizes this "fundamentally essential difference between *being as lived experience* and *being as a physical thing*" in terms of a difference between an intrinsically temporal form of givenness and a spatial form of givenness. Whereas consciousness is given to itself, or lived as an experience, in an intrinsically temporal manner, the manner of givenness of perceptual objects is essentially spatial (Hua III, 95 [89]). We should not take this distinction as equivalent to the distinction between space as the form of external sense (external perception) and time as the form of inner sense (inner perception), in the manner proposed by Kant, since, for Husserl, the spatiality of perceptual objects is, in fact, inseparable from a temporal form of objectivity, as noted above. Moreover, as we shall discover in a subsequent discussion (cf. chapter 6), consciousness itself must be embodied in order to perceive objects. Nonetheless, it is significant that Husserl construes the difference between transcendence and immanence in terms of different manners of givennness. The self-givenness of immanent consciousness is characterized in terms of an immanent *temporality as lived experience*, in contrast to the spatio-temporal givenness of transcendent objects. More accurately stated, Husserl in fact presupposes two forms of temporalization in implicitly formulating, in the quote just cited, a difference between "noematic" (temporal givenness of intentional object) and "noetic" temporalizations (temporal (self)-givenness of consciousness).

There is another possible misunderstanding of the manner in which Husserl draws a distinction between the immanence of consciousness and transcendent objects. Husserl does not mean to draw a distinction between two separate kinds of beings; instead, this distinction is drawn within intentionality, as a distinction between two manners of givenness that are necessarily correlated. Moreover, the example of opening and closing one's eyes, which Husserl employs in order to introduce the distinction between immanence and transcendence, should also not be taken as suggesting a difference between the temporal succession of consciousness and the "non-temporal" identity of the intentional object. As we shall discover shortly, the identity of the object is itself constituted as a form of temporality due to its synthetic character – the different perspectives *of* the object must be synthetically, and thus temporally, unified along with the temporal unification of the perceiving consciousness as such.

Husserl fashioned a sophisticated account of the correlation between what he termed the noematic object – the intentional object as the unity of sense – and noetic acts – acts of intentional consciousness (such as an act of perception) along with a non-intentional, "reell" sensible content, which Husserl also calls "hyletic data."[8] We will take up in greater detail the meaning of this distinction between acts of consciousness and their non-intentional underpinning in immanent content or *hylé*. The details and difficulties with Husserl's theory of intentionality in its mature formulation under the heading of the noetic–noematic correlation need not detain us here, since our principal aim is to situate the significance of time-consciousness within the framework of this centerpiece of transcendental phenomenology.

Husserl distinguishes between the noematic object, the noetic act of consciousness and its "real" ("reell") sensible content, in other words, the sensual basis, or sensing, of my *own* lived experience. The cardinal insight driving Husserl's phenomenological thinking is the claim that intentionality is fundamentally an apriori correlation between a noematic object and a corresponding noetic act (and its hyletic basis). In the earliest treatment of intentionality, in the self-styled "descriptive psychology" of *The Logical Investigations* (first edition), Husserl was only able to provide a one-sided – "act-intentionality" – description of perceptual experience. It is not until the discovery and progressive refinement of the reduction (first articulated in the 1907 *Idea of Phenomenology*) that Husserl comes to understand how to describe within phenomenological reflection both the noematic object and the noetic acts of consciousness in their essential correlation. For the sake of our exposition, what Husserl understands as the noema contains three structural elements: the noematic sense, that is, the sense in which an object is intended; the "thetic character," or mode of belief, in which the object is intended; and the intended object itself, as that determinable "something" or "X."

---

8 I follow here the interpretation of the noema proposed by John Drummond, *Husserlian Intentionality and Non-foundational Realism* (Dordrecht: Kluwer, 1990).

Take this table again. In Husserl's description, the table is intended in its "noematic sense." I perceive the table as a table, that is, as being so-and-so, and it is by virtue of its being so-and-so that this table is given to me as anything at all. As Husserl writes, "each noema has its 'content,' that is to say, its 'sense,' and is related through it to 'its' object" (Hua III, 316 [309]). The noematic sense does not "mediate" between my perceptual act and the object itself – the determinable X that is apprehended as being so-and-so – since, if this were the case, the noematic sense would function as an intermediary, or representation, between my perceptual act and the object itself. However, the object itself can only be given under a definite description of being so-and-so. In this regard, the object is given as its sense; by the same token, the object is intended "through" – in terms of – its noematic sense. The determinable X is that of which it is so-and-so; there is no object without a sense, yet the determination of sense is a determination *of something* as defined by that sense. Within this way of considering the intentional object in terms of its noematic sense, the object considered abstractly as such serves as the "point of unity" (*Einheitspunkt*), or "pole," towards which the various determinations of sense, as different ways in which that object is given, are polarized. Considered in abstraction from its noematic sense, the object itself is a "pure something," an "empty X" (*leeres X*) that serves as the "bearer of sense" (*Sinnesträger*). The sense of an object defines the manner of its givenness as being so-and-so. As Husserl writes, "The 'sense' [*der 'Sinn'*] is this noematic 'object in the how' ['*Gegenstand im Wie*']" (Hua III, 321 [314]).

Within the noetic–noematic framework of intentionality, the constitution of a transcendent object is accomplished through a complex weave of syntheses in both dimensions of the noetic and noematic correlation. The manifold of noematic moments are temporally united in a synthetic consciousness much as, and in correlation with, the manifold of noetic acts and non-intentional sensual manifold. Each dimension of intentional consciousness – noetic and noematic – is unified, and unified with each other, on the basis of a "synthesis of identification" in which consciousness constitutes the identity of an intentional object as meant in and through a manifold of its appearances. The consciousness of intentional objects as meant is fundamentally an activity of *synthesis* with a dual significance of "objective" and "subjective" synthesis, that is, as the synthesis of an objective unity of sense along with a synthesis of consciousness itself, in its synthetic consciousness of an intentional object. As Husserl remarks, "a consciousness and consciousness are not only bound together universally, but they are combined into *one* consciousness the correlate of which is *one* noema which, on its side, is founded on the noemas of the combined noeses" (Hua III, 291).

Especially with an eye towards our discussion of time-consciousness, the synthetic character of consciousness cannot be over-stated. As Husserl writes, "Only elucidation of the peculiarity we call synthesis makes fruitful the exhibition of the cogito (the intentional subjective process) as consciousness-of – that is to say Franz Brentano's significant discovery that 'intentionality' is the fundamental characteristic of 'psychic phenomena' – and actually lays open the method for a descriptive transcendental-philosophical theory of consciousness" (CM, 79 [41]). The intentionality of consciousness is only half the story; without an equal emphasis on consciousness as synthesis, one fails to grasp the full thrust of Husserlian phenomenology. Indeed, it is only when intentionality is taken as a synthesis, and thus, when consciousness is taken as the "movement" of constitution, that the significance of time-consciousness for Husserlian phenomenology can come into view. The synthesis of identification, in terms of which the unity in multiplicity of transcendence in immanence is constituted, implies that "the fundamental form of this universal synthesis [*synthesis of identification*], the form that makes all other syntheses of consciousness possible, is the all-embracing inner time-consciousness" (CM, 81 [43]).

If we recall for a moment Husserl's over-arching "world-problem" regarding the essential connection between "being and reason," that is, as we have learned to formulate, "transcendence in immanence," the argument that the accomplishment of consciousness in its intentional constitution rests ultimately on "the all-embracing inner time-consciousness" assigns to this implied phenomenology of time-consciousness yet to come the task of uncovering the "all-embracing" origin of the difference between reason and being, mind and world. The origin of consciousness in time-consciousness promises to deliver the origin of all possible experience. As Husserl notes, "consciousness is precisely consciousness 'of' something; it is of its essence to bear in itself 'sense,' so to speak, the quintessence of 'soul,' 'spirit,' 'reason.' Consciousness is not a name for 'physical complexes,' for 'contents' fused together, for 'bundles' or streams of 'sensations' which, without sense in themselves, also cannot lend any 'sense' to whatever mixture; it is rather through 'consciousness,' the source of all reason and unreason, all legitimacy and illegitimacy, all reality and fiction, all value and disvalue, all deed and misdeed" (Hua III, 213 [208]).

And yet, despite this promise of time for the project of transcendental phenomenology, Husserl's writings and lecture courses, published and unpublished, are marked by a curious postponement of any extended confrontation with time-consciousness. In the *Ideen*, Husserl explicitly remarks that in this presentation of the "fundamental considerations of phenomenology" he will not enter into "the dark depths of the final constituting consciousness of all experiences of temporality [*dunklen Tiefen des letzten, alle Erlebniszeitlichkeit konstituierenden Bewußtseins*]" (Hua III, 208). As Husserl further remarks, "time is a name for a completely *delimited sphere of problems* and one of exceptional difficulty. It will be shown that in order to avoid confusion our previous presentation has remained silent to a certain extent, and must of necessity remain silent about what first of all is alone visible in the phenomenological attitude and which, disregarding the new dimension, makes up a closed domain of investigation" (Hua III, 198 [193]). This weighty exclusion of time-consciousness (repeated in *Formal and Transcendental Logic, Cartesian Meditations* and *The Crisis of the European Sciences*) means, however, that "the transcendental 'absolute' which we have brought about by the reduction is, in truth, not what is ultimate; it is something which constitutes itself in a certain profound and completely peculiar sense of its own and which has its primal source in what is ultimately and truly absolute" (Hua III, 198 [193]).

In a curious footnote, Husserl observes: "the efforts of the author concerning this enigma, and which were in vain for a long time, were brought to a conclusion in 1905 with respect to what is essential; the results were communicated in lectures at the University of Göttingen" (Hua III, 198 [194]). Husserl's remark that his reflections on time-consciousness "were brought to conclusion in 1905" is meant to reassure us that all is not in vain: the final absolute uncovered in the *Ideen*, and which, indeed, defines the visible framework of phenomenological thinking from the *Ideen* to the *Crisis*, is not compromised through the exclusion of what is truly absolute. As Husserl remarks, "Fortunately we can leave out of account the enigma of time-consciousness in our preliminary analyses without endangering their rigor" (Hua III, 198 [193–192]. In what sense, however, is the absolute that is not truly an absolute truly not *at risk* in the absence of an inquiry into the true absolute of time-consciousness, which, although named, remains as yet invisible? As betrayed in the anxious tone of "fortunately," Husserl's assurance – to himself as much as to his readers – that his investigations into the enigma of time-consciousness "were brought to conclusion in 1905" remains to be confirmed. Indeed, as we begin to explore the entangled beginnings of these 1905 lectures, far from representing a "conclusive finish," these lectures mark the *beginning* of the enigma of time-consciousness for transcendental phenomenology, a beginning that would remain unfinished to the end.

# 2

# A REHEARSAL OF DIFFICULTIES

*Problems that appear small are large problems that are not understood.*

— Ramón y Cajal

## Small change

Of the many ways in which there is truth to Husserl's confession, "without Brentano, I would have never written one word of philosophy," his inheritance of distinguishing between addressing questions and handling problems is especially significant.[1] More pronounced than in Brentano, this distinction is unmistakable in Husserlian phenomenology. On the one hand, substantial questions of philosophy, gathered around the axis of how knowledge is at all possible, feature prominently throughout Husserl's writings. In the 1905 lectures "On Inner Time-Consciousness" (hereafter: ITC lectures), for example, Husserl begins with a stirring evocation of Augustine's *Confessions* and the question *quid est enim tempus* that "nearly brought Augustine to despair." As Husserl is quick to declare, despite "our modern age, so proud in its knowledge, we may still say today with Augustine: *si nemo a me quaerat, scio, si quaerenti explicare velim, nescio*" (Hua X, 3 [3]).[2] In one broad stroke, the sweep of such a question opens a space for

---

1 Quoted in Maria Brück, *Über das Verhältnis E. Husserls zu F. Brentano* (Würzburg: K. Triltsche, 1933), 3.
2 Cf. Hua XXIV, 99. For the relation between Augustine and Husserl, see Nicolas de Warren, "Tempo e memoria in Agostino e Husserl," trans. N. Scapparone in: *La realità del pensiero*, ed. A. Ferrarin (Pisa: ETS, 2007), 93–142; Andrea Staiti, "Il luogo della verità: La presenza di Agostino nella fenomenologia di Husserl," *Quaestio*, 6 (2006), 373–402; Michael Kelly, "On the Mind's 'Pronoucement' of Time: Aristotle, Augustine

reflection; the question of time is decisively raised anew. On the other hand, Husserl attacks philosophical questions by handling specific problems that do not seem proportionate to the burden of the general questions they are meant to shoulder. Under the heading of the question of time, the investigations developed in the ITC lectures as well as in subsequent research manuscripts pursue a circumscribed set of issues, most importantly, the perception of temporal succession and the temporality of consciousness. And even though the constellation of these problems is Husserl's many-pronged attempt to address the question of time, one is often left wondering whether, over and above the unquestionable richness of Husserl's phenomenological reflections, the treatment of such problems amounts to an answer, let alone a theory of time and time-consciousness in the manner traditionally understood in the history of philosophy.

This suspense at the heart of phenomenological philosophy is constantly reinforced by the practice of Husserl's thinking. As Husserl declares in the 1906/07 lectures "Introduction to Logic and Theory of Knowledge" (*Einleitung in die Logik und Erkenntnistheorie*), "it cannot be our task here to solve the most difficult of all phenomenological problems, that of the analysis of time. What matters to me is only to lift the veil a little from this world of time-consciousness, so rich in mystery, that has up to now been hidden from us" (Hua X, 276 [286]).[3] The task is deceptively modest – to lift the veil *a little*, to work out *a few* problems – yet it clearly has the ambition of surpassing Augustine and Aristotle, of succeeding, moreover, where "the ancients" and "the moderns" have failed.[4] This emphasis on the crafting of specific problems gives an unmistakable profile to the breadth of great questions that span and bind Husserl's phenomenological framework into an enterprise of philosophical significance. As Husserl is reported to have once remarked: "I am not interested in large bills, but only in small change."

and Husserl on Time Consciousness," *Proceedings of the American Catholic Philosophical Association*, 78 (2005), 249–262.

3 Cf. Hua XXXIV, 255: "Ich kann hier nur einige rohe Hinweise geben."

4 Cf. Hua X, 394: "Man könnte hier den Anspruch des Aristoteles anführen aus seiner Schrift *De memoria*: 'Das Gegenwärtige ist Sache der Wahrnehmung, das Zukünftige Sache der Hoffnung (oder wie wir auch sagen könnten, der Erwartung), das Vergangene Sache der Erinnerung.' In ähnlichen Sinn versucht Augustinus die drei Modi der Zeit zurückzuführen auf *attentio, expectatio,* und *memoria.* Ob es sich damit nun so verhält oder nicht . . . das liegt ganz auf der Hand."

In introductory comments to the 1904/05 lecture course "Main Parts from a Phenomenology and Theory of Knowledge" (*Hauptstücke aus der Phänomenologie und Theorie der Erkenntnis*), Husserl credits Brentano's "unforgettable" seminar on "Selected Psychological and Aesthetical Questions" (*Ausgewählte psychologische und ästhetische Fragen*) for motivating a set of problems – small change – that first gave direction and impetus to the development of his groundbreaking phenomenology of time-consciousness.[5] As is evident from these lectures, ideas gathered from Brentano's "unforgettable" seminar continued to resonate during the transformative years after the *Logical Investigations* (1900/01) and the definitive break from its residual adherence to "descriptive psychology." Indeed, insights gathered from Brentano's seminar would continue to reverberate in Husserl's analysis of time-consciousness long after having been materially surpassed. At the beginning of the ITC lectures, Husserl announces to his audience that, "we now want to attempt to gain an access to the problems [*of time-consciousness*] by connecting into Brentano's theory of the origin of time" (Hua X, 10 [11]). In another expression of the conviction that without guiding thoughts, one cannot begin to search, Husserl recalls: "my own studies, into which I became more entangled a decade later, led me on different paths on many essential points, but most of all they taught me that the problems are even more entangled and difficult than Brentano had then recognized."[6] The principal aim of this chapter is to understand this entangled beginning by examining what Husserl takes to be the problem of time-consciousness and how he arrives at its initial, yet decisive, formulation by way of his

5 Cf. Hua X, xv: "Die ersten Anregungen zur Beschäftigung mit denselben [*origin of concept of time*] verdanke ich meinem genialen Lehrer Brentano, der schon in der Mitte der achtziger Jahren an der Weiner Universität einer mir unvergeßliches Kolleg über *Ausgewählte psychologische und ästhetischen Fragen* las." While not all of Husserl's notes from Brentano's seminars and lectures have survived, those from lectures attended in 1883/84 can be found in the Husserl Archives in Louvain under the signature Q 9 ("Ausgewählte psychologische Fragen"). In addition, notes exist from the 1887/88 lecture course on "Descriptive Psychology" (Q 10) and from an 1883/84 lecture course on "Metaphysics" (Q 8). Husserl provides an overview of these "unforgettable" 1885/86 lectures in Hua XXV, 307.
6 Hua X, xv–xvi: "Meine eigenen Studien, in die ich mich zumal ein Jahrzehnt später immer <mehr> verwickelte, führten mich freilich in wesentlichen Punkten andere Wege, und vor allem lehrten sich mich, daß die Probleme noch sehr viel verwickelter und schwieriger liegen, als Brentano sie damals geschaut hatte."

critique of "original association" (*ursprüngliche Assoziation*) – the title of Brentano's theory of the origin of time that Husserl takes as his point of departure.

## In praise of old notes

Although much has been made of the significance of Brentano's original association, the details of how it provided a "fountainhead" for a phenomenology of time-consciousness still remain unexplored.[7] Husserl's ITC lectures present an abridged version of Brentano's theory, yet remain by default the only substantial source for its reconstruction, given the paucity of references to original association in Brentano's writings and lecture courses.[8] While in the 1887/88 lectures on descriptive psychology Brentano informs his audience that he has hitherto discussed his idea of original association with only a select group of students – a statement confirmed by Husserl's reliance on a combination of Brentano's lectures and publications by Anton Marty and Carl Stumpf for his own presentation in the ITC lectures – nothing in the expositions of either Marty or Stumpf indicates that Brentano ever developed a fortified theory of original association.[9] Without a robust account of Brentano's original association on its own terms, we are arguably limited from understanding Husserl's criticism completely and, most importantly, the extent to which Brentano's thinking guides, but also haunts, Husserl's investigations. Any attempt to grasp Brentano's original association as a "fountainhead" must therefore supplement

---

7 Herbert Spiegelberg, *The Phenomenological Movement* (The Hague: Martinus Nijhoff, 1960), 44: "There is nothing new about the philosophical puzzle of time. But it has assumed fresh poignancy in the philosophizing of the phenomenologists. The fountainhead for this renewed and intensified interest was again in Brentano's thinking."

8 Elmar Holenstein, *Phänomenologie der Assoziation* (The Hague: Martinus Nijhoff, 1971), 246.

9 Cf. DP, xviii: "Ich selbst habe nie etwas darüber publiziert, und so wird die Lehre in mündlichen Vortrage auf gewissen Kathedern, die Schüler von mir inne haben, vertreten." Husserl bases his discussion of Brentano's theory of original association on the cursory presentation of Brentano's original association in Anton Marty, *Die Frage nach der geschichtlichen Entwicklung des Farbensinnes* (Vienna: Gerold, 1879), 121, Marty's 1889 lectures "Genetische Psychologie," and Carl Stumpf, *Tonpsychologie* (Leipzig: Hirzel, 1890), vol. I, 185–186; 227.

Husserl's own discussion with the available fragmentary material in Brentano's lectures, which allows for a composite picture of greater resolution than the portrait of Brentano's argument offered by Husserl alone.

In fact, Brentano's research on the problem of time constitutes an impressive and largely unexamined body of work, especially his dictations during the final years of his life.[10] According to Oskar Kraus, no other philosophical problem preoccupied Brentano more than the question of time – with the exception of God's existence.[11] On the significance of these writings, Kraus is unequivocal: "never in the history of philosophy and psychology has the problem of time (i.e., the question of the origin of our concept of time and the axioms of time) been so advanced."[12] Despite this importance, or, better, due to this importance, Brentano remained reluctant to publish hastily or complete hurriedly his reflections on the problem of time. On central philosophical issues, Brentano and Husserl both shared a habit of pursuing a program of conceptual experimentation within the (relatively) insular environment of their research manuscripts – a veritable laboratory in writing – without ever arriving at any definitive formulation of their respective views. Over the course of Brentano's thinking, no less than six different theories can be identified among his voluminous manuscripts, dictations, and correspondence with his students, yet these various accounts rarely surfaced in his published writings.[13]

---

10 Franz Brentano, *Philosophische Untersuchungen zu Raum, Zeit und Kontinuum*, eds. R. Chisholm and S. Körner (Hamburg: Feliz Meiner, 1976).

11 Oskar Kraus, *Franz Brentano* (Munich: Oskar Becker, 1919), 39: "Das Zeitproblem hat Brentano seit jeher auf das lebhafteste beschäftigt . . . das Gottesproblem ausgenommen ist Brentano wohl zu keiner Frage öfter und mit unbesiegbarer Geduld zurückgekehrt als zu der Frage nach dem Ursprung unserer Zeitvorstellung und zum Kontinuitätsproblem überhaupt." Cf. Lucie Gilson, *La Psychologie descriptive selon Franz Brentano* (Paris: Vrin, 1955), 154: "L'étude des relations temporelles, qui sont au nombre des relations comparatives, a tenu une place particulièrement importante dans la meditation philosophique de Brentano."

12 PE III, xl: "niemals bisher in der ganzen Geschichte der Philosophie und Psychologie ist das Zeitproblem d.i. die Frage nach dem Ursprung unserer Zeitbegriff und Zeitaxiome so weit gefördert worden, wie es in den unten veröffentlichen Untersuchungen [Brentano's posthumously published: *Von sinnlichen und noetischen Bewußtsein*] der Fall ist."

13 Roderick Chisholm, "Brentano's Analysis of the Consciousness of Time," *Midwest Studies in Philosophy*, 6 (1981), 3–16; Arkadius Chrudzimski identifies six theories in "Die Theorie des Zeitbewußtseins Franz Brentano im Licht der unpublizierten

Of these different accounts, the idea of original association is the earliest, and was fashioned by Brentano during the mid-1880s. In a letter to Marty in 1895, however, Brentano already refers to the theory of original association as his "old view," and informs Marty of the circumstances that provoked him to renounce this early interpretation of the origin of the concept of time.[14] At the time of Husserl's ITC lectures, Brentano had thus abandoned the theory of original association that Husserl takes as *his* critical point of departure.[15] Whether oblivious or indifferent to the progression of Brentano's thinking, Husserl *deliberately* returns to his notes from Brentano's "wonderful lectures" in his effort to develop for his students a

Manuskripte," *Brentano Studien*, 7 (2000), 149–161; Johann Götschl, "Brentanos Analyse des Zeitbegriffes," in *Die Philosophie Franz Brentanos: Beiträge zur Brentano-Konferenz Graz, 4.–8. September 1977*, ed. R. Chisholm and R. Haller (Amsterdam: Editions Rodopi, 1978), 225–248; Kevin Mulligan identifies four stages in "Brentano on the Mind," in *The Cambridge Companion to Brentano*, ed. D. Jacquette (Cambridge: Cambridge University Press, 2004), 66–97. For Brentano's later view of the temporal modification of representations, see Lucie Gilson (*La Psychologie descriptive selon Franz Brentano* (Paris: Vrin, 1955), 154–157.

14 Letter to Anton Marty, March 1895.

15 As noted by both Oskar Kraus and Henri Dussort, Husserl's 1905 critique of Brentano loses some of its critical force when redirected against the position Brentano actually defended around 1905. See Oskar Kraus, "Zur Phänomenognosie des Zeitbewußtseins," *Archiv für die gesamte Psychologie*, 75 (1930), 1–22; Henri Dussort, trans., *Leçons pour une phénoménologie de la conscience intime du temps* (Paris: PUF, 1964), 191: "On voit donc que les critiques husserliennes s'adressent à une théorie abandonnée par son auteur, et *en partie* pour les mêmes raisons: en rapportant le moment temporel aux modes de représentation, Brentano est amené à attribuer à ceux-ci une différenciation (en mode direct et oblique), qu'il luer refusait primitivement. Bien entendu, Husserl ne le savait pas lors de son cours de 1905, mais *dès* 1911 pourtant il aurait pu en être informé, et de même encore en 1919 par l'ouvrage collectif sur Brentano, auquel il a participé, et en 1925 par la publication des inédits." Given these circumstances, assessments of Husserl's Brentano critique have tended to be one-sided in either one of two ways. At the expense of failing to take into account the details of Brentano's theory, certain readings have relied exclusively on Husserl's truncated presentation of Brentano's original association. Alternatively, at the expense of failing to recognize the function of Husserl's critique of Brentano's "old view" as a point of departure for the *phenomenological* problem of time-consciousness, other readings have redressed Husserl's misrepresentation of Brentano's position by acknowledging Brentano's renouncement of his "old view." Yet even if we grant Dussort's suspicion against the trenchancy of Husserl's critique and Kraus' irritation at Husserl's lack of faithfulness towards his mentor's teachings, insisting on the fact that Brentano had abandoned his theory of original association in favor of a position closer to Husserl's in 1905 actually says little about what it was of Brentano's original association that set Husserl's investigation into motion in the first place.

phenomenological analysis of time-consciousness.[16] As Husserl
acknowledges to Brentano in a letter of 1905:

> I looked back to my old notes from your wonderful lectures in Vienna
> from the years 1885/1886 and read from these notes directly to my
> students, using [your] insights as points of departure for further
> analysis. My own proposals for how to handle these extremely subtle
> and difficult problems – problems that nearly bring us to despair –
> remained unsatisfactory.[17]

Husserl's deliberate use of his "old notes" suggests that an assessment
of Husserl's Brentanian inheritance must view Husserl's critique of
Brentano in the ITC lectures in strategic terms, as a staging ground
where the characteristic difficulties engendered by the problem of time
are first rehearsed. This discovery of different and more complicated
paths of questioning *through* Brentano is reflected in the organization
of the opening paragraphs of the ITC lectures. As a first phase, Husserl
begins with a critique of Brentano's original association (§§ 3–6). Once
the question of time has thus been delineated, Husserl translates this
"pre-phenomenological" (i.e., "descriptive psychological") statement
of the problem into the conceptual framework of phenomenological
analysis. In this second phase (§§ 7–9), we are invited to rediscover the
problem of time with new eyes: the intersection of time *and* consciousness

16 Brentano's abandonment of original association in the early 1890s shunted his
understanding on to a track of analysis closer to Husserl's analysis of time-
consciousness in the 1905 lectures. A rapprochement between both authors is
apparent given that Brentano now considers the consciousness of time as grounded
in the temporal modification of acts of presentation (*modo recto/modo obliquo*). This
recognition of temporal modification to acts of presentation (*Vorstellung*) – as opposed
to a modification of sensation – represents a decisive step in the development of his
descriptive psychology. As Brentano acknowledges in an appendix to the second
edition of the *Psychology from an Empirical Standpoint*, "when I wrote my *Psychology from
an Empirical Standpoint*, this [*that there are different modes of presentation*] had not yet
become obvious to me, or at least not entirely so. As a result there are many things that
I not only have to expand upon but also correct. Above all we must designate temporal
differences as modes of presentation" (PE, II, 279).
17 Letter of 27 March 1905: "Meine alten Hefte über Ihre wundervolle Wiener
Vorlesungen aus dem Jahre 1885/1886 hatte ich mir herausgesucht, manche Partien
daraus auch meinen Schülern vorgelesen und zu Quellpunkten weiterer Analysen
gemacht. Unvollkommen genug war, was ich selbst zu bieten vermochte zur
Behandlung dieser allersubtilsten und bis zur Verzweiflung schwierigen Probleme."
The preceding sentences also merit citation: "Besonders durch meine Vorlesungen
[*1904/1905*], in denen ich seit den Weihnachtsferien die deskriptive Psychologie der
Phantansie und der Zeit behandelte, ward ich arg bedrängt. Doch fehlte es inzwischen
nicht an einem nahen geistigen Kontakt mir Ihnen." Quoted in Hua X, xviii.

becomes *less* familiar, stranger, but also more promising, than was previously recognized in the history of philosophy. And yet, the translation of a problem into a phenomenological space of analysis is comparable to the projection of light through a prism: the structure of a problem becomes clarified in revealing a complexity that otherwise goes unseen. The problems inherited from Brentano take on a life and a direction of their own – they obtain their own phenomenological destiny – once seen through the prism of phenomenological reflection.

### "None of my listeners knows what I am truly after"

"Proterästhese" and "ursprüngliche Assoziation" are terms – as Brentano informs the audience to his 1887/88 lecture course on descriptive psychology – that are not to be found in contemporary handbooks and treatises of psychology, for both terms refer to a "fact" (*Tatsache*) of consciousness that has yet to be investigated or even discovered by either psychologists or philosophers.[18] When placed against the broader canvas of Brentano's descriptive psychology, the concepts of *proteraesthesis* and original association presented in these lectures are inconspicuous. Shortly after these lectures, Brentano would in fact discard this proposed characterization of *proteraesthesis* as an original association without, however, renouncing the concept of *proteraesthesis* entirely; subsequent attempts to understand the consciousness of time would retain the heading of *proteraesthesis* without espousing its characterization as an original association. As with other seminal ideas in Brentano's descriptive psychology, the concept of *proteraesthesis* was an idea in the making. Were it not for its role as a catalyst for Husserl's phenomenology of time-consciousness, this early casting of *proteraesthesis* as an original association would have remained what it was for Brentano, a passing stage in the evolution of his thinking, one of numerous attempts to clarify the origin of the concept of time. The concept of *proteraesthesis* as such, however, occupies a notable place in the list of Brentano's conceptual innovations; in contrast to other distinguishing concepts of descriptive psychology

18 DP, xviii: "In der Tat kommt der Ausdruck in keinem Handbuch oder Lehrbuch der Psychologie vor und die damit bezeichnete Tatsache selbst, wird, scheint mir, in keinem, das mir untergekommen, in ihren wahren Eigenheit erfaßt und gedeutet." Brentano employs the terms "original association" and "proteraesthesis" interchangeably until he abandons the theory of "original association" as an appropriate understanding of the phenomena of "proteraesthesis." Unless otherwise noted, I also use both terms interchangeably.

(e.g., intentional relation, immanent object, perception *en parergo*), it is without precedence in the history of philosophy – even Aristotle did not possess a concept comparable to *proteraesthesis*. Perhaps for these reasons, Brentano cautions his audience (and in the same breath reassures himself) that "none of my listeners knows what I am truly after."[19] But even Brentano could not anticipate how *proteraesthesis* would lead his most brilliant student to uncover "the most difficult of all problems of phenomenology."

Brentano's peculiar neologism is composed of the Greek noun αἴσθησις, "sensible perception," and the comparative adjective πρότερος, "before than" or "earlier than." Accordingly, *proteraesthesis* literally means "perception of the earlier than" or "sensation of the earlier than." The choice of the comparative adjective πρότερος is revealing, for it signals Brentano's intention to characterize the perception of an object as *past relative to the present* rather than as past in an unqualified sense, without an immediate and continuous bearing onto the present. At first glance, the significance of proposing a difference between the perception of the past and the perception of the immediate past (or "just-past") is far from evident, even though it is apparent that without a meaningful difference between "the past" and "the just-past" there would be no compelling reason to pursue Brentano's neologism – why invent a new concept rather than adopt and/or adapt the traditional vocabulary of "memory," or some equivalent thereof, if what is meant is simply a consciousness of the past?

Indeed, Brentano's proposal of a neologism is unusual given his established preference for the renewal of concepts from the history of philosophy rather than the fashioning of novel vocabularies without historical reference and responsibility. Brentano's historical doctrine of the cyclical four stages of philosophical development would seem to inveigh against introducing a concept into the history of philosophy that is bereft of precedence and whose origin cannot be traced back to the "Master" of the first historical epoch of Greek philosophy, Aristotle.[20]

---

19 DP, xviii: "daß hier keiner meiner Zuhörer weiß, worauf ich eigentliche ziele."
20 Brentano first published his view of the history of philosophy in "Geschichte der kirchlichen Wissenschaft," a chapter written for J. A. Möhler's posthumously published *Kirchengeschichte von J. A. Möhler*, 3 vols. (Regensburg, 1867–68), vol. II, 526–584. In his 1874 inaugural lecture at the University of Würzburg, Brentano presented a less developed version of his doctrine of the "four stages" of philosophy in "Über die Gründe der Entmutigung auf philosophischem Gebiet," in *Über die Zukunft der Philosophie*, ed. O. Kraus (Hamburg: Felix Meiner, 1968), 83–100.

The establishment of a psychology without a soul, by means of which Brentano seeks to inaugurate a philosophy of the future, exemplifies the first theoretical stage of his doctrine of four stages, but likewise illustrates his understanding of the relationship between historical transmission and experience for the life of concepts. In *Psychology from an Empirical Standpoint*, Brentano develops a psychological science based on a descriptive analysis of phenomena as given in the immanent domain of intentional consciousness. Aside from establishing the conceptual foundation for an empirical psychology, Brentano's method of analysis aims at renewing philosophy itself in the aftermath of its degeneration at the hands of German Idealism. Psychological analysis offers a method for the clarification of philosophical concepts. Concepts are robust with the history of their (re-)interpretation, yet have their origin in experience to which philosophy must periodically return, and reclaim. In the exceptional case of *proteraesthesis*, however, Brentano's neologism designates neither a positivist concept forged exclusively from "the phenomenon" nor the clarification of an inherited concept by way of a descriptive analysis of lived experience. Strictly speaking, from the viewpoint of the history of philosophy, the term *proteraesthesis* remains unrecognizable and, in this sense, unreadable; from the vantage-point of experience, it seems unmotivated and, in this sense, without traction.

If we recall the manner in which Aristotle distinguishes between sensible perception and memory in terms of a temporal difference between the present and the past, the temporal inflection of *aesthesis* in *proteraesthesis* announces a challenge of fundamental significance.[21] In so far as perception, for Aristotle, is defined as the actualization of a sensible form in the material substrate of a sense organ, perception reveals the actualization of change in the present. Although the actuality of the world does not depend on perception, it is only in the present – in the now – that the actuality of the world is given or

---

A relatively more developed exposition is given in "Die Vier Phasen der Philosophie und ihr augenblicklichen Stand," in *Die Vier Phasen der Philosophie*, ed. O. Kraus (Leipzig: Felix Meiner, 1926), 3–32; and in Franz Brentano, *Geschichte der griechischen Philosophie* (Hamburg: Felix Meiner, 1977), 1–23. For a critical reaction to Brentano's reading of medieval philosophy, see Étienne Gilson, "Franz Brentano's Interpretation of Medieval Philosophy," *Medieval Studies*, 1 (1939); for a fuller treatment of Brentano's thesis, see Lucie Gilson, *Méthode et métaphysique selon Franz Brentano* (Paris: Vrin, 1955), 25–67 and Balzs Mezei, "Brentano and Husserl on the History of Philosophy," *Brentano Studien*, 8 (1989), 81–94.

21 *De memoria*, 449b14.

encountered in experience. By contrast, memory is an "affection" (πάθος) or "state" (ἕξις) of either thinking or perception conditioned by a lapse or distance of time. Ambiguously defined as an "in-actual sensation," memory is "a sensation or affection when time has passed in which one has knowledge and sensation but without their being actualized."[22] In *De memoria*, Aristotle further distinguishes between memory (μνήμη) and recollection (ἀνάμνησις), defining the former as the retention of an image, produced like a signet's impression – which exhibits a likeness to that which was once perceived. The distinction between the memory/recollection and perception rests on an underlying difference between a time that has elapsed (an interval of time that is no longer, an elapsed now) and a time that is present, the now. So as to reinforce the basis of this temporal distinction between memory and perception, Aristotle twice remarks in the course of his discussion that "we cannot remember the present while it is present."[23] If perception reveals exclusively the present and memory makes known a present that is no longer, whether by recollection or memory, it is by definition impossible to remember that which has yet to pass completely or perceive directly that which has completely elapsed. It is, however, precisely this impossibility that Brentano calls into question by proposing a concept meant to occupy a yet unopened space between αἴσθησις and μνήμη, and that, by combining the traits of both αἴσθησις and μνήμ, questions the temporal origin of their distinction. Taken as an initial clue, we can read the term *proteraesthesis* as a specific challenge: we are asked to consider the possibility of a sensible perception, that is neither memory nor perception, which reveals the immediate past *along with* the present. We are asked, in other words, to perceive the just-past or "earlier than" (*protero-aesthesis*).

### The motivation of legacy

Brentano's creation of a Greek neologism in German evokes the legacy of Aristotle's seminal definition of time for the history of

---

22 *De memoria*, 449b19.

23 *De memoria*, 449b14; b25. As Richard Sorabji notes: "In English, one can say of something one is seeing that one remembers it, meaning that one recognizes it. But in the Greek it is more natural to stick to the word for recognizing *anagnôrizein*. . . . This is why no one, in Aristotle's opinion, would say he remembered a thing while he was seeing it. He would say instead that he recognized it" (*Aristotle on Memory* [London: Duckworth, 2004], 66). Aristotle appears keen to reinforce his claim (449b15) that "memory is of the past," but not of the immediate past (451b29–31).

philosophy in the form of a challenge directed at his contemporaries: how can we understand what Brentano is truly after, if he employs a concept that does not (yet) exist in the vocabulary of philosophy? What problem motivates the formulation of Brentano's *proteraesthesis*?

In a text written circa 1902, entitled "What Philosophers Have Taught About Time" (*Was die Philosophen über die Zeit gelehrt haben*), Brentano sketches a brief history of time that revolves in plot around the confrontation between metaphysical and psychological definitions of time. Curiously, in this particular account, Aristotle is neither a partisan of a metaphysical (as predominates among Scholastic philosophers and in Descartes) nor a psychological approach (as we find in Locke). In fact, Aristotle's efforts are deemed as failing entirely to address properly the "essence of time." According to Brentano, it is "ridiculous" to think that Aristotle provided an explanation of time in the *Physics*, as ridiculous as thinking that the essence of warmth is explained by a thermometer (RZK, 61). The Aristotelian definition of time as the number of movement with regard to the before and after only speaks to the issue of "when does something happen" and, thus, of how the category of time ("when" or πότε) is to be predicated, or said, of substance.

One could immediately protest that such an appraisal adopts too narrow a focus on the problem of categorical predication, and so fails to do justice to the breadth, but also the complexity, of Aristotle's discussion of time in the *Physics*. Nevertheless, such an over-played attention to Aristotle's framing of the question of time in terms of *when* does something happen (and with an eye to the question of predication and "saying" time) reveals Brentano's underlying commitment to the compact between a "proper" framing of the question of time in terms of *how* we at all perceive temporal succession and a psychological investigation into the *origin* of the concept of time – an inflection of what it means to question time, and of what such a questioning is after, that remains foreign to Aristotle's *Physics*. Aristotle's treatment of time in the *Physics* is situated in an overarching inquiry into the principles of nature and the being of natural entities; this orientation towards natural beings with beginnings and endings, beings of movement and change, translates into the centrality of movement (and, more specifically, the spatial displacement of locomotion) for Aristotle's discussion of time. Time "follows" movement; as Aristotle observes: "we would not recognize or sense the passage of time in our own soul without some kind of

movement."[24] This relationship between time and movement liberates time from the skeptical impasse of time's non-being, or non-
existence, raised in the opening of Aristotle's discussion. Time is not
movement, yet without movement of some kind, there is no time.
Time does not itself have being, but neither is time "nothing." We
must therefore look to beings in time in order to grasp the essence
of time that eludes, in passing through, being and non-being.

In the *Physics*, it is on the basis of an analogy between magnitude,
movement and time that Aristotle fashions his influential definition of
time as the number of movement with regard to the before and the
after. Temporal determinations are predicates of beings undergoing
movement of some kind, the implicit geometry of which can be
compared to a linear unfolding from a beginning to an end, with
earlier and later states of change separated in the now. As Aristotle
argues, the now divides the before from the after while simultaneously
uniting both into a continuous whole (the continuity of change),
much as a mathematical point divides and unites the continuum of
a line. The now is both a point and a limit, a "one and two." This
image of time as a line connects with the form of simultaneity that
excludes other nows other than the now that is itself present, or a form
of temporal givenness other than the form of the now. Nothing of
time can be actually given but the now. This conception of time as the
number of movement with regard to the before and after implicates
the soul directly as that which counts; yet, Aristotle only touches on
the difficult question of whether time therefore depends on the
counting activity of the soul. As Aristotle notes: "Whether if the soul
did not exist time would exist or not, is a question that may fairly be
asked."[25] Aristotle's silence on the question of the soul's relation to
time provoked, as is well known, debate among generations of commentators, Greek, Scholastic and Modern. For Brentano, Aristotle's
silence on the relation between time and the soul is further evidence
of the degree to which Aristotle's definition of time presupposes the
perception of time familiar to all of us, on the basis of which further
determinations, such as that of time as the number of movement, are
possible.

And yet, Brentano is not completely ignorant of Aristotle's other
fragmentary pronouncements on the question of time. In a set of

---

24 *Physics*, 218b21–24; 219a5.     25 *Physics*, 223a21.

dictations entitled "Towards an Understanding of Aristotle's Theory of Time" (*Zum Verständnis der Aristotelischen Lehre von der Zeit*) from 1915, Brentano admits that Aristotle indeed addressed the perception of time, not in the *Physics*, but in various psychological writings: *De anima, De sensu et sensibili, De memoria et reminiscentia.*[26] These writings "contain something significant about the origin of the presentation of time" (RZK, 146). Yet, despite such "a close relation" between "the conceptual determination of time" and "the origin of the presentation of time," the origin of "specific temporal differences" nevertheless remains obscure. As Brentano comments:

> Above all, the question of *from which intuition* does the concept of time emerge is not clearly established. He [*Aristotle*] at one point says that one *senses* [*empfinde*] something as earlier or later. How? Is it, however, not the case that sensation can only refer to something that is present? But perhaps what he means is that memory must be brought in to help. Nevertheless, how does what is earlier show itself? . . . The appearance of the present [*Gegenwart*] is characterized as a double-limit, yet what is not investigated is what must also appear as not-present [*Nichtgegenwärtigen*]. Nowhere is it established that the difference between present, past, and future can only be intuitively given, and nowhere are the different modes of presentation, which are here required, mentioned.[27]

In these remarks, the problem that motivates the concept of *proter-aesthesis* is succinctly formulated. Aristotle's writings contain few references to the perception of time, and Brentano's criticism deliberately exploits Aristotle's ambivalent characterization of how the common

---

26 Brentano begins by noting: "In den Darstellungen der Lehre des Aristoteles über die Zeit wird meistens nur das berücksichtigt, was er darüber im vierten Buche der *Physik* sagt; um seine Gedanken zu diesem Gegenstande gründlicher kennenzulernen, müssen aber auch die Äußerungen in den anderen Schriften herangezogen werden" (RZK, 138).

27 RZK, 148: "Vor allem ist die Frage, *aus welcher Anschauung* der Zeitbegriff stamme, nicht klargestellt. Er sagt einmal, man *empfinde* etwas als früher oder als später. Wie aber? Scheint doch die Empfindung auf Gegenwärtiges allein zu gehen. Und so könnte einer meinen, es müsse die Erinnerung zuhilfe genommen werden. Doch wie zeigt sich in ihr das Frühere? . . . Die Erscheinung der Gegenwart als doppelte Grenze ist zwar so charakterisiert, aber es wird nicht untersucht, was dabei von Nichtgegenwärtigen mit erscheinen müsse. Nirgends ist gesagt, daß die Differenzen von Gegenwart, Vergangenheit und Zukunft das einzige uns anschaulich Gegebene sind, und nirgends wird der Verschiedenheit von Vorstellungsmodis, die hier gefordert sind, Erwähnung getan." The latter part of this quote refers to Brentano's view of temporal differences as modifications of acts of presentation – a later view that replaced the earlier idea of original association.

sensible of time is discerned by the soul. In *De anima* (in the quote above, Brentano's remark "at one point" is most likely a reference to *De anima* 448a19), Aristotle writes that when common sense "takes in and senses" (νοεῖ καὶ αἰσθάνεται) an object, the soul discerns (or judges) that the object is now in a manner that is indivisible; the now is sensed instantaneously (426b22). Aristotle also argues in *De sensu* that a continuous magnitude is grasped "not by any of the nows in it," that is, not instantaneously, but in continuous time (448b6). Whether the soul perceives an instantaneous now or the now as an interval (and in the *Physics*, the intellect must grasp the now as both one and two), it remains ambiguous, on either account, how the soul "senses" or "perceives" an object in the now *with regard* to its before and after (τὸ πρότερον καὶ ὕστερον), given Aristotle's restriction (already signaled above) of perception (αἴσθησις) to the now; sensible perception can only reveal what is in the present. An appeal to memory does not resolve the issue at hand. On Aristotle's own definition, memory presupposes an elapsed duration or interval of time: an object must already have been perceived in a now with regard to its before and after in order to be remembered as having once been experienced in a now. Moreover, Aristotle explicitly notes in *De memoria* that the immediate past cannot be the object of memory, and should be considered, instead, as part of the now, since a now possesses "a certain span, and includes within itself experiences which one has just had."[28] On *this* view, Aristotle appears to embrace a notion of the now as a specious present that is "taken in and sensed," i.e., perceived; on the view noted earlier, however, Aristotle appears to embrace the now as an indivisible point, without temporal breadth.

Here is not the occasion to debate which conception of the now one should take as Aristotle's considered view. What is significant for our interests is Brentano's attention to the Aristotelian characterization of the now, or the present (*Gegenwart*), as a "double-limit," and its implication of the "appearance" or "phenomenon" (*Erscheinung*) of that which is not present (*Nichtgegenwärtigen*). As Brentano is keen to note, Aristotle's definition of time presupposes the perception of the "before" or "earlier-than": the now is "taken in and sensed" *with regard*

---

28 Sorabji, 66. As Sorabji further remarks, "the present which Aristotle envisages here has something in common with the specious present. For at any given instant, it has a certain span, and includes within itself experiences which one has just had" (91). Cf. *Physics*, 221a20–24.

to the before and the after, that is, *in view of* the before and the after.
To perceive the now, one must look two ways at once. Aristotle's claim
that the soul "takes in and senses" (νοεῖ καὶ αἰσθάνεται) the now
therefore requires the appearance of the "just-now," and thus, by
implication, a *proteraesthesis*, which, however, the assumptions
governing his thinking render impossible. This insistence on how
the "earlier-than" (*das Frühere*; πρότερος) appears, that is, in what
kind of sensation (*Empfindung*; αἴσθησις) is the "earlier-than" given,
allows Brentano to align the problem of time towards the givenness of
*absence* in the specific form of "earlier-than" or "just-past." In this
manner, Brentano's reservations against Aristotle already delineate a
basic orientation towards the givenness of the past that will guide
Husserl's phenomenological reflections.[29]

In an aside, Brentano notes that, "the appearance of succession in a
melody, or speech, etc., is nowhere discussed" in Aristotle's discussion
of time (RZK, 148). This remark – introduced without further
commentary – can be taken as shorthand for the basic thrust of
Brentano's critique of Aristotle's definition of time. Ever since
Augustine's appeal to reciting a line of poetry in the *Confessions*, music
and the spoken word have often served as preferred examples
for temporal succession. As Henri Bergson explains, music provides
"la plus pure impression de succession que nous puissions avoir."
In contrast to spatial displacement in which "il y a distinction nette
de parties extérieures les une aux autres," with temporal succession,
"la vérité est qu'il n'y ni un substratum rigide immuable ni des états
distincts qui y passent comme des acteurs sur une scène."[30] As
opposed to a random set of sounds, a melody is heard as an organization of sounds in time and, in this regard, is based on the consciousness of a distinct temporal order. Each note does not instantaneously
disappear from the aural grasp of consciousness once a subsequent
note enters the scene of hearing. With the arrival of a new note,
previous note(s) glide into the immediate past and cede their place,
yet are still heard in relation to the new note in the now, but in a
modified way, as earlier than the note heard in the now. To hear
musical form, we must still hear earlier notes in relation to the note

---

29 Brentano never entertains the idea of an "hysteroaesthesis" as a complement to his
   proposed "proteraesthesis."
30 Henri Bergson, "Le perception du changement," in *Oeuvres*, ed. A. Robinet (Paris:
   PUF, 1959), 1384.

currently heard. Every note is a tone-interval, a relation of "pure in-betweenness"; it is through the constant play of grasping the "in-betweenness" of notes that we hear musical structure.[31] If each of the notes in a melodic form were heard simultaneously, it would prove impossible to discern musical form in the articulation of notes as earlier-than and later-than. Instead of a melodic form, we would hear a chord, a tone cluster or simply noise. In hearing a melody, consciousness must accomplish what at first glance appears paradoxical: it must at the same time hold apart a succession of tones yet grasp this succession of tones as a present whole, as belonging to a melodic unity that is irreducible to any single note or all of the notes taken singularly.

Brentano's reference to the "appearance of succession in melody" reinforces his critique that the temporal (*zeitlich*) differentiation of before and after cannot be derived from a difference between prior and posterior states of movement. For Aristotle, temporal succession is not primary, but is derived from the prior and posterior in movement, which, in turn, is based on the prior and posterior of magnitude. The definition of time as the number of movement thus presupposes a temporal horizon against which the states of movement can at all be recognized as before and after *in time*. Aristotle's definition of time (*Zeit*) presupposes what Brentano identifies as temporality (*Zeitlichkeit*), as exemplified in the perceptual experience of a melody that does not follow any movement other than its own intrinsic temporal unfolding.

### The train of time and consciousness

This Aristotelian failure to uncover the origin of the concept of time in perceptual experience becomes partly resolved with Locke, whom Brentano recognizes as *the* decisive figure in the history of the problem of time. In the *Essay on Human Understanding*, the traditional formulation of the question "what is time?" is recast into the question of the origin of the concept or presentation (Brentano switches between *Begriff* and *Vorstellung* as translations of Locke's "idea"). As Brentano observes, "the conceptual determination of time is brought [*with Locke*] into close connection with the question of the origin of this concept" (RZK, 67). This significant transformation of approach

---

31 Roger Scruton, *The Aesthetics of Music* (Oxford: Clarendon Press, 1997), 21.

to the question of time goes hand in hand with a methodological reliance on a reflective analysis of our consciousness of time. As Brentano argues: "In so far as Locke looks for the clarification of the concept of time in certain intuitions [*gewisse Anschauungen*], in which the concept of time appears concretely, he announces an entirely correct methodological conviction" (RZK, 68).

Brentano's endorsement of Locke's "correct" methodological approach entails an implicit endorsement of Locke's "correct" formulation of the problem of time that matches Shaun Gallagher's recent designation of the "Lockean, phenomenological paradigm" and its paramount significance for a clear statement of what is problematic about the problem of time. Locke's introspective description of "certain intuitions in which the concept of time appears concretely" is inseparable from a general characterization of consciousness as a "train of ideas" or flow of consciousness. This to an unspecified degree metaphorical characterization of consciousness describes the manner in which consciousness possesses an intrinsic temporal form. As Gallagher notes, "the concept of the succession of ideas or flow of consciousness introduces time itself into the very heart of subjectivity."[32] But this introduction of time itself into consciousness is as much the introduction of consciousness into time; indeed, the demand to render time itself visible mutually entails the rendering visible of consciousness itself. Rather than look to the nature of beings in time as we find in Aristotle, Locke institutes a Copernican Revolution of his own in turning inwards to consciousness itself – to the consciousness of time in the time, or train, of consciousness. As Gallagher observes, the "radicality of this notion [*consciousness as a stream*] may be measured by the fact that philosophers have been able to explicate two problems that, prior to Locke's analysis, were never clearly defined as problems."[33] Of the two problems identified by Gallagher, Brentano explicitly addresses the first in his critique of Locke and comes to address the second, in connection with the first, in his efforts to understand the origin of time, albeit not in the same terms as either Locke or Gallagher.

Locke's characterization of consciousness as a train of ideas allows for a perspicuous articulation of what is entailed in the perception of temporal succession. As soon as consciousness becomes inherently

32 Shaun Gallagher, *The Inordinance of Time* (Evanston: Northwestern, 1998), 8.
33 Gallagher, 9.

defined through the notion of succession, as a train of succeeding
ideas, it becomes evident that the consciousness of succession cannot
be accounted for merely on the basis of the succession of consciousness itself. As repeatedly argued by any number of Brentano's contemporaries, each inspired by Locke's seminal insight, successive objects
(e.g., notes in a melody) can only be represented as successive if
apprehended simultaneously within a single act of consciousness;
the consciousness of succession is not the same as the succession *of*
consciousness. Yet, this insight generates a "cognitive paradox," to
invoke Gallagher's term, that defined the landscape of debate for
Brentano and his contemporaries at the end of the nineteenth
century.[34]

The paradox inherent in the proposition that the succession of
consciousness is not equivalent to the consciousness of succession
comprises two related issues. The consciousness of succession
depends on a momentary act of consciousness in which successive
notes in a melody, for example, are brought together and united into
the representation, or consciousness, of succession. The paradox here
is that either the consciousness of succession is based on a momentary,
and thus temporally not-distended act of consciousness (in which case
the consciousness of time is itself not temporal), or the apprehending
act of consciousness is temporally distended, in which case, one runs
the risk of equivocating the succession of consciousness with the
consciousness of succession, not to mention the further problem that
a succession of consciousness would imply *another* act of consciousness, in which succession is apprehended, which would in turn require
a further act, and so on . . . Moreover, if consciousness must represent
successive objects *at the same time,* this implies that objects of the
immediate past must somehow still remain, or persist, in the presence
of consciousness; but how does consciousness retain the past as past,
yet in the present?

This first problem of "objective synthesis" (Gallagher's term) is
connected to a second problem, namely, that of personal identity,
or, in other words, the unity of consciousness itself and its own
"subjective synthesis" of constituting itself a unified train or flow of
"ideas." In Locke's classic formulation, "we must consider what a
Person stands for; which, I think, is a thinking intelligent Being, that

---

34 Cf. Milic Capek, *The New Aspects of Time: Its Continuity and Novelties* (Dordrecht: Kluwer,
     1991).

has reason and reflection, and can consider it self as it self, as the same thinking thing in different times and places."[35] Consciousness is inseparable from reflection, and thus, from its self-manifestation in the form of self-consciousness: to be conscious is to be oneself conscious. An experience can be ascribed as mine to the degree that I can be aware of myself at the same time as that experience. As Locke argues: "When we see, hear, smell, taste, feel, meditate, or will anything, we know that we do so. Thus it is always as to our present sensations and perceptions: and by this every one is to himself that which he calls *self*."[36] Simultaneous experiences belong to me on the condition that I am reflectively aware of myself as now seeing, hearing, touching, etc. Reflective self-awareness can, moreover, extend to experiences of the past; the reach of remembrance extends the reach of my personal identity. Locke's theory of personal identity and self-consciousness are not, as noted above, the immediate concern of Brentano's critique; we shall shortly recognize how Brentano combines the consciousness of time with self-consciousness in a manner that brings together the "objective synthesis" of temporal succession with the unification of (self-)consciousness.

Locke begins his discussion of time in the *Essay* by referring to "the Answer of a great Man," St. Augustine, whose conclusion "might perswade [*sic*] one, that Time, which reveals all other things, is it self not be discovered."[37] Locke does not develop further this critical view of Augustine, but proposes instead to reveal the "abstruse" nature of "duration, time, and eternity" by tracing each of these concepts to "their Originals" in either sensation or reflection. Locke distinguishes between duration (or succession; Locke does not strongly distinguish between duration and succession) and time (or an interval of duration); the now is the unit of duration's measure. According to Locke, the idea of duration is a simple mode, in other words, a complex, yet unified idea containing different simple ideas. A simple mode is a "variation of a same idea," an idea modified through its own repetition (the same idea "added together"), as opposed to a mixed mode in which different ideas are compounded together.[38] In contrast to simple ideas produced in the mind by either sensation or reflection, simple modes are not passively received; "out of its simple Ideas, as the

---

35 John Locke, *An Essay Concerning Human Understanding* (Oxford: Clarendon Press, 1975), 335.
36 *Essay*, 335.     37 *Essay*, 181.     38 *Essay*, 163ff.

Materials and Foundations of the rest [*of ideas in the mind*]," the mind
fashions complex ideas through the exercise of its own powers. As a
simple mode, Locke therefore denies that duration is a relation, a
setting together of two ideas, or a simple sensation. The idea of
duration is an aggregate unity that has its source in a combination
of sensation and reflection.

Through introspection, we discover a constant train of fleeting
ideas on the basis of which our mind forms the idea of duration,
defined as "another sort of distance," from the "fleeting and perpetually perishing parts of succession," provided, however, that we remain
awake and attentive to the succession of each idea. As Locke remarks:
"That we have our notion of *Succession and Duration* from this Original,
viz. from Reflection on the train of *Ideas*, which we find to appear one
after another in our Minds, seems plain to me, in that we have not
perception of *Duration*, but by considering the train of *Ideas*, that take
their turns in our Understanding."[39] Locke thus clearly recognizes
that the succession of ideas is not equivalent to the idea of succession;
the consciousness of succession emerges from a *reflection* on the train
of ideas. I am aware of temporal succession only to the extent that
I am conscious of myself as a train of ideas; it is as if reflection
establishes a vantage-point over the succession of ideas, much like
an observer on the bank of a flowing stream. In this manner, an act
of reflection must bring together, or "compound," numerically discrete ideas into the simple mode of the idea of duration. The idea of
duration thus arises not from "the Observation of Motion by our
Senses," but from observation of the motion or flow of ideas in the
mind that is independent from the duration of motion.[40]

In his brief remarks, Brentano takes issue with Locke's account.
According to Locke, the idea of duration is based on the perception of
successively appearing discrete ideas (let us take the first two notes –
E and D# – of Beethoven's *Für Elise*) as well as the apprehension that
E occurred before D#. Consciousness of succession requires not only
that I perceive distinctly both E and D#. I must observe E as earlier
than D#. Yet, I can only perceive D# as following E on the basis of
retaining E as earlier than D# – but this already presupposes apprehending E as earlier than D#, and thus, already presupposes the idea
of duration. As Brentano notes, I cannot form the idea of succession

from the succession of ideas without tacitly presupposing duration, i.e., the interval between E and D#. Locke, however, speaks of "retention" as that "Faculty of the Mind" by which simple ideas, of either sensation or reflection, are kept before the mind and, moreover, distinguishes between this kind of retention from the "other way of retention" by which ideas, "which after imprinting have disappeared," are "revived again in our Minds." Yet, the thrust of Brentano's critique is less the issue of Locke's tacit assumption of duration (which an appeal to retention could account for) as it is the issue that merely retaining ideas of the past in the present is still insufficient to account for the consciousness of succession.[41] This point is trenchantly formulated by Husserl in his own critical discussion of Brentano: "If, in the case of a succession of tones, the earlier tones were to be preserved *just as they had been* while at the same time new tones were to sound again and again, we would have a simultaneous sum of tones in our representation but not a succession of tones" (Hua X, 12 [12] emphasis mine). What is required is not simply that past tones are retained in the consciousness of the present; those tones must be retained in a *modified* form of givenness; the mere retention of past tones delivers a "sum" of tones, but not the temporal relation of succession. In Brentano's terminology, Locke, much as Aristotle, fails to recognize the need for a *proteraesthesis*; but where Aristotle's failure is complete, Locke fails to grasp that the "retention" of an idea entails a modification of that idea such that it is specifically apprehended *as past*, and not, *as still present*.

## Temporality and sensibility

Aristotle and Locke each originate two different approaches to the problem of time that shaped the history of the concept of time. For the sake of brevity, whereas Aristotle orients his reflection on time towards *nature*, with a clear accent on what it is for beings to be in time, Locke orients his reflection towards *subjectivity*, in its essential manifestation as *consciousness*, with a clear accent on what it is for consciousness itself to be in and of time. From Brentano's point of view, Aristotle and Locke each assume the origin of temporal differences (the temporal difference and distance of earlier-than

41 *Essay*, 149.

and later-than) in the intuitive apprehension of temporal succession, which depends on understanding how consciousness can intuitively apprehend – perceive – the "just-past," or "earlier-than," along with the apprehension of the now. The question posed to Aristotle of how does the "earlier-than" *appear concretely* is equally posed to Locke: in both instances, the question of time's presence falters on the question of time's passing away. Before, however, we move directly to Brentano's answer to the problem of time in his account of *proter-aesthesis*, in the form of an "original association," we first need to locate its place within the framework of Brentano's descriptive psychology.

Brentano's Aristotelian-inspired discovery of a defining intentional correlation between objects of experience (physical phenomena) and acts of consciousness (mental phenomena) propelled a theoretical program of investigation in which diverse manifestations of consciousness could be studied descriptively. As Brentano remarks in his 1888/89 lectures "Descriptive Psychology," mental phenomena can be described from two complementary points of view:

> When we compare different mental activities along with their correlations, we find that differences exist among mental activities: either with regard to the object to which they refer or with regard to the manner in which they refer to an object, in which case differences can be more or less fundamental, often allowing for subordinating view-points to emerge. (DP, 131)

More than simply a psychological thesis, the characterization of consciousness as an intentional relation provides the guiding clue for an ontological determination of consciousness and its unique "texture" or "composition" (*Verwicklung* or *Verwebung*). This emphasis on the ontological tenor of Brentano's definition of mental phenomenon allows for a greater appreciation of the degree to which a "psychology without a soul" challenges the assumption that "the soul [is] something strictly unitary and completely simple" (DP, 10). In rejecting a metaphysical conception of the soul as a simple and indivisible substance, Brentano proposes that his descriptive psychology remains true to what he terms the "common man's view" that consciousness is a "bundle of ideas." Yet, Brentano is quick to reject Hume's construal of consciousness as a "bundle of ideas," which misleadingly suggests a "bond" or "string" that would hold together the "ebb and flow" of presentations (impressions or ideas, in the case of Hume), each given in a discrete and atomistic manner. Brentano's reservation against Hume (and by extension also Locke, despite their differences) recalls directly William James' objection against "associationist Philosophy." As James remarks:

> The chain of distinct existences into which Hume thus chopped up our "stream" was adopted by all of his successors as a complete inventory of the facts . . . Somehow, out of "ideas," each separate, each ignorant of its mates, but sticking together and calling each other up according to certain laws, all the higher forms of consciousness were to be explained, and among them the consciousness of our personal identity.[42]

Brentano amplifies the temporal constitution of consciousness through his characterization of consciousness as a stream or flow, as opposed to the "atomistically" laden conception of "train" or "chain," and designates each individual presentation within the life of consciousness as a "wave" within a continuous stream. Consciousness is a multiplicity of presentations that do not appear "next to each other" (*nebeneinander*) in a spatial relationship, but rather, in a purely temporal form in which each event of mental life (each presentation, i.e., mental phenomenon) is entwined into the whole of an actual human consciousness (*das Ganze eines wirklichen menschlichen Bewußtseins*). In addition, consciousness is also defined as an origin, since all knowledge, whether directly or indirectly, is based on sensible representations (*sinnliche Vorstellungen*) or sensibility (*Sinnlichkeit*). Equal in weight to Brentano's intentional thesis and its ontological implication, the conception of consciousness as an origin emphasizes the fundamental connection between "sensibility" (*Sinnlichkeit*) and the "sense" (*Sinn*) of concepts; all concepts have their origin in sensible experience. Stated differently, all knowledge is based on "being-affected," or *givenness*, in the primordial form of sensibility.

Brentano's intentional thesis articulates in a new key the function of sensibility for the foundation of knowledge. Brentano pursued the ramifications of his intentional thesis for an analysis of sensibility only after the first edition of *Psychology from an Empirical Standpoint* (1874). Indeed, the *Psychology* presents an incomplete portrait of Brentano's psychology since, in addition to its unfinished publication, it omits an explicit treatment of sensibility, but also fails to distinguish clearly between sensible presentation and perception (which Brentano considers a form

---

42 William James, *Principles of Psychology* (Cambridge, MA: Harvard University Press, 1981), 334.

of judgment). Brentano's investigations into sensory and noetic forms of consciousness flowered during the 1880s in the context of his programmatic distinction between "genetic" and "descriptive" psychology.[43] Under the title of descriptive psychology, Brentano understands an analysis of the basic elements and structural connections of consciousness. A reflective method of description leads to the classification of the different "parts" and "relations" that structure consciousness, and, in this fashion, establishes "an ontology of mind."[44] In contrast, genetic psychology studies the physiological basis of mental phenomena as well as the conditions under which mental phenomena emerge (i.e., the origin of mental life in the physiology of the human organism).

Brentano divides presentations (*Vorstellungen*) into two kinds: fundamental or sensible presentations and "superimposed" presentations (judgments and affections of pain/pleasure) that are based on fundamental presentations in which an object is at all presented, or given, to consciousness. In the ontology of parts and wholes that shapes Brentano's psychological classification, these three kinds of presentations are defined as "separable [*ablösbare*] parts" of consciousness since any mental phenomenon can be separated from each other (i.e., the givenness of one kind does necessarily entail the givenness of another kind) as well as from consciousness as such, in so far as every presentation is an occasional event, a "wave" in the stream of consciousness. As with every kind of mental phenomenon, a sensible presentation is characterized as an intentional relation: "the sensing is real, the object of sensation is not-real."[45] A sensation is any aspect of an object's

---

43  Cf. Franz Brentano, *Meine letzten Wünschen für Österreich* (Stuttgart: Cotta, 1985). L. Gilson *La Psychologie descriptive selon Franz Brentano* (Paris: Vrin, 1995), 77: "Ainsi, quand on entre dans le detail des analyses, la *Psychologie* de 1874 paraît parfois juxtaposer purement et simplement ce qui relèvera plus tard de l'une et de l'autre branche de la psychologie. . . . Elle prepare cette distinction, elle ne la formule pas. De là son caractère à certains égards paradoxal: son exposé de la method paraît surtout viser à la constitution de ce qui s'appellera ensuite psychologie génétique; les autres etudes seraient des preliminaires necessaires, l'essentiel étant la determination des lois de succession des phénomènes, et pourtant les analyses descriptives y sont de beaucoup les plus nombreuses puisqu'elle doivent venir en premier et que l'oeuvre est demeurée inachevée."

44  Cf. Kevin Mulligan and Barry Smith, "Franz Brentano on the Ontology of Mind," *Philosophy and Phenomenological Research*, 45 (1985), 627–644.

45  Here is not the occasion to enter into Brentano's theory of intentional relation and its changing fortunes. Significant for our concerns is the distinction between the "sensing" of red and the sensed red-content as an immanent (hence: "not-real") content *or* object of consciousness.

sensible appearance that is immediately "experienced" (*erlebt*) and given in an intuition (*Anschauung*); due to its intentional character, a sensation is not a "simple state" or "feeling," but a "manifold of parts" (*eine Mehrheit von Teilen*) or a "plurality" (*Vielheit*) (DP, 135; 137 [144; 146]). Within this manifold of parts, Brentano further distinguishes two distinct, yet non-separable parts or "moments" belonging to sensible presentations: spatial location and quality. Under the aspect of quality, Brentano understands a relative degree of brightness or darkness as well as a relative degree of what Brentano terms *Sättigung* or *Ungesättigung*. Spatial location and quality are distinctive moments of sensation in the sense that they "contribute reciprocally, and interpenetrate, to the individualization of sensations" (DP, 17 [20]).

In addition to spatial location and quality, sensations also possess a temporal character; each sensation appears in a temporal continuum, or finite stretch of time (*Zeitstrecke*). Yet, the temporal character of a sensible presentation is not a distinctive part of sensation, but a modification of the manner in which a sensation is given ("in der Art verändert auftreten")[46] (DP, 16 [19]). Expressed in Brentano's terminology, a sensation's temporal duration is an "enriching attribute" (*bereichernde Bestimmung*) in contrast to the "determining attributes" (*determinierende Bestimmungen*) of spatial location and quality. Determining attributes pertain to the content of sensation whereas enriching attributes, or what Brentano also calls "modifying attributes" (*modifizierende Bestimmungen*), modify in a total manner the way in which a sensation is given. The temporal character of a sensation, for example, a musical note, as "just-past" does not change the content of the sensation: the note of middle-C remains as much middle-C through its duration. Moreover, Brentano argues that the temporal determinations of "earlier than" and "later than" cannot be considered as true predicates of an object since a sensation that is past cannot, by definition, be said to exist. *Zeitlichkeit* is neither a sensation

---

46 Cf. Brentano's rejection of Ernst Mach's argument in his *Beiträgen zur Analyse der Empfindungen* (and in *Erkenntnis und Irrtum*) that experience is composed of sensations from which complex objects are constructed. These sensations include color, sounds, etc., but also "spaces" and "times" (*Räumen* and *Zeiten*). Mach argues for the existence of "time-sensation" or "time-moment" (*Zeitempfindung* or *Zeitelement*) as the sensible ground for the intuition of time (*Zeitanschauung*). Brentano rejects this idea of a "time-sense" (*Zeitsinn*) and its "time-sensation" (*Zeitempfindung*). As Brentano remarks: "Eine Zeitempfindung kann es nicht geben" (*Über Ernst Machs "Erkenntnis und Irrtum*," ed. R. Chisholm and J. Marek [Amsterdam: Rodopi, 1988], 143).

nor a distinctive part of sensation; yet temporality is always given along with, and in this sense, follows, the actuality of sensation, since an actual sensation is given with a temporal duration. On the one hand, Brentano argues that there is no sensation of time; as Hume already recognized: "The idea of time is not derived from a particular impression mixed up with others, and plainly distinguishable from them, but arises altogether from the manner in which impressions appear to the mind . . ."[47] On the other hand, the idea of time is not, as Hume argued, based on the *succession* of impressions, since this presupposes the duration or "temporality" of each individual sensation.[48] The upshot for Brentano is that not every "sensible presentation" (*sinnliche Vorstellung*) is composed solely of a "sensation-presentation" (*Empfindungsvorstellung*). The concept of sensibility (*Sinnlichkeit*) is not exhausted by the concept of sensation (*Empfindung*); in the example of hearing a temporally extended sensation (e.g., a musical note), the duration of the note is composed of the actual sensed content as now as well as the sensed content as "just-past." The sensible presentation of the sensed content as "just-past," however, is itself not derived from, or based on, an actual sensation (*Empfindung*); yet this sensible presentation is likewise not itself given independently of an actual sensed content as now, to which it is attached, and in this sense "follows," much as time "follows" the mobile (the moved object) in Aristotle's account. This discovery of the origin of time in sensible presentation that is, on the one hand, not derived from a sensation, yet, on the other hand, not itself given without a sensation also informs Brentano's consideration of self-consciousness or "inner consciousness," to which we shall now turn as the final implement in our approach to the theory of original association or *proteraesthesis.*

### Inner or self-consciousness

Brentano's definition of a mental phenomenon as an intentional relation remains incomplete without an examination of the accompanying thesis that every act of consciousness is also implicitly conscious of itself, but only to the extent that it is a consciousness of something other than

47 David Hume, *Treatise of Human Nature* (Oxford: Clarendon Press, 1965), 65.
48 Hume, 65: "[*Time*], since it appears not as a primary distinct impression, can plainly be nothing but different ideas, or impressions . . . disposed in a certain manner, that is, succeeding each other."

itself. Brentano's understanding of self-consciousness (in Brentano's terminology: inner consciousness) influenced Husserl's phenomeno-logical treatment of self-consciousness to a comparable degree as to his theory of original association, and not without reason: in Brentano's psychology, time-consciousness and self-consciousness are intrinsically connected through their shared reference to sensibility. Against the position of self-consciousness in Kant and German Idealism, Brentano situates self-consciousness within sensibility, or sensible presentations, which, as noted above, constitute the class of fundamental presenta-tions. By contrast, Kant does not consider self-consciousness to be *directly* connected to time, even though Kant recognizes time as an apriori form of intuition; rather, self-consciousness is the basic principle of the understanding, in its synthetic capacity of judgment. Self-consciousness is "non-temporal" as well as "non-sensible." As Man-fred Frank notes, this Kantian inability to bring together time and self-consciousness ultimately points back to the two sources of knowledge in sensibility and the understanding: "Both have their apriori: Time (but of course also space) is the apriori of sensibility; the apriori of the understanding is the capacity of judgment that springs forth from the categories."[49] Among the numerous ways in which Brentano's descrip-tive psychology is a direct response to, and rejection of, German Idealism, inner consciousness (which should not be conflated with Kantian inner sense) is situated within sensibility, not, however, as its apriori form, but as a feature inherent to the intentional structure of sensible presentation as a mental phenomenon. For Brentano, self-consciousness is not derived from a sensation – it is not an *Empfindungsvorstellung*), yet it is nonetheless manifest as a sensible presentation (*sinnliche Vorstellung*), much as the temporal character of sensible givenness is not derived from a sensation, yet equally not manifest in an absence of sensation.

Every mental phenomenon, as an intentional relation, is com-posed of two distinct presentations: the presentation of the sound (the immanent content or object) and the presentation of the act itself (the perceptual act of hearing), both of which Brentano char-acterizes as perceptions, external and internal. As Brentano observes, "along with the presentation of a sound, we have a presentation of the presentation of this sound at the same time" (PE, 171 [121]).

---

49 Manfred Frank, *Zeitbewußtsein* (Tübingen: Neske Pfullingen, 1990), 7.

When taken at face value, if every intentional act of consciousness is itself the "object" of consciousness (i.e., perceived internally), an infinite multiplication, or what Brentano calls an "infinite complication," of mental acts would seem unavoidable. If every act of consciousness contains an intentional object other than itself yet also relates to itself, and thus takes itself as its own "object," would this not require a separate act of consciousness, namely, another act of consciousness through which consciousness becomes for itself the object of its own consciousness? Notoriously, this complication in any "reflection theory" of self-consciousness leads either to an infinite regress or to the positing of an unconscious act – a door-stopping mental act that lacks its own consciousness. Brentano, however, rejects both alternatives on the perceived conceptual advantage of an essential ramification of his intentional definition of mental phenomena.

According to Brentano's diagnosis, defusing the paradox of a reflection theory of self-consciousness turns on the issue of whether an act of consciousness is ontologically simple – composed of one single presentation – or ontologically complex – composed of several presentations as a unified whole. In other words, does self-consciousness divide consciousness from itself, and in this fashion, provoke an infinite regress (as with a reflection theory and its subject–object relation) or does self-consciousness coalesce along with a presentation as a whole?

Brentano argues that the threat of an infinite regress depends on the assumption that only a one-to-one relationship obtains between an act of consciousness and its intentional object; that for each object of consciousness, there corresponds one distinct act of consciousness. If the perceptual act of hearing a sound is itself the "object" of consciousness, this would require, on this assumption, another distinct act, other than the original act of hearing the sound, and so on, *ad infinitum*. It is precisely this assumption, however, that is challenged by Brentano's intentional characterization of mental phenomenon. In questioning the assumption that an act is defined by a relationship to a single object, Brentano considers that an act of consciousness may have different objects *simultaneously*. As an example, Brentano argues that "the presentation of the sound and the presentation of the presentation of the sound (i.e. hearing) form a single mental phenomenon; it is only by considering it in its relation to two different objects, one of which is a physical phenomenon and the other a mental phenomenon, that we divide it conceptually into two presentations" (PE, 179 [127]). On this view, the division between the *object* of consciousness (physical phenomenon) and the *act* of consciousness (mental phenomenon), as two objects, is a *conceptual* distinction introduced through an act of reflection which divides the primordial and pre-reflexive unity of a mental phenomenon (as a *single* presentation) into "primary" and "secondary" objects of consciousness. The sound is the *primary object* of the act, whereas the act of hearing is a *secondary object*, not, however, in the sense of being a second primary object, but in a secondary or incidental sense of taking itself "on the side," or in Aristotle's language, as a perception *en parergo*.

This forked structure of a mental phenomenon attests to its specific temporal form: "in the same mental phenomenon in which the sound is present to our minds we *simultaneously* apprehend the mental phenomenon itself" (PE, 179 [127] my emphasis). The simultaneity of dual apprehension is here stressed in order to underline a basic decision regarding the temporal constitution of an act of consciousness. It is the simultaneity of the primary and secondary objects that structurally prevents an unwarranted multiplication of acts. Both presentations (primary and secondary) are necessarily contained in the unity of *a* consciousness, in the same indivisible time of the act. And yet the sound is nevertheless primary since "a presentation of the sound without a presentation of the act of hearing would not be inconceivable, at least *a priori*, but a presentation of the act of hearing without a presentation of the sound would be an obvious contradiction" (PE, 179 [128]). Brentano clearly reveals his Aristotelian heritage; by allowing for the possibility that a primary object could be presented *without* a presentation of a secondary object (a possibility that Husserl will dismiss in his own investigations), Brentano accepts that inner or self-consciousness is not a condition of possibility for the presentation of a primary object. Self-consciousness is given only to the extent that consciousness relates to an object other than itself; consciousness is only actualized in so far as a primary object dwells within it. Self-consciousness is a pre-reflexive "self-feeling" or affection, in other words, an original self-manifestation of consciousness for itself. This pre-reflexive self-consciousness can subsequently become an explicit theme of reflection through an act of self-observation by which the secondary object of an *earlier* presentation is turned into a primary object of reflective consciousness.[50] To a significant degree, the "ego" for Brentano is a "thing" (*ein Ding*), a transcendent object of consciousness, as opposed to its pre-reflexive (self-)consciousness. Importantly, Brentano does not consider incidental self-consciousness as a relation of identification nor as an act of introspection but rather as an accidental modification of substance, even if inner consciousness enjoys an epistemological seniority that remains alien to Aristotelian psychology. Expressed in Brentano's Scholastic terminology, self-consciousness is an accidental and modifying attribute of the substance of the soul that is actualized in intentional consciousness.[51]

A secondary object can never at the same time serve as a primary object (that is, I cannot observe myself observing in the same act) since this act of consciousness would have to be observed by means of a second, simultaneous act of (self-)consciousness – it is impossible for there to be two different acts of consciousness at the same time within the same unity of consciousness, given Brentano's adherence to an Aristotelian view that potentiality can only become *one* actuality at a time. Brentano dismisses an inner consciousness of a secondary object given the impossibility of consciousness to be actual *at the same time* in two separate acts (a condition that is assumed if we argue

50 In *Psychology from an Empirical Standpoint*, Brentano has yet to formulate a distinction between explicit and implicit self-awareness, which he develops in his lectures on descriptive psychology during the 1880s.

51 A more extensive analysis of Brentano's psychology would have to include a discussion of Brentano's metaphysics and his novel interpretation of the relationship between substance and accident. Let us note briefly, however, that according to Brentano, an accidental property "contains" the substance in which it inheres. In Brentano's writings, the German "angehören" translates Aristotle's *hyparchein*, which in turn was translated into the Latin *inesse*. Every intentional act of consciousness is an accidental modification of the substance of the soul. Inner perception suggests indirectly that consciousness is of a substance distinct from the life of intentional accidents. God is a substance without accidents. An intentional accident (e.g., any intentional act of consciousness) is not a necessary feature of the soul since the soul endures even in the absence of any actualized state of consciousness (e.g., in sleep). The absence of any intentional activity does not terminate the subsistence of the soul, provided that the soul does not succumb to a substantial change (i.e., death). But even though accidents belong to the substance of the soul (as contingent modifications or attributes), Brentano considers these accidents as a modal enrichment (*eine Bereicherung*) of substance, as a whole that embraces the substance in a modal fashion (*modalumfassendes Ganzes* or *modal befassendes Ganzes*) (Franz Brentano, *Kategorienlehre* [Leipzig: Meiner, 1933], 269; 272) For a discussion of Brentano's view of substance and accident, see Enrico Berti, "Brentano and Aristotle's Metaphysics," in *Whose Aristotle? Whose Aristotelianism?* ed. R. W. Sharples (Aldershot: Ashgate, 2001), 135–149.

that a second and separate act is required to account for how the presentation of hearing is itself the "object" of consciousness).[52] A secondary object does not imply a second simultaneous act, since it is a second simultaneous object of one single and temporally indivisible, yet forked act. An act of consciousness does not possess duration; it is not divisible into parts that exist separately in time. Difficulties remain with Brentano's innovative account of self-consciousness; these difficulties notwithstanding, Brentano points in the direction of rethinking the unity in multiplicity of an act of intentional consciousness – "die eigentümliche Verschmelzung" of consciousness – that would influence and haunt Husserl's own struggles with self-consciousness and time-consciousness.[53]

### Original association or *proteraesthesis*

Under the heading of original association or *proteraesthesis*, Brentano identified the origin of the concept of time in what he termed *Zeitlichkeit*, which refers to the temporal extension of sensation: sensation

---

52 Franz Brentano, *Über Aristoteles*, ed. R. George (Hamburg: Felix Meiner, 1986), 279 ff. Brentano also addresses the following concern: "When a person is aware of seeing and hearing, he is also aware that he is doing both at the same time. Now if we find the perception of seeing in one thing and the perception of hearing in another, in which of these things do we find the perception of their simultaneity? Obviously, in neither of them. It is clear, rather, that the inner cognition of one and the inner cognition of the other must belong to the same real unity" (PE, 226 [160]).

53 Among more recent criticisms, Dan Zahavi (*Self-Awareness and Alterity: A Phenomenological Investigation* [Evanston: Northwestern, 1999]) has lucidly argued for the limitations of Brentano's theory. In Zahavi's assessment: "But although his [*Brentano's*] account of how this self-awareness is to be explained avoids the problem confronting the version of the reflection theory which takes reflection to be a relation between two different acts, his own proposal is, as Cramer and Pothast have shown, faced with an equally disastrous problem. An act which has a tone as its primary object is to be conscious by having itself as its secondary object. But if the latter is really to result in self-awareness, it has to comprise the entire act, and only the part of it which is conscious of the tone. That is, the secondary object of the perception should not merely be the perception of the tone, but the perception which is aware of both the tone and of itself" (30). Yet, Zahavi's demand that the having of itself as a secondary object must "comprise the entire act" is in fact what Brentano argues, since, as an accidental modification of substance – as a *modifying atttribute* – the secondary object modifies the entire intentional relation; this is precisely what Brentano means when he writes that "the presentation of the sound and the presentation of the presentation of the sound (i.e. hearing) form *a single mental phenomenon*; it is only by considering it in its relation to two different objects, one of which is a physical phenomenon and the other a mental phenomenon, that we divide it conceptually into two presentations" (PE, 179 [127] my emphasis).

appears, or is given, in a finite duration, or time-segment (*Zeitstrecke*). Along with the actual content of presentation, earlier sensed content (e.g., earlier notes in a melody) are still contained within the grasp of consciousness, albeit in a modified manner. A note as just-past is retained in a modified manner since it is no longer actually given, whereas the note actually heard as now is given in an unmodified manner. This modified content of what has just been perceived attaches itself by way of an original association (or *proteraesthesis*) to the perception (*aesthesis*) of the actual content in the now, and thereby produces, or adds, the dimension of the past (*Vergangenheit*) to the experienced dimension of the present. As Husserl reports, Brentano proposes a "general psychological law" according to which "a continuous series of presentations is fastened by nature to every given presentation such that each presentation belonging to this series reproduces the content of the one preceding it, but in such a way that it always affixes the moment of the past to the new presentation" (Hua X, 11 [12]).

The original association between the modified content of the "earlier than sensation" and the unmodified sensation of the present is cast in the mold of a relation between a primary and secondary object; in other words, the structural form of an intentional relation (primary object and secondary object) is applied to the relationship between the sensible presentation of content in the now (*aesthesis*) and the sensible presentation of an "earlier than" content.[54] For example, an actual note in the now is a primary object of consciousness, whereas an "earlier than" or "just-past" note is a secondary object. The secondary object (the earlier note) is attached to the primary object by way of an original association. This association is "original" in two senses: it is an association that is produced or created, and thus not derived from an actual sensation; the associated content is *reproduced*, since past content cannot in any manner actually exist. It is therefore not an association of two contents, each of which is actually present, and consequently not a species of what Brentano calls "association of ideas" (*Ideenassociation*), which describes the brand of association in Hume. Instead, what is distinctive of original association is that one of the associated elements, namely, the secondary object, no

---

54 Here, as elsewhere, the terms "content" and "object" are interchangeable, for the sake of discussion, but also in keeping with Brentano's own ambivalence at the time of his experimentation with the idea of original association.

longer exists as a real sensation; yet it – the sensation just-past – is nevertheless still sensed as earlier than (i.e., associated with) the actual sensation in the now so as to produce the consciousness of duration, i.e., of a sensation as temporally extended. This secondary object is not the object as past, but rather my having once sensed the object as present; in this fashion, *proteraesthesis* (or original associ-ation) is inseparable from inner consciousness. I cannot be aware of an immediately elapsed note without an awareness of myself as having just perceived the note as present. The "fusion" (*Verschmelzung*) of primary and secondary objects in an original association is not an additive relation nor is it based on an act of reflection (as is the case with Locke); the unity of present and past, of *aesthesis* and *proteraesth-esis*, is an original unity that cannot be reduced to its parts.

In order to illustrate this mechanism of original association, Brentano devised a diagram that would serve as the basic template for Husserl's own time-diagrams. A horizontal line represents the succession of "nows"; vertical lines, each placed at ninety-degree angles at each "now-point," represent the consciousness of succession, i.e., the unity of a primary object (*aesthesis*) and a secondary object (*proteraesthesis*) (Figure 1).[55] As a sequence of four notes in a melody unfolds (A, B, C, D), one and only one note is actually the content of perception in the now (i.e., a primary object). At point B along the horizontal line of successive notes, consciousness apprehends note A as earlier than B while at the same time perceiving note B as now; I hear note B as following note A. With each successive note along the horizontal axis, earlier notes are continually pushed back, increasing their relative distance to the new now, and further "stretching" the durational unity of the melody as a whole. Whereas the horizontal line represents the succession of consciousness, the vertical line represents the consciousness of succession, ever encompassing, in this example, the whole melody of notes with the progression of each note. Along each vertical line, earlier sensations are reproduced, and so "added" to the new now, thus further expanding a unified consciousness of succession. For example, at note D, the segment A-B is reproduced with the added (earlier) moment C, so as to form the row of previous tones in the original order of succession relative to the new

55 Stumpf presents Brentano's diagram in *Franz Brentano: Zur Kenntnis seines Lebens und seiner Lehre*, ed. O. Kraus (Munich: Oskar Beck, 1919), 136; For Husserl's version of Brentano's diagram, see Hua X, 399.

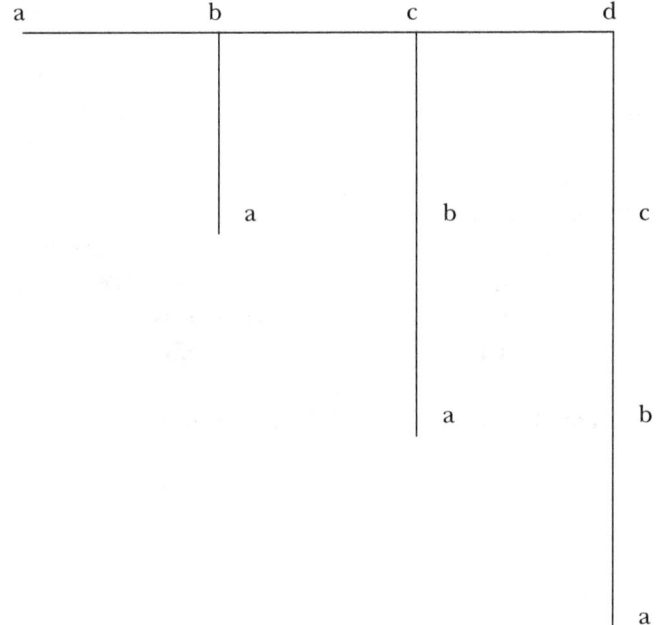

**Figure 1**

moment D. At the moment of hearing D, I am conscious of the ordered succession A-B-C-D. The reproduced notes along the vertical line do not have a relationship of "next to each other" (*Nebeneinander*); the objective succession of notes along the horizontal axis is neither mirrored nor stamped onto the vertical line or consciousness of succession. Whereas the notes along the horizontal line are "punctual" or "point-like," each vertical line represents a continuum of elapsing content (earlier notes) attached to each punctual now, much like the tails of a comet, as Husserl will say.

According to Brentano, the imagination (*Phantasie*) is responsible for this unique modification of content and the corresponding production of *Zeitlichkeit. Proteraesthesis* is a sensible presentation of the imagination. In contrast to both perception and memory, imagination is a form of sensible presentation that creates its content.[56] Both

---

56 At the beginning of his comments, Husserl mistakenly interprets Brentano's original association as a "memory presentation" (*Gedächtnisvorstellung*), even though he continues by correctly discussing original association in terms of the imagination.

memory and perception are non-productive forms of givenness: memory re-produces or re-calls past objects in the present whereas perception receives objects in the present. The service of the imagination, however, is to create the appearance of an appearance: the modified content appears as still given, as having just been, even though it is no longer. Yet, the imagination is not productive in the sense that it reproduces the original content of a perception in the form of an image, by way of a decaying of sense or a slackening of intensity. Brentano rejects Hume's association model (imagination as decaying sensation) as well as elements of Aristotle's understanding of imagination as a mental image. Instead, as Brentano writes, "the presentation of imagination contains, so to speak, an intuitive core, that is, it approximates the core of intuitions."[57] The reproduction of the imagination, on the one hand, contains an "intuitive core" of presence, yet, on the other hand, produces a form of subtraction by creating a distance – Locke's "perishing distance" – from the core of presence.[58] Reproducing loses its meaning of producing again by repetition of what has already been given and takes on the meaning of an original type of "de-presentification."

This "de-presentification" of sensation by way of the imagination results in what Brentano calls an "inauthentic presentation" (*uneigentliche Vorstellung*), since it is a sensible presentation emptied of intuitive givenness (*Anschaulichkeit*). And yet it is an empty presentation that still refers to intuitive givenness from a distance, no longer to be regained but only to be sensed as passing away. For these reasons, the presentation of the imagination (and its content, a "phantasm") is not a genuine presentation since it appears to present what is not, that is, no longer, actually given. Imagination is the instigator of an original association that produces an "irreal" or "quasi-sensation" of the earlier than; it is not a species of memory, which Brentano defines as an "inauthentic" (*uneigentliche*), yet "intuitive presentation" (*anschauliche*

---

57 Franz Brentano, *Grundzüge der Ästhetik*, ed. F. Mayer-Hillebrand (Hamburg: Felix Meiner, 1959), 84: "die Phantasievorstellungen enthalten sozusagen einen anschaulichen Kern, d.h., sie nähern sich den Anschauungen." Cf. Husserl's comment, "the imagination thus proves to be productive in a peculiar way here. This is the sole instance in which the imagination creates a truly new moment of representation, namely, the temporal moment [*das Zeitmoment*]" (Hua X, 11 [12]).

58 Chisholm, 12: "These attributes [*modifying attributes*] (rather than adding to the subject) subtract from the subject."

*Vorstellung*).[59] Even though the presentation of imagination is neither authentic nor intuitive, it nevertheless functions as an essential supplement to an authentic, intuitive presentation, since an *aesthesis* of the present and *proteraesthesis* of the immediate past form a single unity of consciousness. The past exists in the present as a ghost or "phantasm" of its former self. There is no perception of the present without an incidental grasp of the immediate past, just as every primary object is given along with a secondary object of inner consciousness. But likewise, just as it is only the primary object that is perceived, there is no perception of the immediate past, but only a perception of the momentary present. Only the point-like-now, for Brentano, is authentically perceived. By a curious turn, Brentano's *proteraesthesis* names what it cannot have: an *authentic* perceptual apprehension of the immediate past or "earlier than."

With every intentional relation, an original association is always attached to a primary object, since sensation is always given in a temporally extended manner. Is an original association also attached to self-consciousness, that is, to the secondary object that is always given along with the primary object of an intentional consciousness? According to Brentano, it is only the primary object that is given in a temporal extension and modified by an original association. Were a secondary object also temporally extended, this would lead to the "shocking implication" (*monströsen Annahme*) of a continuum of infinite dimensions, and promptly reanimate the threat of an infinite regress.[60] Not only would we stumble again upon the problem of an infinite regress, it also would suggest that my past self could become the object of inner consciousness. Granting temporal extension to inner consciousness would entail what Brentano refers to as a "double row" (*doppelte Reihe*) of temporality in which every primary object would have to be presented in a double fashion: directly in the mode of the immediate past and indirectly as the present object of a past consciousness. That is to say, both the sound and the act of hearing would have to be presented as just past. However, even though the sounds are perceived in temporal succession and benefit from the continuous original association of immediately past content and present content, the act of hearing, in Brentano's view, does *not* appear as a successive experience. In every moment of experience, I am *now*

59 *Grundzüge der Ästhetik*, 85.
60 The expression is Carl Stumpf's in a letter of June 16, 1906 to Oskar Kraus (PE, lxxxix).

implicitly aware of hearing yet I am not aware of having been aware of having heard. As Brentano notes: "I do not appear to myself as someone who earlier sensed something as present, but rather, as someone who now senses something as past [*erscheine ich mir nicht als einer, der früher etwas als gegenwärtig empfunden hat, sondern als einer, der etwas jetzt als vergangen empfindet*]."[61] My past self can never be the object of an inner consciousness. It is only with an act of remembrance, by which the incidental self-consciousness of a previous act is raised to the visibility of a primary object that I can become aware of myself as having (just) heard the (earlier) sounds. It is only because I am always implicitly aware of myself as now perceiving, that I am aware of perceiving things in the now, and as fading away.

## Husserl's critique

Husserl launches his critique of Brentano's original association with the laudatory comment that "psychologists until Brentano struggled in vain to locate the authentic source of this [*of time*] presentation" due to their "confounding of subjective and objective time, which misled the psychological investigators and completely prevented them from seeing the real problem before them" (Hua X, 11 [12]). As examined above, Brentano's analysis of original association avoids the conflation of subjective and objective time by arguing that *Zeitlichkeit* is neither an objective feature of sensation nor the subjective succession of presentations. When placed against the vignettes of Aristotle and Locke, Brentano's originality is first and foremost to have rejected the traditional terms in which the origin of the concept of time had been sought. Although every sensation is necessarily temporal in character (*zeitlich*), the duration of sensation is not reducible to a particular part of sensation, even though it is nonetheless a sensible presentation (*sinnliche Vorstellung*). Whereas the concept of color, for example, can be traced back to a distinctive part of sensation, as Brentano argued in his lectures on "Descriptive Psychology," *Zeitlichkeit* is not a distinctive part of sensation; instead, the temporal duration of sensation is the manner in which sensation at all appears in consciousness that necessarily "accompanies" or "follows" the actual givenness of sensation, much as self-consciousness is implicitly given

61 PE, Bd. III.

along with, and taken on the side of, the consciousness of an intentional object.[62]

In Husserl's view, the unity of primary and secondary objects of original association (the unity of *aesthesis* and *proteraesthesis*) delineates the phenomenological nucleus of time-consciousness and, in this manner, reveals the "real problem" that thus far has eluded philosophers and psychologists. Indeed, much of Husserl's attention in the earliest stages of his reflections will be devoted to fashioning a proper phenomenological description of the relation between an "original impression" and its "retentional modification," or "retention," to already invoke Husserl's own concepts. Husserl thus continues along a trajectory of investigation first broached by Brentano; indeed, if we recall the thrust of Brentano's critique of Aristotle, the "real problem" of time-consciousness resides in understanding how the now is a "double-limit," including both the appearance of the present along with the appearance of that which is no longer present. As Husserl remarks, "the unity of the consciousness that encompasses in an intentional manner what is present and what is past is a phenomenological datum" (Hua X, 16 [16]). But as we noted above, in a curious turn of thinking, despite the intention driving the term *proteraesthesis*, Brentano ascribes the consciousness of the immediate past to the "inauthentic" sensible presentation of the imagination (*Phantasie*), and thus refuses any perceptual character to the immediate past. Thus, even though Brentano *begins* to challenge the traditional coincidence of the present and perception, he remains at the end of the day faithful to Aristotle's definition that a sensation cannot reveal the past – even the immediate past; as Brentano notes, "a sensation that exists is something present."[63] The modified and reproduced sensation of *proteraesthesis* is a "phantasm" or ghostly presence, an "irreal" sensation, disembodied of any authentic givenness. Duration (*Zeitlichkeit*) is

---

62  "That we have our notion of succession and duration from this original, viz. from reflection on the train of ideas, which we find to appear one after another in our own minds, seems plain to me, in that we have no perception of duration but by considering the train of ideas that take their turns in our understanding" (*Essay*, 239). Husserl must have Locke in mind when he writes, "Naturally we must make precisely the same objection against those who wish to trace the representation of duration and succession back to the fact of the duration and succession of psychic acts" (Hua X, 12 [12–13]).

63  Franz Brentano, *Über Ernst Machs "Erkenntnis und Irrtum,"* 151: "Eine Empfindung, die ist, ist ja etwas Gegenwärtiges." Cf. Aristotle, *De memoria*, 449b14.

not perceived; only the momentary now is "authentically" perceived – a now that, following Aristotle, is not a "part" of time, but rather a division or limit in an "irreal" temporal continuum of past and future nows.

As a first difficulty, Husserl takes issue with Brentano's characterization of the immediate past as imaginary or "irreal." Is Brentano able to distinguish between the past as remembered and the past as reproduced in an original association? What differentiates the "irreality" of the immediate past from the "irreality" of the remote past? Brentano is indeed mindful of a difference between the imagination and memory; the former is an inauthentic presentation lacking any intuitive givenness (*Anschaulichkeit*), whereas the latter is an inauthentic presentation with intuitive givenness. Nevertheless, the problem remains that any *temporal* difference between a remembered past and an imagined past becomes erased. The consciousness of the immediate past slips through the distinction between the imagination and memory; both share the *same* defining boundary against the "real" presence of the now. On the one hand, Brentano's designation of the immediate past as "irreal" is motivated by the recognition that the immediate past is no longer present, and thus, no longer contained within the actual now; but on the other hand, the "irreal" character of the immediate past amputates its *intuitive* continuity with the actual now. If the modified content (phantasm) of *proteraesthesis* is "irreal," yet combined in an intentional *unity* with an actual sensation, in what sense can this mixture of "irreal" and "real" give rise to the *real* continuity of temporal passage? Husserl suggests that Brentano's failure to distinguish between the content of apprehension, the objectifying act, and the intentional object, is the source for his inability to make descriptive sense of how the past *transcends* the present. As Husserl trenchantly formulates this issue: "The being-present of an A in consciousness through the annexation of a new moment, even if we call that new moment the moment of the past, is incapable of explaining the transcending consciousness: A is past" (Hua X, 18 [19]). This transcendence of the past vis-à-vis the present cannot be interpreted as a form of memory. Although memory does indicate the transcendence of the past vis-à-vis the present, memory already presupposes (as Brentano's analysis makes apparent) the original temporality of the unity of (immediate) past and present. An event must have elapsed in order to be remembered as past; whereas memory is based on a discontinuity between the past and the present, the sinking-away of the present is continuous with the now. Brentano thus falls short of explaining what becomes central to Husserl's own analysis, namely, the transcending character of the immediate past and its intentional character. As Husserl asks: "Where do we get the idea of the past" (Hua X, 18 [19])? The phenomenon of "de-presentification" at the core of Brentano's insight is betrayed by its interpretation as the product of the imagination.

This failure to clarify the transcendence of the immediate past vis-à-vis the present is due to Brentano's claim that the content of the immediate past (i.e., the secondary object) is contained within the immanence of consciousness, i.e., an immanent content that is itself present, or contained, *in* consciousness. As presented earlier, the primary and secondary objects of time-consciousness – the now and the just-now – are encased within the simultaneity of one act of consciousness. The actual now-sensation and the "inactual sensation" (or "phantasm") of the earlier-than are both contained within a single act of consciousness. Despite its two-fold object, an act of consciousness is itself not temporally divisible, since admitting temporal distension within the act of consciousness itself would lead to an infinite regress. As noted above, Brentano understands the unity of the immediate past and present to be a relation between a secondary and primary object. The unity of this two-fold structure of original association is grounded in the simultaneous "givenness" of both contents in the same act of consciousness. As Husserl remarks, both the primary and secondary object "are there now, enclosed within the same consciousness of an object; they are therefore simultaneous. And yet the succession of time excludes simultaneity" (Hua X, 18 [19]). Even though primary and secondary objects are connected together in a unity of consciousness by the feat of original association, the encompassing unity of consciousness has the temporal footprint of simultaneity. But, how can the succession of two sounds be based on their simultaneity in consciousness – if both contents are simultaneous, how can consciousness discern one content as having been earlier than the other?[64]

Looking at Brentano's diagram, the horizontal line represents the succession of sensations, whereas the vertical line represents the consciousness of succession that takes the form of an original association

---

64 In the example of hearing a sound, the primary object of the sound must be "earlier" than my inner consciousness of hearing: "Denn zeitlich treten sie [*primary and secondary objects*] zugleich auf, aber der Natur der Sache nach ist der Ton das frühere" (PE, 180 [128]).

between the now-sensation (A) and the re-produced earlier-than sensation (B). On the one hand, Brentano's diagram clearly illustrates that the succession of consciousness is not the same as the consciousness of succession; yet, on the other hand, this diagram also shows that the consciousness of succession depends on the simultaneity of represented content in consciousness, and thus the representation of succession in simultaneity (the simultaneity of contained content along the vertical line). The consciousness of succession is based on a momentary or duration-less act of consciousness; within this momentary act, re-produced sensations of the immediate past are contained as immanent to consciousness. Husserl's two objections against Brentano's original association thus go hand in hand: the perceptual apprehension of the momentary now depends on the sensible apprehension of the just-past that, itself, for Brentano, is not authentically, and thus, not perceptually given, but only given as an "irreal" sensation, as the phantasm of the imagination; this means that the apprehension of the past is based on a content that is itself present; moreover, this content is contained within the immanence of consciousness, a momentary act of consciousness. Brentano thus reproduces the two assumptions defining the "cognitive paradox" of the perception of time: the perception of succession depends on a momentary act of consciousness (the simultaneity of content in consciousness); the perception of succession depends on the persistence or retaining of past content *in* the present, in the form of "phantasms," as in the case of Brentano's original association.

### Mental presence-time

After presenting his critique of Brentano's original association in the ITC lectures, Husserl turns to William Stern's critique of the dogma of the "momentariness" of acts of consciousness in order to complete the rehearsal of difficulties with which to propel his own phenomeno-logical analysis of time-consciousness.[65] As Husserl remarks, "there

---

65 L. William Stern, "Psychische Präsenzzeit," *Zeitschrift für Psychologie und Physiologie der Sinnesorgane*, 13 (1897), 325–349; English translation: "Mental Presence-Time," trans. N. de Warren, in *The New Yearbook for Phenomenology and Phenomenological Philosophy*, vol. V, ed. B. Hopkins and S. Crowell (Seattle: Noesis Press, 2005), 325–351. For a more detailed treatment, from which the presentation of Stern in this chapter draws, see Nicolas de Warren, "The Significance of Stern's *Präsenzzeit* for Husserl's Phenomenology of Time Consciousness," *The New Yearbook for Phenomenology and Phenomenological Philosophy*, vol. V (2005), 81–122.

works as a driving motive in Brentano's theory an idea that derives
from Herbart, was taken up by Lotze, and that played an important
role in the entire following period, namely, the idea that in order to
grasp a succession of representations (A and B, for example), it is
necessary that the representations be the absolute simultaneous
objects of a knowing that puts them in relation and that embraces
them quite indivisibly in a single and indivisible act" (Hua X, 19
[21]). This "driving motive" is recognizable in Brentano's original
association as the simultaneous unity of a consciousness encompassing primary and secondary objects, the now-sensation (*aesthesis*) as
well as the earlier-than sensation (*proteraesthesis*). Husserl's cursory
discussion of Stern's significant concept of "mental presence-time"
forms a seamless extension of his Brentano critique and culminates a
process of establishing "the genuine problem of time-consciousness."
By identifying how Stern's critique of "the dogma of the momentariness of consciousness" challenges a basic assumption in Brentano's
thinking – an assumption that Brentano shares with the history of
philosophy – Stern leads Husserl beyond Brentano's preoccupation
with time-consciousness towards the yet uncharted territory of *inner*
time-consciousness.

The dogma of momentary consciousness reflects the assumption
that "only those contents can belong to a whole of consciousness
[*Bewußtseinsganzen*] that are present in consciousness simultaneously."[66] At any given moment in the life of consciousness, a temporal cross-section (*Querschnitt*) of the stream of consciousness must
contain every content belonging to a single unity of consciousness.
This dogma is based on the presumed momentary character of an act
of consciousness: a plurality of content contained in an act of consciousness must be simultaneously contained in consciousness given
the temporally indivisible unity of an act. Stern argues, however, that
numerous examples can be provided in which an act of apprehension
is produced on the basis of temporally distributed content. In these
cases, the unity of an act of consciousness is based on what Stern calls
the "non-simultaneity" (*Ungleichzeitigkeit*) of content.

Stern proposes that, "mental events that play themselves out [*sich
abspielende psychische Geschehen*] within a certain stretch of time can
under circumstances form a unified and complex act of consciousness

66 Stern, 326 [313].

regardless of the non-simultaneity of individual parts."[67] This stretch of time in which mental acts "play themselves out," or what Stern calls "mental presence-time" (*psychische Präzenzzeit*), should not be confused with the specious present or "the duration block." Mental presence-time is not the extension of the sensed or perceived *content* of consciousness, but the temporal distension of the *act* of consciousness itself, the *sensing* of content. Indeed, one may grant the notion of a specious present that is intuitively apprehended without thereby recognizing the temporal distension of the act of consciousness, as, for example, in William James, who assumes a momentary act of consciousness despite his insistence on the intuitive givenness of the specious present.[68]

For Stern, in experiences where content is experienced discretely – individual tones in a melody or written letters on a page – we have, on the one hand, the discrete succession of content and, on the other hand, a unifying "connection of consciousness" (*Bewußtseinsband*), an act of apprehension. Even though individual notes are given discretely in succession, the entire sequence of notes shapes a single act of apprehension, played out within a single and unified presence-time. Yet the act of apprehension does not have a fixed location within the successive unity of hearing a melody as Stern denies that the act of apprehension coalesces into its discernible form in a momentary act that apprehends the series as a whole at the end of succession. On the contrary, a cross-section (*Querschnitt*) in the stream of consciousness does not contain every element (e.g., notes) in a series, which, however, are contained in the "length-section" (*Längsschnitt*) of presence-time – a cut that runs through the distended unity of consciousness along the trail of the immediate past's fading away. Whereas the cross-section of presence-time represents a momentary apprehension, the length-cut within the distension of presence-time as a whole represents the apprehension of non-simultaneity (*Ungleichzeitigkeit*). The concept of presence-time encompasses both kinds of temporal relations (*Successiva, Simultanea*) and lessens any sharp distinction between the two: successive content can give rise to simultaneous acts

---

67 Stern, 327 [315].

68 Gallagher has instructively clarified the conflation of Stern's "presence-time" with James' "specious present" that is often in the secondary literature. As Gallagher correctly notes, "the *Präsenzzeit* is precisely the temporal extension of the psychical act, not the temporal extension of the sensed content that James calls the 'specious present" (35).

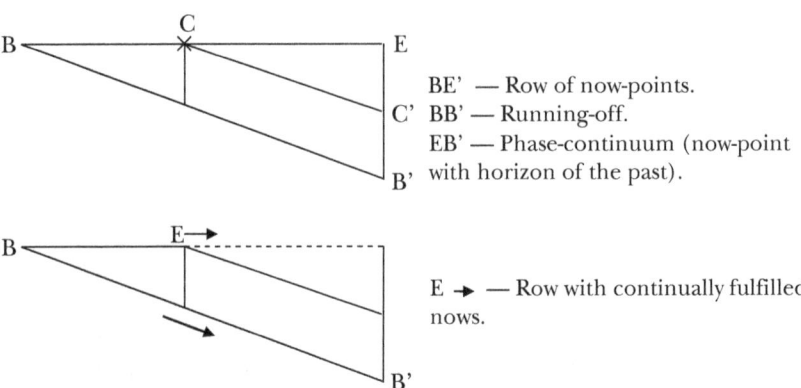

BE' — Row of now-points.
BB' — Running-off.
EB' — Phase-continuum (now-point with horizon of the past).

E → — Row with continually fulfilled nows.

Figure 2

of apprehension just as simultaneous content can give rise to successive acts of apprehension.

Although Stern did not employ a time-diagram in his seminal paper, even the most cursory of glances at one of Husserl's diagrams illustrates the thrust of Stern's proposal. Recall the diagram introduced by Brentano (Figure 1): a horizontal line represents the succession of nows; a vertical line represents the consciousness of succession. When expressed in this visual idiom, Stern's critique of the momentariness of consciousness translates into introducing a third line of angular distance vis-à-vis the horizontal and vertical lines. In Brentano's diagram, the content of the immediate past stretched along the vertical line is entirely simultaneous with the actual content situated on the horizontal line; both are contained immanently within a momentary act of consciousness. The consciousness of the succession of notes A, B, C is based on the apprehension of A, B, C along the vertical line as constituted by the unity of *aesthesis* and *proteraesthesis*. Stern's argument that the length-cut of presence-time encompasses non-simultaneous content can be represented by rotating the angle of incident of the vertical line to an angle *less than* ninety degrees (Figure 2), as we find in Husserl's diagram. The continuous running off of the immediate past along this "length-wise cut" is accordingly no longer entirely simultaneous with the apprehension of the now since it "cuts across" or "intersects" previously elapsed, and still elapsing, segments of earlier notes (as represented by vertical segments). The "length-wise cut" represents the "presence-time" of the act of apprehension, its inner temporal distension, on the basis of which temporal succession is apprehended. In Stern's manner of speaking, temporal relations are perceptually grasped within mental presence-time.

In addition to the perception of temporal relations *within* presence-time, Stern further argues that mental presence-time is itself perceptually given: "in addition to the temporal relations contained within it, this presence-time that belongs objectively to the act of consciousness can also immediately become a subjective content."[69] This significant claim introduces a distinction between the temporal relations that are played out within presence-time (or what Stern calls immediate time-consciousness) and presence-time itself, not the consciousness of time but the temporality of consciousness itself. As Stern, however, observes, the introduction of an inner temporal distension to an act of consciousness runs the risk of reverting to an equivocation of the consciousness of succession with the succession of consciousness. Yet, the notion of mental presence-time introduces a distinction between the temporal *extension* of the object of consciousness (specious present) and the temporal *distension* of the act of consciousness itself (mental presence-time). The perception of time within presence-time is not equivalent to the inner consciousness, or perception, of presence-time itself. The "peculiarity of time-intuition" (*Eigenart der Zeitanschauung*) consists in the fact that time can only be the object of our apprehension when we also recognize that time belongs to our own consciousness in the form of mental presence-time: the experience of something happening in time depends on the fact that consciousness is aware of itself as existing in time even though it does not depend on the explicit registering of that fact, but only, that the apprehending act of consciousness is itself temporally distended, but not in the same manner as its extended temporal content.[70]

As with other insights in Stern's argument, the concept of mental presence-time does not receive the sustained analysis that it requires, as Stern remains satisfied with the remark that "here we arrive at a psychological concept of time that has up to now received scant recognition and which merits more precise analysis."[71] This invitation, to which Husserl will respond, consists in developing how the presence-time of an act of consciousness itself "appears" or is "manifest"

69 Stern, 332 [323].    70 Stern, 332 [323].    71 Stern, 333 [323].

("erscheint") in addition to ("nebst") the consciousness of temporal relations, or what Stern calls "immediate time-consciousness." The combination of Brentano and Stern bequeath to Husserl two novel, yet underdeveloped insights into the consciousness of time, providing two keys that must be cut and turned in unison: the perceptual apprehension of the past *as past*, as no longer present, and the temporal distension of the apprehending act of consciousness.

# 3

# THE GHOSTS OF BRENTANO

*Die Leichname werden wieder lebendig und grinsen uns hohnlächelnd an.*

— Husserl

## A compelling beginning

The ITC lectures are an exception among Husserl's abundant, yet diffuse writings on time-consciousness. Whereas other writings are fragmentary and/or assume familiarity with earlier gains in Husserl's investigations, with individual manuscripts often continuing *in media res* without clear indication of orientation or trajectory, the ITC lectures – in the form of their 1928 publication – present a structured exploration of "the hidden world of time-consciousness, so rich in mystery" that fully set into motion "the most difficult of all phenomenological problems" (Hua X, 276 [286]).[1] In preparing these lectures for publication as per Husserl's instructions, Edith Stein deliberately attempted to retain the original path of thinking in the lecture-course while also accommodating Husserl's fastidious reworking of his seminal insights after 1905. Despite her efforts, the patchwork character of these reconstructed lectures generates a number of difficulties for grasping Husserl's argument, in both its substance and its

---

1 Gerard Granel is right to stress, as have other commentators, that "[c]'est donc ce texte [*the 1905 lectures*] qu'il faut interroger pour connaître la pensée de Husserl sur les problèmes de la temporalité, quel que puisse être par ailleurs l'intérêt des manuscrits inédits qui portent sur le même sujet. Non point tant parce que les *Leçons* bénéficient de l'autorité de la chose publiée, que parce que l'acquis de ce travail, et d'abord son langage, sont indispensables à la compréhension des inédits" (*Le Sens du Temps et de la Perception chez E. Husserl* [Paris: Gallimard, 1968], 11).

development.[2] The 1928 edition is composed of textual material – from the original lecture manuscripts, subsequent research manuscripts, and other lecture courses – spanning the critical years during which the entire phenomenological enterprise, and especially the investigations of time-consciousness, underwent significant transformation. Not surprisingly, the progress of Husserl's treatment of time-consciousness during these decisive years is obscured by the 1928 edition's indiscriminate juxtaposition of different stages in Husserl's thinking between 1904 and 1911. For these reasons, our interpretation of Husserl's analysis does not rely on the 1928 edition alone, and includes other manuscripts in *Husserliana* X, yet follows the path of questioning first broached in the ITC lectures, which provided the original template for Husserl's subsequent revisions, improvements and corrections.

As reflected in the deficiencies of Brentano's theory of original association, two inseparable problems steered Husserl's nascent phenomenology of time-consciousness. As argued in chapter 2, the consciousness of temporal succession remains inexplicable without a perceptual grasp of the immediate past in tandem with the perception of the present – the unity of the just-past and the now inadequately described under the heading of original association. In turn, this binocular perception of time implies the temporal distension of the perceptual act of consciousness, or what Stern dubbed "mental presence-time." A momentary act of consciousness remains blind to the passage of time. The restriction of the perception of time to the perceptual apprehension of a now bereft of temporal breadth goes hand in hand with the restriction of a perceptual act of consciousness to the form of simultaneity, and thus to an act of consciousness bereft of temporal distension. Armed with the decisive recognition that "the perception of a temporal object itself possesses temporality," Husserl closes his rehearsal of difficulties with a cue for his own phenomenological efforts: "A deeper analysis must begin here" (Hua X, 23 [25]).

The way forward, however, is neither a continuation of Brentano's descriptive psychological approach nor an attempt to craft solutions to Brentano's statement of the problems by other, phenomenological

---

2 For Edith Stein's editorial work, see Roman Ingarden, "Edith Stein on her Activity as an Assistant of Edmund Husserl (Extracts from the Letters of Edith Stein with a Commentary and Introductory Remarks,)" *Philosophy and Phenomenological Research*, 23 (1962), 155–175.

means. An unmistakable trait of a philosopher is to make the problems of others problems of one's own. Phenomenological philosophy neither accepts – naïvely – problems from the past without question nor claims to discover – naïvely – entirely new problems without historical precedence. Instead, old problems are sought out in new ways – given renewed life – through the methodological operations of suspension and reduction, both of which facilitate the transposition of established philosophical concerns into an uniquely phenomeno-logical register of analysis, predicated as much on the "seeing of things themselves" as on the seeing of the problems themselves, clearly and distinctly, as exhibited in the phenomenon under question.[3] The transition to a phenomenology of time-consciousness proceeds hand in hand with a transition to a phenomenological way of handling problems that launches the ark of questioning time anew, but not from where we just started.

## Phenomenological suspense

"Naturally," as Husserl remarks, "we all know what time is; it is the most familiar thing of all" (Hua X, 3 [3]). For most of us today, time is understood in terms of what it is to be on, behind, or ahead of time, where punctuality serves not only as the basis for time's measurement but also reflects an elaborate form of socialization: to be on time, be it for people or events, is a distinctly modern urgency elevated to the rank of a social virtue. Time is equally familiar to us through the periodic cycles of nature that pace the course of daily life – the alternation of night and day, the yearly march of seasons. And time is also familiar to us with reference to the sacred and the profane, especially in the secular distinction between unproductive (free or play-time) and productive time (work or commodity-time). Without minimizing the importance of these various forms of understanding time, we can nevertheless recognize that such diverse ways in which time is indeed "the most familiar thing of all" centers on a pheno-menon without which we could not meaningfully speak of time in each of the instances mentioned above. As captured with expressions such as "time flies" or "the river of time," the irreversible passage of time is

3 Eva Braun puts it well when she remarks: "No writer on time will bring the questionableness of the question under closer scrutiny than Husserl" (*What, Then, Is Time?* [New York: Rowman & Littlefield, 1999], 126).

inseparable from its ubiquitous significance for human affairs. This conspicuous feature of time announces what is most puzzling about it: time is the form of presence in which all experiences occur that cannot remain present. Familiar and inescapable, this passage of time has significance for us in many ways; it is only given to us as experienced.

In calling for a phenomenological clarification of time, Husserl's principal aim is to retrieve the consciousness of time's passage from the veil of naïveté and the web of over-laying conceptions, without which any conception of time is without a meaningful basis. Accordingly, Husserl's challenge consists in providing a "description of the naïvely given as well as scientifically presupposed time-consciousness in its meaning [*nach seinem Sinn*]" (Hua X, 188 [194]). As established in chapter 1, a phenomenological description is inseparable from the method of suspension and reduction. In the case of time-consciousness, the suspension and reduction jointly offer a methodological circumstance in which the "sense" (*Sinn*) and "validity" (*als was gilt uns die Zeit*) of time is made into a theme of reflection; the suspension renders the "sense" of time into a central phenomenological concern, whereas the reduction turns this concern into a workable transcendental problem. What does Husserl understand by the "sense" and "validity" of time-consciousness, which, once made into a theme of reflection, provides the focal point of his investigations? An answer – or answers – to this question will emerge over the course of our study. Yet, we can already signal that such framing of "sense" and "validity" serves notice of the overall strategy of approaching time-consciousness from the framework of intentionality. As indicated in chapter 1, the significance of intentionality does not merely reside in the claim that consciousness is the consciousness of an intentional object. Taken in its specifically phenomenological expression, the significance of intentionality reveals consciousness as an intrinsic reference to "sense." An intentional object is always intended, or meant, *as being so and so*; for Husserl, the problem of consciousness is the problem of sense.

Is time-consciousness an independent domain of investigation – an independent form of intentionality – as suggested in § 81 of the *Ideen* or is the intentionality of time-consciousness inseparable from other forms of intentionality – perceptual experience, the imagination – much as the perception of time is inseparable from the perception of objects in time?

Phenomenology is a method of rendering the familiar strange in transforming what is taken for granted into a theme of reflection. As a first step towards the recovery of what we take for granted of time, Husserl argues that a phenomenological analysis must abstain from any reference to the "existence" or "reality" of time. This demand for the suspension of time should not be conflated with a denial of time's existence. As Husserl's choice of terminology indicates (*epoché; Ausschaltung*), the suspension calls for a learned form of naïveté – a metaphysical agnosticism – in which we remain indifferent, or neutral, towards any presumed understanding of time's "existence" or "reality" until the *givenness* of time in consciousness is exposed to phenomenological scrutiny. This inaugural indifference forces a displacement rather than an abandonment of a theoretical interest in determining the "existence" or "reality" of time – for Husserl clearly sets his sights on an understanding of "objective time" (he anticipates providing a foundation for chronometry), even though he stresses that "through a phenomenological analysis, one cannot find the least traces of objective time" (Hua X, 6 [6]). The act of phenomenological suspension replaces the invocation of skeptical paradoxes regarding the being or existence of time that traditionally framed discussions of time, as in Aristotle's *Physics* or Augustine's *Confessions*. In this manner, Husserl reveals "existence" as itself a determinate "sense" and "validity" for consciousness, with its appropriate transcendental story of constitution. As a direct consequence of this shifting of terrain, the problem of time becomes refreshingly liberated from the combined metaphysical and psychological assumptions, exemplified in Brentano's original association, that prevented a genuine rendering visible – and thus questioning – of time itself. But the suspension and reduction equally liberate consciousness, in its fundamental form as time-consciousness, from its own "self-forgetting." Consciousness comes to see itself through the suspension and reduction as an original condition that has always been the ground of any sense, including its own.

## Clock and metaphor

The force of Husserl's suspension of objective time is aptly summarized with the remark that "there is no now, past or future in objective time" (Hua X, 189 [195]). What Husserl means is exemplified by considering the function of clocks. Our common as well as scientific understanding of objective time is mirrored in the service of clocks in terms of which time is represented as a linear succession of fixed "now-points" marching stepwise in ordered regularity. Each now-point excludes every other; there is no inter-penetration of now-points, but a steady progression of exclusion, each now-point dividing the past from the future, but also uniting the past and the future into a continuous whole, in terms of which a metric of comparative measure is possible. At the basis of this conception of time is the image of a line and its generation through successive points. Along this line, two now-points cannot be given simultaneously without being indiscernible from each other. The line of time is thus coupled to the form of simultaneity, as the form of presence that excludes a presence other than itself or a form of givenness different than its own.

A clock is an object in the world, yet an object unlike other objects with which we have daily commerce. We do not just perceive clocks, we use them; and we are capable of using them to the extent that we can read them. Knowing how to read a clock is learned and inseparable from asking certain kinds of questions such as "how much time do I have for X?" or "what time is it?" When I ask myself "what time is it?" and look to my watch, I must interpret what I perceive in the specific manner of reading what I see. Reading time from the face of a clock is not identical with counting or registering the progress of a clock's movement since neither the act of counting nor the perception of spatial displacement (or any type of change) inherently represents temporal succession. On the contrary, both the act of counting and the perception of change presuppose a temporal horizon against which prior and posterior states of change (or numbers) can be distinguished as prior and posterior states *in time*. If I register the movement of a clock, I grasp its change of position as exhibiting the advance of time on the basis of a temporal difference, implicitly understood, between before and after. At the moment in which I notice the hand on "7," I must grasp that the *same* hand *was* at "6" and relate where the hand once was to where the hand now is. It is not enough to register a difference in spatial positions (or the progression of numbers in the case of digital displays). I must understand what it means that something just was, yet no longer is, that another now occurred earlier than this now. Yet the past lacks a visible face; a clock can never show me the time that just was, but only the time that is now. Indeed, it is significant that we commonly speak of *reading* a clock: much as when I read different letters of a word in succession, and connect each letter into the composition of the word as a whole,

I must also take in each mark ticked-off by the movement of the clock's hand, and relate earlier ticks with later ticks into an articulate whole, so as to grasp an interval of time between 6:00 and 6:15. Moreover, the proficiency of distinguishing between past, present and future depends on the indexical self-awareness of reading a clock *in the now*. If I read a clock that shows me that it is now 3:00, the clock does not show me that it is *now* 3:00 unless I am also aware that reading the clock as 3:00 coincides with my awareness of now reading it. In other words, I must be aware that hearing the alarm clock coincides temporally with my awareness that I am now hearing the alarm clock in order to know that it is now 3:00. A clock tells me the time that is *now* only because of the consciousness that I now have.

If I must lend my own consciousness of time to the clock, what awareness of time does a clock give me in return? Here is not the occasion to pursue in detail the function of time-measuring devices for practical and scientific enterprises or to engage with one of Husserl's stated intentions of providing a foundation for the scientific laws of time (chronometry). One aspect of objective time deserves, however, particular mention. A clock permits a form of abstraction and ordering that facilitates the ascription of dates to events as well as their arrangement into sequences that reaches beyond, and encompasses, the time-spans of our lived experiences. Clocks establish an inter-subjective framework of chronological order, comparative measure and, most importantly for the modern world, increased precision and prediction. This abbreviated truth of clock-time – that time is an order of things – becomes recuperated in Husserl's transcendental exposition of the constitution of time as a fixed order of temporal positions.[4]

The suspension of objective time can also be examined from another vantage-point as the suspension of time's master metaphor. The metaphor of time as a "flow" or "stream" captures a deep-seated intuition that time is an embracing container or form in which events happen irreversibly. We can never "step in the same river twice," as Heraclitus once remarked. But as Merleau-Ponty observes, "the fact

---

4 The point of Husserl's suspension of clock-time is not to deny its accuracy or usefulness, but, on the contrary, to understand the basis of its possibility. As Husserl notes in the margins of his copy of *Sein und Zeit*: "Als ob die 'vulgare' Zeitauffasung nicht ihr ursprüngliches Recht hätte, das durch die konstitutive Analyse nicht in mindesten verschwindet."

that the metaphor based on this comparison [*of time like a river or flowing substance*] has persisted from the time of Heraclitus to our own day is explained by our surreptitiously putting into the river a witness of its course."[5] The metaphor of time's flow cannot be taken to mean that time is a movement or becoming independent of consciousness. But neither can the metaphor of time's flow, once transplanted into consciousness as *its* master metaphor, be taken to mean a movement or becoming of consciousness independent of *self-consciousness*. As a metaphor for either objective or subjective time, the flow of time presupposes a consciousness of temporal passage smuggled into its implied metaphorical landscape. We can imagine two different scenarios for where to place our surreptitious consciousness *within* this metaphor. We can either place an observing consciousness on the bank of the stream or in a boat gliding along the stream's flow. In the first case, the consciousness of time is itself not "in" time; as with Kant's transcendental apperception, the ego organizes the temporal manifold of its own experiences without itself being determined by time. In the second case, consciousness is submerged in time, caught in the flux of all things. In either case, the metaphor of time's flow presupposes a consciousness of time even though the metaphor's intuitive appeal remains immune to the ambiguity of whether consciousness resides within the stream or at its edge. In fact, each option leads to an impasse. If the consciousness of time is itself "non-temporal," how can a non-temporal consciousness grasp time? If the consciousness of time is itself in time, how can we avoid the infinite regress of positing another consciousness to grasp the stream of consciousness itself? Setting these issues aside for the moment, this master metaphor of time as a stream resonates with our vivid sense of time's irreversibility and its continuous, unbroken passage. We literally *can* turn back a clock, but we cannot thereby regain time past nor undo what we have since become.

If we survey the ground just covered, a common denominator emerges: the phenomenological suspension is shorthand for the argument that the linear succession of time (the succession of "nows" in clock-time; the stream of time) presupposes a consciousness of time. Expressed in terms harkening back to Brentano's critique of the philosophical tradition, the suspension of objective time and its

5 Maurice Merleau-Ponty, *Phenomenology of Perception*, trans. Colin Smith (London: Routledge, 1962), 411.

master metaphor allows Husserl to recover the origin of the concept of time (*Zeit*) in the "temporality" (*Zeitlichkeit*) of time-consciousness. Yet even though the suspension of objective time tacitly reiterates Brentano's historical critique of Aristotle and Locke, it does not thereby repeat the same understanding of what is in play with a phenomenological analysis of temporality.

## Reduction and immanence

As discussed in chapter 1, Husserl developed over the course of his phenomenological thinking three different paths to the reduction, each fashioned in various steps of accomplishment and methodological significance. Husserl's presentation of phenomenology in *Ideen I* was largely dominated by a Cartesian formulation according to which the reduction discloses the residuum of pure consciousness as the domain of absolute givenness. In addition to the Cartesian path, Husserl further identified a Kantian path that takes its bearings from the problem of constitution and the apriori conditions of possibility for experience, and a Brentanian path that progresses from psychology to phenomenological psychology to the science of transcendental subjectivity. Each of these paths leads to the discovery of transcendental subjectivity as the foundation for the possibility of experience, and all three paths are employed in various degrees in the investigation of time-consciousness from the 1905 lectures to the C-manuscripts. In any given writing, however, the methodological status of Husserl's reflections is often difficult to discern and rarely made explicit; in certain moments of the Bernau Manuscripts, for example, it is even questionable whether Husserl's reflections actually operate within the suspension of the natural attitude. In both collections of texts (*Husserliana X* and the *Bernau Manuscripts*) treated in this study, Husserl moves freely (and at times unthinkingly) between his paths to the reduction. To complicate matters, the 1905 lectures and supplementary manuscripts included in *Husserliana X* represent a primarily *noetic* analysis of time-consciousness, in accordance with the "act-phenomenology" of the *Logical Investigations*. The noematic expansion of time-consciousness only begins in earnest with the Bernau Manuscripts in which we find a fully transcendental analysis, where different constitutive forms of time-consciousness are explicitly connected with different regions of objectivity.

As discussed in chapter 1, the neutralization of the natural attitude performed by the suspension frames Husserl's interest in the problem of givenness and sense. The theme of "givenness" is exploited through the phenomenological reduction, the purpose of which is to disclose the immanence of pure consciousness, as the "infinite field" of concrete phenomenological inquiry that survives the figurative "destruction of the world" (to invoke the controversial Cartesian formulation). As Husserl writes, "We therefore do accept 'an existing time' [*eine seiende Zeit*], however, it is not the time of the world of experience but rather the immanent time of the flow of consciousness" (Hua X, 5 [5]). It is, Husserl notes, "meaningless" (*sinnlos*) to deny the consciousness of time since to doubt the consciousness of time would require that consciousness deny itself the temporality intrinsic to its own act of doubting. The reduction to immanent time-consciousness investigates how time is given *in* consciousness, yet it does not thereby in turn naïvely presuppose how consciousness itself is temporally given. Indeed, a significant ambivalence in Husserl's term "time-consciousness" already makes itself felt in the opening of the reduction: "time-consciousness" can (and must) be read from two directions at once, as the consciousness of time *and* as the time of consciousness. This apparent circularity lodged at the core of time-consciousness points to an issue of fundamental significance: is a non-circular account of time-consciousness possible? As we shall discover, this issue implicates the essence of Husserl's phenomenological determination of consciousness and its "auto-manifestation" as absolute time-consciousness.

The methodological strength of the reduction allows Husserl not only to exploit positively the suspension of objective time but also to formulate the suspension of subjective time. Stated differently, the suspension refuses to accept the succession of consciousness (e.g., Locke's train of ideas) as a brute given; the reduction of time is also a reduction *of consciousness* to the immanent temporality of consciousness.[6] The ambivalence of time-consciousness thus articulates a *dual* question, where each question implicates the other; this dual question shapes Husserl's phenomenology of time-consciousness into two directions of inquiry. Husserl thus retains the basic characterization of consciousness as a stream or flow, yet purges it of any psychological

---

6 As Husserl notes: "Jedenfalls ist es klar, dass die Sukzession des Bewusstseins dasselbe Problem in sich birgt wie jede andere Sukzession" (Hua XXX, 97).

or "naturalized" meaning, as exemplified in Locke and Brentano. Another way to express the radicalization of Husserl's pursuit is in the form of questioning the extent to which Locke and Brentano each implicitly relied on the *metaphorical* meaning of time as a train or stream. Viewed from this angle, Husserl's ambition is nothing less than the attempt to think through the metaphor of time as a stream, replace it with concrete phenomenological description, and in this manner, reach beyond the metaphor to the things themselves, on the basis of which the metaphor finds traction and life. The ultimate test for Husserl's account resides in whether it can genuinely bring to intuitive givenness the phenomenon of time-consciousness in *conceptual* descriptions whose meaning does not entirely depend on metaphors; or, if metaphors are required, whether phenomenology can reactivate the origin of the concept of time in time-consciousness on the basis of which such metaphors are at all "alive."

The scope of Husserl's suspension of subjective time is not limited, however, to the "Lockean–Brentanian" paradigm that construes consciousness as an immanent succession of "ideas" or "presentations," but also includes accounts of the perception of time based on the interplay of the three separate faculties of *memoria, perceptio, expectatio,* as in Aristotle and Augustine. Both Aristotle and Augustine operate on the assumption that an act of consciousness (or the "soul") does not possess an inner temporality. The distribution of the different aspects of temporal extension – past, present, future – onto a parallel distribution of separate faculties of the soul tacitly adheres to the psychological dogma of momentary consciousness and a metaphysical assumption of the coincidence of the perceptual present – narrowly defined to a momentary apprehension – and the punctual now. Husserl, however, notes that, "An analysis of the consciousness of perception, of the imagination, of memory and of expectation is not complete as long as temporality is not included; conversely, that an analysis of these acts presupposes an analysis of time-consciousness, that goes without saying" (Hua X, 394). As with his reservations against Brentano's original association, Husserl once again insists on the temporality of consciousness (and once again revealing Stern's impact) on the basis of which the perceptual apprehension of temporal extension is rooted. This emphasis on an investigation of the temporality of consciousness directs the phenomenological demand for a description of lived experience. Yet, as Husserl cautions, "lived experiences" (*Erlebnisse*) or, more explicitly, "lived experiences of time" (*Zeiterlebnisse*), should not be taken in an ordinary sense (*im gewöhnlichen Sinn*) within a phenomenological context of analysis; "lived experience" (*Erlebnis*) does not possess a "pyschological-empirical sense" in Husserl's proposed analysis. As Husserl notes, "With respect to the problem of time, this means that we are interested in *experiences* of time. That these experiences are themselves fixed in objective time, *that they belong in the world of phsyical things and psychic subjects,* and that they have their origin in this world does not concern us and we know nothing about it" (Hua X, 9–10 [10]).

## Different lines of inquiry

Little is achieved for any particular field of phenomenological inquiry by stating abstractly what Husserl's reduction is meant to accomplish. In phenomenological philosophy, method is performance, and the performance of method is always situated in the midst of handling problems. The method of phenomenological philosophy must be applied concretely and refashioned over the course of working through specific lines of inquiry that both motivate and challenge it.

Although Husserl retains Brentano's focus on the perception of time by continuing to pursue the question of how consciousness apprehends what Husserl calls a "time-object" (*Zeitobjekt*), the reduction repositions this problem within a phenomenological framework of intentionality, according to which, intentional objects are apprehended by self-transcending acts of consciousness on the basis of immanent, non-intentional content. Throughout his investigations, this orientation towards the constitution of time-objects is continually maintained, and reflects the centrality of the relationship between temporality and objectivity for a phenomenology of time-consciousness. As Husserl remarks, "We apprehend the moment of temporality along with the apprehension of a perceived real entity" (Hua X, 274 [284]). We may here profitably recall an Aristotelian insight (taken, however, in appropriately modified form) that although time is distinct from perceptual objects (do we ever perceive time as such?), and, indeed, should not to be conflated with objects in time, time is nonetheless only given along with perceptual objects. The consciousness of time is inseparable from the consciousness of objects in time. As Husserl states, "a phenomenological analysis cannot clarify the constitution of time without taking into consideration the constitution of time-objects; because objective temporality is constituted phenomenologically and only appears for us as objectivity or a moment of objectivity on the basis of such constitution" (Hua X, 22/23 [24]).

The term "time-object" has a broad and a narrow meaning in Husserl's analysis.[7] In its general meaning, the term refers to any perceptual object – a tree in the yard, a bird in flight – that appears in time. A perceptual object that appears in time occupies a definite now. The now in which an object appears defines an object's temporal individuation as *this* object in a fixed temporal position (along with a definite spatial position of "there" relative to the absolute "here" of my lived-body). Moreover, as discussed in chapter 1, a perceptual object is a system of appearances, or, in other words, a synthetic unity of different perspectives or adumbrations; different sides of an object must be brought together into a unitary form of apprehension. This synthetic form of objectivity inserts temporality into the constitution of objectivity and, in this sense, all perceptual objects – as the synthetic accomplishment of a constituting consciousness – can be regarded as time-objects. In its narrow meaning, "time-object" refers to an individual perceptual object that intrinsically contains temporal extension, i.e., whose parts are necessarily distributed over time (e.g., a melody or a spoken word). Due to the constituted character of all intentional objects, the constitution of a time-object (in both senses) must be based on an objectifying act of consciousness. This implication, that a corresponding act of intentionality underlies the givenness of time-objects, allows Husserl to consider time-consciousness as a *form* of apprehension (as the form of objectivity) rather than as a kind of sensation (Aristotle's common sensible), a modification of sensation (Brentano's original association), or a relation (association) that unites disparate content in consciousness. An analysis of the intentional accomplishment of time-consciousness thus requires a description of how time-objects are constituted in perceptual acts of consciousness, and Husserl first tackles this task in the ITC lectures by turning to the conception of intentionality formulated in the Fifth Logical Investigation, and by looking at time-objects in the *narrow* sense.

This line of inquiry recapitulates the traditional problem of the perception of time (time-object in the narrow sense), but also expands its scope to the form of objectivity as such (time-objects in the broader

---

7 Husserl uses interchangeably the terms *Zeitobjekt* (borrowed from Alexis Meinong) and *dauernde Objekt*.

sense). In addition, Husserl proposes another direction of investigation, which, as with the first line of inquiry, responds to Brentano's failed original association. As Husserl asks: "How, in addition to 'time-objects,' immanent and transcendent, does time itself, the succession and duration of objects, become constituted?" (Hua X, 22 [24]). At first glance, this formulation is potentially misleading as it could be taken to suggest a separation between, on the one hand, "time-objects" and, on the other, "succession/duration," as if we were speaking of two independent phenomena. The reference to "succession and duration" is indeed ambiguous in the context of this introductory section of the ITC lectures since it invites two possible readings. Husserl could have in mind the relation of succession between time-objects and their respective time-positions. In due course, Husserl comes, however, to distinguish between the constitution of time-objects and the constitution of objective time, the latter construed as a fixed order of temporal positions, as structured by laws of chronometry. In this specific passage, Husserl's invocation of "succession/duration" hinges on the term "in addition" (*neben*) which immediately recalls Brentano's "incidental perception" or "secondary object" of inner consciousness. Indeed, in speaking of "succession and duration" *next to* (or on the side of) time-objects, Husserl identifies a second direction of inquiry within the problem of time-consciousness that remained structurally excluded from Brentano's original association, due to Brentano's rejection of an inner temporality to acts of consciousness.[8] As Husserl notes immediately after the question raised above, "it is certainly evident that the perception of a temporal object has itself temporality, that the perception of duration itself presupposes duration of perception, that the perception of any temporal form itself has its temporal form." The expression "succession and duration" *next to* time-objects refers to the temporality of perception (Stern's mental presence-time) in which time-objects are given, but which is also itself given, or manifest, along with – yet not as the same – time-objects. The consciousness of an object in time is at the same time a consciousness of oneself as experiencing time. We recall that this is precisely what Stern flagged as the "peculiarity of time-intuition" (*Eigenart der Zeitanschauung*) – that time can only be the object of our apprehension when we also recognize time as a property of our own

---

8 Husserl is rarely consistent with his use of "duration" and "succession."

consciousness. Husserl's second direction of inquiry responds directly to Stern's passing remark: "here we arrive at a psychological concept of time that has up to now received scant recognition and which merits more precise analysis."[9]

If the perception of a time-object is based on the temporality of perception, a complete analysis of time-consciousness must address both directions of inquiry: the constitution of time-objects in perceptual acts *and* (in light of) the temporality of this constitution, i.e., the temporality of (constituting) perceptual acts. The relation between the constitution of time-objects and the temporality of consciousness must furthermore be clarified by describing how the temporality of perception provides the foundation for the temporality of objects in the specific form of being given "next to" time-objects. Looking back to chapter 2, Brentano's failure to acknowledge the temporality of perceptual acts prevented him from recognizing that an answer to the first problem (the apprehension of time-objects) depends on acknowledging the second direction of inquiry *as a problem* (the temporality of the perceptual acts). Looking forward to Husserl's own analysis, the distinction between the consciousness of time-objects and the temporality of consciousness delineates different lines of inquiry within the "hidden world of time-consciousness," and so establishes the trajectory of Husserl's analysis in the ITC lectures from the constitution of time-objects (intentionality of time-consciousness) to the temporality of consciousness (temporality of intentionality). This trajectory of investigation also delivers the underlying plot to the drama of Husserl's struggle with "the most difficult of all phenomenological problems." Both directions of consciousness are to be described in terms of intentionality; whether intentionality (and in what form) provides an unambiguous description of the perception of time and the temporality of consciousness, and whether the relationship between both dimensions of consciousness can be captured through intentionality remains to be seen.

## Intentionality, again

A phenomenological analysis of consciousness, we recall, is descriptive to the degree that it is based on reflection in the strict phenomenological sense outlined in chapter 1. Since the appearance of objects

9 Stern, 333 [323].

is grounded in intentional acts, an analysis of the intentional accomplishment of consciousness must inhabit the view of objectifying acts. Consciousness does not look upon itself at the expense of its intentional reference, but rather looks through its intentional accomplishment at itself in the act of reflection. Yet, an intentional act of consciousness is open to investigation only by way of an act of reflection – itself of an intentional character – that discloses the entire structure of an intentional experience, in which, prior to its thematization, consciousness merely "lives" in a form of "self-oblivion" or naïveté. Prior to reflection, consciousness lives in its intentional experience, directed towards its object, and not towards itself. The transcendence of an object – its appearance for consciousness – necessarily implicates a corresponding constituting immanent act of consciousness; every form of transcendence contains a hidden story of constitution that in plot turns on distinguishing constituted objectivity from constituting act and, in this light, describing the givenness of transcendence *for* immanence.

In once again recalling the basic features of intentionality, our present aim is not to revisit the entire framework of intentionality detailed in chapter 2. Instead, our aim is to return to intentionality from the angle of time-consciousness, and thus pursue Husserl's effort to describe time-consciousness – along the two lines of inquiry formulated above – as a type of intentionality.

Listening to music is often described as an experience of complete absorption. Immersed within the music, we are focused exclusively onto the musical sounds, drawn into their movement and mood. We are temporarily detached, as it were, from ourselves. The experience of listening to music, of course, need not always saturate our consciousness. Muzac in shopping malls is an example of "music" that does not captivate consciousness, but which lingers like an uninvited guest in the backdrop of a remote horizon. In both instances, musical objects – a melody, a song, etc. – are intentional objects of consciousness. As an intentional object, the musical sounds appear "out there" in the world, not as something *extra mentum*, entirely divorced and independent from consciousness, nor as something merely in my mind – I do not hear a representation of music, but the sounds themselves. Although my act of apprehension (the perceptual act of hearing) is directed towards the intentional object, and intends the object itself as so and so, I can always turn my attention towards my act of listening and make my listening a theme of reflection at the expense of no longer giving myself directly to the sounds heard.

As with any intentional object, a musical object transcends my consciousness. The transcendent character of an intentional object is inseparable from its constitution as a noematic unity of meaning that is never exhausted in any given performance or appearance; it is also inter-subjectively available, there for others to hear, and enjoy. Bach's *Partita Nr. 4* is an intentional object that is neither adequately presented in a single experience (even the most perfect performance remains imperfect) nor in the accumulated form of my entire biography of listening (or, indeed, in the entire history of its performance). Bach's *Partita Nr. 4* is – to adopt Husserl's technical vocabulary – a system of appearances, centered on an identity of meaning that admits various degrees of intuitive fulfillment, depending on the proficiency of any given performance. As discussed in chapter 1, an intentional object is continually intended as the same, as an identity, through a changing landscape of profiles as exposed in the materiality of immanent, sensible content or, as Husserl remarks, "However we may call those founding contents at the foundation of apprehension" (Hua XIX.1, 399 [568]). Although it is the same *Partita* that I hear today as I heard yesterday, the sensible contents on the basis of which I hear the *Partita* today are not the same as those experienced yesterday. Likewise, while listening to Bach's *Partita*, the experienced sensible content underlying my apprehension of its notes is continually different. Many notes of the same kind may be apprehended, but each time the same note is heard, the note is given to me in a qualitatively different sensation, exhibited in a "lived content" unique to its moment of givenness.

In Husserl's tripartite schema of explication (commonly referred to as the "apprehension/content of apprehension schema") that characterized his earliest conception of intentionality in the *Logical Investigations*, an objectifying act of consciousness apprehends its transcendent object as so and so, namely, in whatever manner that object is meant, on the basis of non-intentional sensible content. Husserl variously speaks of intentional acts of apprehension as "interpreting" (*deuten*) or "animating" (*beseelen*) sensible content or, in Husserl's later terminology from the *Ideen*, "hylétic sensations or data." For its part, such a sensible underpinning of consciousness exposes or exhibits correlating sensible qualities of the intentional object; in this regard, Husserl also speaks of "exhibiting content" (*darstellende Inhalt*) in order to specify that such immanent content exhibits objective qualities of an intentional object without being those qualities. In other words, experienced immanent content is not to be conflated with the traditional notion of sensation (e.g., Brentano's physical phenomenon) that belongs to an object of experience and impresses itself onto consciousness. Sensible qualities of an object that I perceive (the green of the tree, etc.) are objectified qualities of the object; the corresponding experienced content of what it is like to perceive green is itself not green. Nevertheless, Husserl does argue for a variable degree of resemblance between immanent sensible content and the sensible qualities of an intentional object, and, thus, still leaves room for the charge that he has yet to liberate himself convincingly from a muted form of representationalism. And yet, the primary motivation for Husserl's fashioning of immanent sensible content stems from the dynamic structure of intentionality and its constitutive distinction between "empty" and "fulfilled" intentions. A correlation exists between changes in the givenness of the intentional object and changes in its sensible underpinning; increased fulfillment of an intentional object's presence or intuitive givenness – its plenitude (*Fülle*) – corresponds to a change in sensible content. Immanent sensible contents allow for a precise phenomenological characterization of the varying degrees of intuitive fulfillment, or "intuitivity," of an intentional object's givenness to consciousness. In this manner, the concept of sensation undergoes a significant phenomenological transformation; immanent contents are not "sensible parts" of the intentional object (as with Brentano's notion of sensation), but neither is it the case that immanent contents, despite their "real" (*reell*) inclusion within consciousness, can be straightforwardly placed, so to speak, on the side of consciousness since Husserl also characterizes hylétic content as an intrinsic *alterity*, or "non-ego," *within the immanence of consciousness.* This suggestive ambivalence in the concept of hylétic content eventually leads Husserl to distinguish between "noetic" and "noematic" hylé. In this fashion, the materiality of hylétic content, in terms of which consciousness lives through its experience of the world, can no longer be seen as a "medium" or "representation" in between consciousness and the world. On the contrary, this hylétic dimension of consciousness, when developed under the heading of genetic phenomenology and the theme of affectivity, designates the pre-given "facticity" and "situatedness" of consciousness in the world.

As Husserl repeatedly stresses, immanent hylétic content as well as intentional acts of consciousness are "experienced" (*erlebt*), but do not appear as objects to consciousness. As Husserl writes: "Sensations and 'apprehending' or apperceptive acts are experienced, but they no do appear as objects ... Objects, on the other hand, appear and are perceived, but they are themselves not experienced" (Hua XIX.1, 399). This choice of the expression "lived experience" (*Erlebnis*) to characterize immanent sensations and acts of consciousness already implies a temporal form to immanent consciousness, even before Husserl has explicitly turned to consider the temporal constitution of consciousness as such. The verb *Erleben* (from which the noun *Erlebnis* is derived) means "to still be alive when something happens" or "to undergo an event," and suggests the immediacy "with which something is grasped." Both shades of meaning (immediacy and acquisition) are compressed in the English rendering of *Erlebnis* as "lived experience." The reference to life (*Leben*) is at once significant and indefinite in the term *Erlebnis*, and undoubtedly contributed to its appeal for Husserl. As Gadamer remarks, the term *Erlebnis* entered into the vernacular of German philosophy towards the end of the nineteenth century in order to supplant the traditional terminology of sensation and thus "more sharply define the concept of the given."[10] As Husserl's preferred characterization of immanent consciousness, the term *Erlebnis* – not to be conflated with Husserl's more ordinary use of the term *Erfahrung*, which refers to the experience of constituted objectivities – more sharply defines the temporal givenness of consciousness for itself, as is also evident in the operative expression of "living" in "lived experience" (*Erlebnis*), which also surfaces in another key phenomenological concept: the "living present" (*lebendige Gegenwart*). The intrinsically temporal nature of what it is for consciousness to experience is readily apparent in the type of examples most often used by Husserl to illustrate this crucial phenomenological distinction between experienced immanence and transcendent objects, between self-givenness and other-givenness. In the ITC lectures, Husserl picks up a piece of chalk, and observes: if I interrupt the act of perceiving the chalk by

---

10 Husserl is not always consistent and employs "Erlebnis" to characterize intentional acts as well as non-intentional content. As Gadamer notes in *Truth and Method*, the noun *Erlebnis* entered into common philosophical currency only in the 1870s and primarily in the context of biographical writings. Hans Georg Gadamer, *Truth and Method*, trans. J. Weinsheimer and D. Marshall (London: Continuum, 2004).

closing my eyes, then I open my eyes, I still perceive the same chalk. The intentional object remains numerically identical, yet the lived content is numerically different. We cannot step into the same stream of immanent consciousness twice even though we can perceive the same object twice.

The intentional acts of consciousness, along with their immanent basis in sensible content, allow for the presentation of intentional objects, and are themselves present, or lived and experienced, by consciousness; the consciousness of an intentional object is also conscious of itself – self-conscious – in a non-objective manner. Husserl speaks of experienced content as intrinsically "conscious" (bewußt), not in the sense of an "object" of reflection or "secondary object" of inner consciousness, but in the form of an immanent non-appearance or "non-object." Immanent consciousness is itself given; there is no *distance* between its appearance and its givenness. And yet, insofar as acts of consciousness and their sensible underpinning are "self-given," that is, experienced in a pre-reflexive and non-objective manner, they are also distributed, and so constituted, temporally, and thus, "time-objects" in *some* sense yet to be determined. Despite their non-intentional status, experienced content and intentional acts can under appropriate circumstances become "objects" in the sense that we can attend to them in reflection and provide for them a definite phenomenological description. As objects of phenomenological reflection, these immanent objects are without exception immanent time-objects in the specific sense of an object whose parts are distributed over time. In this regard, Husserl applies the term "time-object" to the temporal unity of immanent content and acts as objectified through phenomenological reflection; he can therefore speak of both "transcendent and immanent time-objects," depending on the context of his discussion.

In light of these remarks, an idiosyncratic feature of Husserl's various descriptions of tones, melodies, clocks, and postal whistles deserves an aside. When examining an immanent time-object in phenomenological reflection, Husserl routinely speaks of "turning his look" in different directions as if he were visually inspecting an object in his hand: we are asked to look at the tone; our look is asked to follow the trail of the tone's sinking away; we are asked to look at its mode of givenness. In what sense can I direct my look in phenomenological reflection? In adopting such a manner of speaking, Husserl is keen to substantiate his over-arching claim that phenomenological descriptions are rigorous in the sense of clarifying invariant structures of consciousness, as revealed in *eidetic* intuition – but rarely is the process of eidetic variation, implied throughout the analyses of time-consciousness, made explicit in his research manuscripts. In addition, the presentation of examples (e.g., hearing a tone) tacitly presupposes the structural accuracy of remembrance. We must continually appeal to our power of recall when dissecting in our mind's eye the brief life story of a tone's passing away; but, we also require the imagination in order to review frame-by-frame a musical note's destiny in time-consciousness – all of which Husserl performs, when he exclaims, "One can say nothing further here than 'look'" (Hua X, 77 [82]).

## The phenomenon of "running-off"

When seen through the prism of intentionality, hearing a musical object is an experience composed of an intentional object (a melody, a single note, etc.), a continuous apprehending act of consciousness (the act of hearing), and experienced immanent content in which the changing perspectives of the intentional object are exhibited. As already stipulated, a musical object is a time-object in both a broad and narrow sense; as such, it has a beginning and an end, and the span between these limits defines the time-object's duration. Indeed, that a time-object necessarily possesses a beginning *and* an end means that the beginning and ending phases must be separated by an interval of duration: every beginning implies duration. Within this temporal duration, Husserl abstractly distinguishes a series of "now-points" or "now-phases," each of which "flows-away" or "runs-off" (*ablaufen*) into the immediate past while giving way to a renewed, and different, now-phase, which, in turn, runs-off – this drama of the expiration and renewal of now-phases recurs continuously over the course of the time-object's duration as a whole. As Husserl declares, "it belongs to the essence of perception, that it not only has in its regard a now with the character of a point, and not only releases from its regard something that has just been of which it is nevertheless 'still conscious' in the appropriate mode of 'just having been' (primary memory), but also that it passes over from the now to now and meets the now half-way in its regard" (Hua XXIII, 258 [314]).

Husserl argues that the "running-off" of each now-phase is perceived, in the form of a primary memory (to which we shall shortly return), along with the perceptual grasp of every succeeding and renewed now-phase. Yet, the thrust of Husserl's observation is not merely to affirm the perceptual apprehension of the "phenomenological datum" of the now and the just-now. A subtle, yet significant shift of accent away from the now as such, in isolation from the phenomenon of running-off, to the now *as it is* running-off, is notable in Husserl's description of the essence of perception. With this shift of emphasis to the transitional character of the now, Husserl's focus becomes more finely calibrated to perceptual apprehension of temporal "in-betweenness," or the difference between the now and the just-now as an intuited difference. We perceive the "in-betweenness" of the now-phase, as now and just-now, as one and two, as unity *in* difference. Consciousness looks two ways at once: back along the spine of an earlier now-phase's running-off and forward towards the next arriving now, which is met half-way, and caught in "mid-flight," so to speak. As Husserl notes: "the regard [*der Blick*] from the now toward the new now, *this transition*, is something *originary*, which first paves the way for future experiential intentions" (Hua XXXIII, 259 [314]; my emphasis). Time-consciousness is a consciousness of original temporal difference, and it is as a consciousness of difference that consciousness is itself awake. As Husserl further remarks: "the waking consciousness [*wache Bewusstsein*], the waking life, is a living towards [*Entgegenleben*], a living from the now towards the new now" (Hua XXIII, 259 [315]). And yet, the waking arc of time-consciousness should not be conflated with attention (*Aufmerksamkeit*); to be awake is the foundation for the attention and inattention of the ego towards objects – actual and possible – in its surroundings. The characterization of time-consciousness as awake defines the *openness of consciousness as such*, not to any particular object of experience, but as the condition for any possible experience. To be awake is to be conscious of time; an openness for what is yet to come in the wake of a past that is still not beyond our grasp. Waking consciousness is an "originary intentionality" that "transitions from now to now," centered on the perpetual *renewal* of the now. The term "Entgegenleben" at the center of this phenomenological insight is difficult to render precisely into English, as it contains different threads of meaning, each of which plays on conjugations of "Leben" (life) and "Entgegen" – "to meet half-way," "to approach or ride towards," "to face or await," "to run contrary to." As with many of

Husserl's self-fashioned terms, an effective translation and understanding can only emerge over the course of its interpretation.

As I hear a musical note, I am conscious of an identical note with a definite duration. I hear a middle C for a certain – however brief – duration. Over the course of its duration, its identity as a note does not change. The note remains as much a middle C in its beginning phase as in its final phase of manifestation. In the same vein, Husserl argues that the temporal form of the time-object's givenness in the now does not change, nor does, in this respect, what Husserl further designates as the time-object's "fixed temporal position," in terms of which the ascription of dates and comparative measurements are possible. The same intentional object is intended in its identity across its shifting profiles; and much as the identity of the time-object (its being so and so; in this example: a middle C) appears across shifting profiles, so does the fixed temporal position, as well as the form of a time-object's givenness as now, appear across shifting profiles. The consciousness of the note's duration as a whole is a consciousness of a beginning to an end. Throughout, consciousness of the note is both a consciousness of continuity, a consciousness of a beginning to an end, *and* the consciousness of self-differentiating now-phases. The stable identity of the note stands in contrast to the manner in which the note is "given" in consciousness, which is always other than as just given (*ein immer anderer*), always in a new "now-phase" caught in a constant flux (*in einem beständigen Flusse*) in the mode of "running-off" (Hua X, 25 [26]).

Husserl thus implicitly draws a distinction between the temporal duration of the time-object as a whole, its form as now, and within this form, different now-phases, each of which is grasped within the horizon of earlier and later now-phases. The unchanging form of temporal givenness, as now, in its changing now-phases is nothing other than the form of the time-object itself. The living present is thus characterized as both "standing" and "streaming." Although all experience happens in the form of the now, it is impossible for experience to remain in the form of the now. We can speak of the now in an extended manner, as a unified whole, as when I say that a melody is now playing; but we can also speak of each now-phase, or slice, of the melody as it unfolds. Husserl's description countenances both; following Eugen Fink, we can speak of "an authentic entwinement of a double-constituted present of experience [*ein eigentümliches Ineinander einer doppelten konstituierenden Erlebnisgegenwart*]" which ultimately, when further developed in later writings, embraces the noetic and noematic components of time-consciousness.[11]

During this early stage of analysis, Husserl remains guided by the "apprehension-content of apprehension" schema of intentionality and construes the constitutive interplay of changing temporal profiles and intended temporal form along the dynamic lines of the changing perspectives of one and same pole of identity. As Husserl writes: "The points of temporal duration recede for my consciousness in a manner analogous to that in which the points of a stationary object in space recedes for my consciousness when I remove 'myself' from the object" (Hua X, 25 [26]). This analogy might provoke a Bergsonian suspicion of a spatialization of *durée*. Importantly, however, Husserl does not speak of "time-points" as "moving away" from me, for such a characterization would imply that the running-off of each time-point resembled the passage of individual train cars as I stand, watching motionless, on a platform. Instead, Husserl speaks of "time-points," or "now-phases," as receding from me *as when I remove "myself."* This qualification around "myself" is here crucial since my temporal point of view in the midst of temporal passage does not in any meaningful sense move as I would move in space. And yet, my temporal outlook, or vantage-point, does incessantly "change," so to speak, in the precise sense that the now-phase is always renewed and replenished in its absolute proximity and originary givenness. In spatial movement, distance and perspective are correlated; every perceptual object within my spatial surroundings is seen from an outlook centered on the absolute "here" of my body. If I move away from a pen placed on a table, the pen increasingly appears smaller to me as a function of the increasing distance between myself and the pen. What has changed is not the actual spatial configuration of the pen, but my perspective as a function of a distance between the pen and myself. I continue to perceive the pen at its own place (on the table) and as the same object (it has not changed into a pencil). But as I move away from the table, my perspective is continually changing. At a certain vanishing point, the pen has disappeared entirely; it has slipped beyond the horizon of my field of vision, and I can no longer make out that black point as the pen that I still recognized a moment ago. In the case of temporal displacement, distance and perspective are also correlated; however,

11 Eugen Fink, *Studien zur Phänomenologie* (Den Haag: Martinus Nijhoff, 1966), 21.

THE GHOSTS OF BRENTANO                                    121

not through the dimensions of my embodiment, but through the dimension of consciousness itself, and its axis on the absolute renewal of the now. Time-consciousness is both a consciousness of proximity *and* a freedom of perspective, a consciousness of distance within this proximity.

In thinking about Husserl's analogy, it is important to keep in mind that a comparison is drawn between the "points of time-duration" (*die Punkte der Zeitdauer*) and the "points of a spatial object at rest." The analogy, in other words, is between "time-duration," or the specious present, and the spatial object at rest, and therefore *not* between a spatial object and time-constituting consciousness, i.e., the temporality (*Zeitlichkeit*) of the act of consciousness. As Husserl writes, "The object keeps its place, just as the tone keeps its time. Each time-point is fixed, but it flies into the distances of consciousness [*Bewußtseinsfernen*]" (Hua X, 25 [27]). The running-off that constitutes the "distances" of time-consciousness is also typified as "a kind of perspective" (*eine Art zeitlicher Perspektive*) that becomes "compressed" or "contracted" (*Zusammenziehen*) within the original temporal perspective (*innerhalb der originären zeitlicher Perspektive*).[12] In order to further describe the "running-off" into "distances of consciousness," Husserl fashions two complimentary pairs of concepts – distance–proximity and clarity–darkness – both of which add texture to the phenomenon of time-consciousness. Such distinctions render more sharply what Husserl dubs alternatively as the "phenomena of running-off" (*Ablaufsphänomene*) or "modes of temporal orientation" (*Modis der zeitlicher Orientierung*). With this focus, Husserl sets a further course of analysis: "It is now a matter of investigating more closely what we are able to find and describe here as the phenomenon of time-constituting consciousness in which time-objects with their temporal determinations become constituted" (Hua X, 26 [28]).

### The double continuum of time-consciousness

Husserl's phenomenological description of the continuous "running-off" of time-consciousness is crystallized in a diagram (Figure 3), offered as a "complete image" of what is discovered as the "double

---

12 Cf. William James, *Principles of Psychology*, 593: "There is thus a sort of *perspective projection* of past objects upon present consciousness, similar to that of wide landscapes upon a camera-screen."

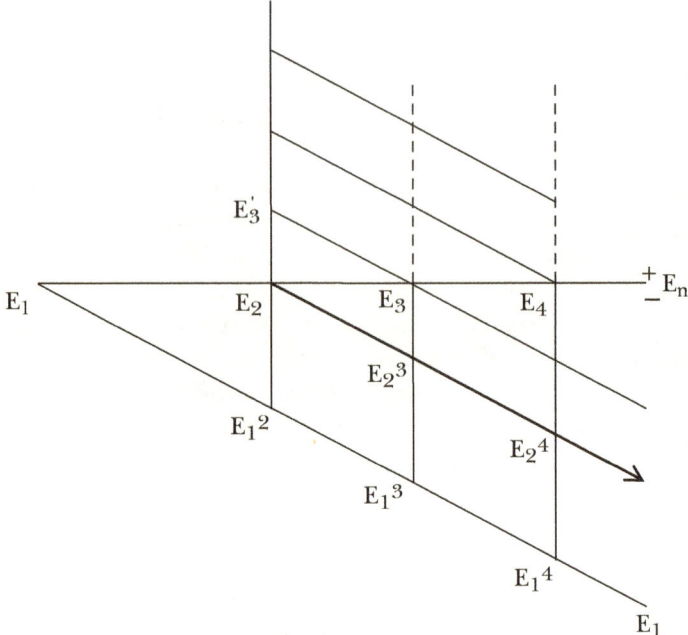

**Figure 3**

continuity of running-off modes." With this diagram as the focal center of renewed reflection, Husserl calls attention to the "remarkable circumstance that each later running-phase is itself a continuity, a continuity that constantly expands, a continuity of pasts [*Vergangenheiten*]" (Hua X, 28 [30]).

A horizontal axis $E$–$E_n$ represents the perceived duration (*Zeitdauer*), or objective temporal succession, of a time-object from its beginning phase, or "now-point," ($E_1$) to its final phase, or "now-point" ($E_4$). As Husserl expresses himself, "as we walk along" the continuum of $E_1$–$E_4$, we encounter a continual transformation (*Abwandlung*) of each now-phase, as each now-phase runs-off, or sinks away, into the immediate past. Each of these now-phases, taken in abstraction from the continuous temporal progression of the time-object as a whole, is a momentary slice or cross-section (*Querschnitt*) of the time-object's duration. Despite such terms as "now-point," "now-phase," and "momentary now-phase," each momentary now-phase is in fact *not* experienced as a point without temporal breadth, though this does not preclude the useful abstraction of considering each now-phase as a point for the purpose of analysis.[13] As Husserl remarks: "Each perceptual phase has intentional reference to an extended section of the temporal object, and not merely to a now-point necessarily given in it and simultaneous with it" (Hua X, 232 [240]). Indeed, each momentary phase encompasses an intentional relation to an earlier now-phase in its mode of running-off as well as to an imminent now-phase yet to come; each now-phase is situated within a perceptual horizon, or "time-halo" (*Zeithof*), of before and after. In this manner, each momentary now-phase can be characterized as what Husserl calls a "phase-continuum," shaped in the wedge of punctuality. Expressed in the visual idiom of Husserl's diagram, the vertical segment $E_2$–$E_1^2$ designates the "phase-continuum," or now-phase, within the consciousness of time in and through which the objective temporal sequence of now-points (horizontal axis $E_1$–$E_n$) is constituted.

Let us introduce a further degree of complexity by identifying a third now-point within the temporal unfolding of a time-object – for example, a three-note melody – in between now-phase $E_1$ and now-phase $E_3$. At now-phase $E_2$, the earlier now-phase B is no longer apprehended as now, yet it is still apprehended as just-now in its mode of "running-off" relative to the apprehension of the newly arrived now-phase $E_2$ (segment $E_2$–$E_1^2$). With the subsequent arrival of now-phase $E_3$, now-phase $E_2$ as well as now-phase $E_1$ are *both* displaced into the immediate past, relative not only to now-phase $E_3$, but just as importantly, relative to each other *as past*. Both now-phases ($E_1$, $E_2$) are still continuously running-off relative to an ever renewed, yet different, now-phase. The earlier of the two now-phases (now-phase $E_1$) does not cease to run-off once its immediate proximity to the actual now-phase has been displaced by the running-off of another now-phase. From the vantage-point of now-phase $E_3$, the continuous running-off of now-phase $E_1$ is further removed relative to the running-off of now-phase $E_2$. That is to say: as the third and final note of a three-note melody enters in the perceptual grasp of consciousness, I am still conscious of the earlier notes $E_1$ and $E_2$ as running-off, and, in this manner, apprehend the sequence of three notes in a determinate order of temporal succession. The melody as a whole is perceptually grasped in the living present. Yet, the earlier notes ($E_1$, $E_2$) are not heard as

13 "The parts that we single out by abstraction can exist only in the whole running-off; and this is equally true of the phases, the points that belonging to the running-off continuity" (Hua XI, 27 [29]).

*equally* just-past relative to the final note $E_3$. Rather, note $E_1$ is given to me as having just been (heard) *earlier than* note $E_2$, which in turn, is still given to me as just having been (heard) earlier than note $E_3$. The original sequence among *earlier* now-phases remains preserved in the consciousness of the immediate past. Earlier now-phases, in their continued mode of running-off, become sedimentated, or layered, upon each other, thus giving depth and distance to the consciousness of the past. As Husserl writes: "Since a new now is always entering on the scene, the now changes into a past; and as it does so, the whole running-off continuity of pasts belonging to the preceding point moves 'downwards' uniformly into the depths of the past" (Hua X, 28 [30]). Once the third note has elapsed, the melody is no longer actually heard, as it has sunken into the remote past, beyond the horizon of the living present and its originary temporal field. I no longer hear the melody; it has become "dead" to me, but still remains an object of experience to which I can return, albeit in the form of remembrance (to which we shall return more extensively in chapters 4 and 5).

Whereas the segment $E_3$–$E_2^3$–$E_1^3$ designates the "phase-continuum" of a now-phase $E_3$ (or *Querschnitt*), the segments $E_1$–$E_1^2$–$E_1^3$ and $E_2$–$E_2^3$ each designate what Husserl calls a "stretch-continuum" of the consciousness of the immediate past, or a "length-cut" (*Längschnitt*), which, as we have just described, trails every apprehension of a now-phase (i.e., segment $E_3$–$E_2^3$–$E_1^3$), akin to the tails of a comet – to note one of Husserl's more vivid metaphors (Hua XI, 30 [32]). Husserl construes this running-off continuum of a now-phase as "stretched" in order to denote the "stretched" character of its consciousness as past, its increasing distance and depth, as well as its increased loss of original intuitiveness until it has slipped into the realm of the "dead," or what Husserl will eventually call "far retention" (cf. chapter 5). As Husserl observes: "Every accomplishment of the living present, that is, every accomplishment of sense or of the object becomes sedimented in the realm of the dead, or rather, [*sleeping*] horizonal sphere, precisely in the manner of a [*steady, continual*] order of sedimentation" (Hua XI, 178 [227]).[14]

Husserl's diagram is meant to provide a "complete image" of the double continuum of time-constituting consciousness as containing both "phase continuums" and "stretch continuums." As Husserl writes:

---

14 An implicit distinction is thus drawn between what Husserl will explicitly call "near" and "far" retentions; see chapter 5.

"We know that the running-off phenomenon is a continuity of constant changes. This continuity forms an inseparable unity, inseparable into extended sections that could exist by themselves and inseparable into phases that could exist by themselves, into points of continuity" (Hua X, 27 [29]). In every now-phase within the consciousness of a time-object, we have a double form of consciousness, or double-consciousness: the consciousness of the succession of now-phases belonging to the time-object $(E_3-E_2^3-E_1^3)$ *as well as* a consciousness of the running-off continuity of each now-phase, in relation to the actual now-phase of consciousness, but also in relative relation to each elapsing now-phase within the immediate past as a whole $(E_1-E_1^2; E_2-E_2^3)$. Importantly, time-constituting consciousness cannot be identified with any *single* segment in the diagram, but rather with the structured field as a whole (Husserl calls this field an "orthogonal manifold"), or what Merleau-Ponty calls an "intentional weave."

The continued running-off of the time-object as a whole, in each of its now-phases, represents the contraction of perspective (the perspective of the immediate past), the increasing of distance, and the lessening of the intuitive vivacity of the constitution of the past in the distances of consciousness. This ordered iteration of temporal displacement is inseparable from the eternal recurrence of a now-phase; each renewed now-phase is *not* the return of the same now-phase, but the renewal of a now-phase in its difference from previous phases. It is as if with the irruption of each now-phase, the entire history of previous now-phases, in their respective modes of running-off, is reinvigorated as well as repositioned relative to a renewed center of gravity that the past could only anticipate but not itself prescribe. The double continuum of original time-consciousness reveals not only that every now has a past, but that every now has its *own* past. The past does not exist in the singular; the continuum of "running-off" is a continuum of "pasts" (*Vergangenheiten*).

## Original time-consciousness

Original time-consciousness is composed of three forms of intentional apprehension, which Husserl calls "primary memory," "original impression," and "primary expectation," and each intends a respective temporal profile of a time-object's now-phase. Each of these three intentional declensions is not an independent act of consciousness, but contributes to the constitution of a perceptual act as a whole. This three-fold intentionality is *internal* to a perceptual act of consciousness; it is an "operative intentionality" (*fungierende Intentionalität*) that renders possible the act of consciousness as such, in its intentional directedness towards an object as well as its own intrinsic consciousness as a lived experience.[15] The operative three-fold intentionality of primary memory, original impression, and primary expectation describes original declensions of time-consciousness; *not* the temporal determinations of the time-object, but determinations of time-constituting consciousness on the basis of which time-objects are objectively constituted. As Husserl argues, "the perceiving of a melody is in fact a temporally extended, gradually and continuously unfolding act, which is constantly an act of perceiving. This act possesses an ever new 'now'-point. And in this now, something becomes objective as now (the tone heard now), while at the same time some one member of the melody is objective as just-past and others are objective as still further past; and perhaps also something or other is objective as 'future'" (Hua XI, 167–168 [171]). When transposed into the visual idiom of Husserl's diagram (Figure 3), each vertical segment (e.g., $E_3^1$–$E_2$–$E_1^2$) represents a three-fold "phase-continuum" within original time-consciousness, and through which the objective succession of the time-object is constituted. As I hear note $E_2$, an original impression provides in an intentional manner the consciousness of note $E_2$ as now; a primary memory provides in an intentional manner a consciousness of note $E_1^2$ as just-now; a primary expectation provides in an intentional manner a consciousness of note $E_3^1$ as yet to come, as almost now. Whereas the phases of the time-object along the horizontal axis are constituted as past, present and future with respect to each other, the intentional declensions of primary memory, original impression and primary expectation are neither past, present or future with regard to each other (within each vertical segment) *nor* with regard to the phases of the time-object. In earlier stages of development, Husserl's diagram noticeably omits any indication of primary expectation. Although Husserl acknowledges the function of primary expectation within the three-fold intentionality of time-consciousness, this omission is revealing of Husserl's preoccupation with the transformation of an original impression into a primary memory. It is only in the Bernau Manuscripts that the dimension of

15 Merleau-Ponty, *Phenomenology of Perception*, 418.

"protention" is accorded its conceptual due (as reprinted in Figure 3), and with consequences that we shall explore in chapter 5.

In this emphasis on the continual running-off of the now, Husserl steadily pursues what Brentano was after with his law of original association. But, whereas Brentano denied any perceptual or intuitive (*anschaulich*) consciousness of a note as just-past, Husserl insists on its perceptual character in the form of primary memory. Each of the three temporal declensions of intentionality is perceptual in character, yet not in the same manner: I *hear* the *entire* melody as a structured relation of before and after. The continuous perceptual character of time-consciousness is centered in the continuous renewal of the original impression and its function as the "source-point" (*Quellpunkt*) and "well-spring" (*Urquelle*) of "intuitivity" as such and its required temporalization. Every original impression is caught in a "constant state of change" – caught in mid-flight so to speak – insofar as every original impression succumbs to its transformation into a primary memory (Hua X, 29 [30]); every original impression necessarily becomes other than itself in "running-off."

Here as elsewhere in his writings on time-consciousness, Husserl's choice of terminology uneasily compresses different strands of meaning into a single conceptual figure, the meaning of which varies over the course of Husserl's thinking, much as a colored object varies in hue over the course of the day. "Original impression" (*Urimpression*) combines two expressions, each of which might at first glance appear redundant in light of the other's meaning. Approached from its historical provenance in Hume, "impression" designates the "vivacity," "immediacy" and "force" with which all perceptions first make their appearance in consciousness; "impression" indicates the fundamental *passivity* of consciousness in the givenness of experience.[16] In Husserl's appropriation, however, "impression" does not refer to an aspect or quality of an intentional object that "strikes" or "impresses" itself on consciousness. Nor is Husserl's employment of "impression" meant to convey the idea of a "time-sensation" or other kind of sensation. Instead, Husserl's "original impression" must be grasped within its argumentative context as a description of the *now-phase of consciousness*, or what Husserl also calls an "impressional consciousness" (*impressionale Bewußtsein*) (Hua XI, 29 [31]). An original impression is

---

16 Cf. Hume, *Treatise*, 1ff.

the renewal of consciousness itself, that form of impressional consciousness that intends the actual now-phase of a time-object. As Husserl notes: "The impression ... must be taken as a primary consciousness that has no further consciousness behind it in which it would be intended" (Hua XI, 90 [94]). Time-consciousness is "primary," "impressional" or "originary" (to string together three equivalent expressions) in the sense of *being* consciousness through and through, as intrinsically conscious of itself as the event of its own lived experience. An original impression is the "null-point" or "zero-degree" (*Nullpunkt*) of temporal orientation in two, related senses: as the proximity of consciousness for itself (the now of consciousness is the consciousness of now); as the absolute reference for the temporal constitution of time-objects: a clock does not show me that it is *now* 3:00 unless my awareness of reading the clock as 3:00 coincides with my awareness of now reading the clock.

The passivity of consciousness denoted by the term "impression" is offset, however, by its further qualification as "original" (*Ur-*), a predicate that infuses the meanings of "creative," "productive" and "spontaneous" into the "passivity" of impression, and thus blends activity and passivity into a self-affection more properly described as medial in the sense of "opening itself" or "self-showing."[17] The self-affection of an original impression is the event of subjectivity in its openness to that which is other than itself. As Husserl specifies: "The original impression is the absolute beginning of this production (*i.e.* temporalization), the primal source, that from which everything else is continuously produced. But it itself is not produced; it does not arise as something produced but through *genesis spontanea*" (Hua X, 100 [106]). And yet, this spontaneity of consciousness – the irruption of original time-consciousness – does *not* create or construct anything. Consciousness perpetually creates itself anew from nothing and yet creates nothing. An original impression is the incessant impression of an original renewal, a repetition of original *difference* (or "in-betweenness"). As Husserl further remarks: "The peculiarity of this spontaneity of consciousness, however, is that it creates nothing 'new' but only brings what has been primarily generated to growth, to development ... It is

---

17 The term "original" (*Ur-*) also harkens back to "original association," which Brentano explicitly distinguished from Hume's "association of ideas" (*Ideenassoziation*); the former is not an association between two ideas, but an association between an impression (in Brentano: a sensation) and its being sensed as just-past.

what is primarily produced – the 'new,' that which has come into being alien to consciousness [*das bewußtseinsfremd Gewordene*], that which has been received, as opposed to what has been produced through consciousness's own spontaneity" (Hua X, 100 [106]). The subtle claim here is that an original impression is an "opening" or "disclosure" of consciousness for itself such that consciousness opens itself or discloses that which is not-consciousness. When Husserl states that "consciousness is nothing without impression," the self-affection of consciousness is neither internal nor external, but the opening of the distinction between internal and external insofar as consciousness differentiates itself from itself. We will return to the significance of this original double differentiation within original time-consciousness.

One cannot address the significance of "original impression" in its numerous meanings – as self-affection, as the origin of intuitivity (or "visibility"), as self-presence, etc. – without considering its necessary transformation into primary memory and, furthermore, without a full consideration of the three-fold intentionality of the living present. Indeed, an original impression is itself an abstraction since an original impression is itself only given through its necessary transformation, or modification, into a consciousness of its running-off, or primary memory. In this manner, the axis of the original impression is "ecstatic," since, in Lévinas' insightful formulation, an original impression "throws its center of gravity outside itself."[18] As we remarked upon earlier, Husserl locates the axis of temporal constitution on the consciousness of an original temporal difference. We perceive the "in-betweenness" of the now-phase, as now and just-now, as one and two, as unity *in* difference; as exemplified with music, every note is a tone-interval, a relation of "pure in-betweenness"; it is through the constant play of grasping the "in-betweenness" of notes that we hear musical structure.

Husserl adopts the term "primary memory" (*frische Erinnerung; primäres Gedächtnis; primäre Erinnerung*) as a designation for the perceptual consciousness of a note as just-past.[19] As Husserl remarks: "We have, then, characterized the *past* itself as *perceived*. In point of

18  Emmanuel Lévinas, *En découvrant l'existence avec Husserl et Heidegger* (Paris: Vrin, 1988), 153.
19  For the sake of expository and interpretative clarity, I reserve the term "primary memory" (as well as "primary expectation") when dealing with Husserl's early interpretation, that is, from 1904–1907, which relies on the schema of apprehension/ content of apprehension. When Husserl's mature view enters, which we discuss in chapters 4 and 5, I follow Husserl in an appropriate switch of terminology, and speak

fact, do we not perceive the passing, are we not directly conscious in the cases described of the just-having-been, of the 'just-past' in its self-givenness, in the mode of being given itself?" (Hua X, 39 [41]). Primary memory is distinguished from remembrance, or "secondary memory." In remembrance, an object already past is brought back into the present, and thus given again, albeit as past; primary memory does not retrieve an object from the past, but instead, "still holds onto" (*noch im Griff*), or retains, the object as past, as just-now. With this conception of primary memory, Husserl arrives at what Brentano was truly after, a *proteraesthesis* in the genuine sense of a perceptual grasp of the just-now.[20]

### "What does immanence here mean?"

Husserl exemplified the ideal of philosophical self-critique that Nietzsche championed in his remark that one has convictions in order to challenge them. Indeed, no sooner had Husserl's phenomeno-logical analysis of time-consciousness settled into the form examined above, that it began to unravel under the pressure of inconsistencies and paradoxes that progressively provoked a significant revision of its central insights, with far-reaching consequences for Husserl's concep-tion of subjectivity as a whole.

As we have thus far witnessed, the drama of Husserl's phenomeno-logical analysis gravitates around the plot of whether, and in what sense, time-consciousness – defined by the guiding problems of the perception of time and the temporality of perception – can be under-stood as a form of intentionality. A phenomenological analysis of the

of "retention" and "protention." However, it should be noted that even in his mature stage, Husserl often slips back and speaks of "primary memory."

20 It is important, however, to signal that the retention of earlier-now phases in primary memory, and thus the constitution of the continuity of temporal passage and time-objects in their duration, is not phenomenologically sufficient as an account of the constitution of objective time and the fixed temporal positions of objects. As Husserl observes: "with the preservation of the individuality of the time-points as they sink back into the past, however, we still do not have the consciousness of a unitary, homogeneous, objective time" (Hua X, 69 [72]). In order to attain a full constitution of objective time, reproductive memory is required, much in the same fashion that the identity of an object is constituted through repetition (re-cognition) and what Husserl terms "Bewährung." This also means that objective time and fixed temporal positions are inter-subjective; the objectivity of time refers to world time of an inter-subjective community.

intentionality of time-consciousness is meant to provide a "description of the naïvely given as well as scientifically presupposed time-consciousness in its meaning [*nach seinem Sinn*]" (Hua X, 188 [194]), and, in this manner, reveal the origin of the concept of time. In sum, Husserl argues that "the time-object becomes constituted in a continuously unfolding act in such a way that, moment by moment, a now of the time-object is perceived as the object's present point while at the same time a consciousness of the past is connected at each moment with the consciousness of the present point, allowing the portion of the time-object that has elapsed up to now to appear as just-past. Apprehension contents are *there* [*da*] *at each moment*: sensations for the now and phantasms for what is past, to the extent that the past was actually intuitive [*anschaulich*]" (Hua X, 234 [241–242]; my emphasis). The constitution of a time-object, in turn, presupposes a perceptual act of consciousness that is itself temporally distended (i.e., mental presence-time) in terms of the three-fold intentionality of primary memory, original impression and primary memory (original time-consciousness). As Husserl reiterates: "an act claiming to give a time-object itself must contain in itself 'apprehensions of the now,' 'apprehensions of the past,' and so on; specifically, as originally constituting apprehensions" (Hua X, 233 [241]).

On this account, each temporal declension of original time-consciousness intends a determinate now-phase of a time-object (as either "now," "just-now," or "not yet now") on the basis of immanent, sensible content; this sensible footing for the apprehension of time-objects, along with their immanent kin – acts of consciousness – is immanent, in the sense of "self-given" and "itself present," in the continuous flow of consciousness itself. As Husserl notes in the passage just cited, immanent contents are "there" or "presently given" (*da*) in consciousness "at each moment" within, or better: *as* the flow of immanent consciousness itself. An original impression, as the impressional consciousness of the now-phase, is based on an immanent sensation (*Empfindung*), whereas the apprehension of primary memory finds its footing in an immanent "phantasm" (*Phantasma*) – Husserl's designation for a modified sensation experienced as just-now, but which, however, cannot truly be described as a sensation, *as a sensing* of a now, since an (immanent) sensation can only "correspond to the consciousness of perception as in a now-consciousness" (Hua X, 234 [241]). This Husserlian adaptation of the term "phantasm" allows for a differentiation *within* immanent time-consciousness between an immanent content specific to a now-apprehension and an immanent content specific to a primary memory. For its part, primary expectation is devoid of any immanent content; primary expectation intends the now-phase yet to come in an empty manner, that is, as devoid of any intuitive "fullness." It is only with the irruption of an original impression that the empty intention of primary expectation becomes fulfilled with "intuitivity," as Husserl will further develop in the Bernau Manuscripts. More accurately stated, original impression is the well-spring of "intuitivity" or "visibility" as such, and, in this light, the source of immanent content as such. In contrast to Brentano's employment of the term "phantasm" for the modified sensation in original association, Husserl, as we have seen, emphatically insists on the perceptual, or intuitive (*anschaulich*) character of primary memory: we *hear* the note as just-past; we *see* the now in its running-off. Moreover, whereas the modification of sensation in original association is, for Brentano, produced by the imagination (*Phantasie*), Husserl's "phantasm" is not an imaginary semblance of the now, but its ghostly presence, as animated by the apprehension of primary memory.

A first difficulty with Husserl's thinking resides in the argumentative burden placed on the tripartite distinction of transcending intentional object, objectifying act of apprehension, and non-intentional immanent content. In this framework, intentionality is composed of an intentional object that transcends consciousness, an objectifying act of apprehension, and its underpinning non-intentional experienced content; whereas it is the musical note itself that appears to me, and which I hear, its underpinning sensible contents are not perceived as such, but, instead, lived (*erlebt*) as the "real" (*reell*) content of imma-nent consciousness. Since Husserl argues that the temporal determin-ations of a time-object – the meaning of "past," "present" and "future" as determinations of a transcendent object – become constituted in and through forms of temporal apprehension, immanent content must therefore lack any *intrinsic* relation to the now-phases of a time-object. Immanent contents are, in this regard, "non-temporal" or "extra-temporal" (*unzeitliche Materie* or *außerzeitliche Materie* (Hua X, 417), neither past, present nor future. But, if immanent content is "non-temporal," it would be entirely arbitrary whether *this* immanent content exhibits *this* intended now-phase of a time-object as either now or just-now. As John Brough wryly notes: "That it [*sensible content*] is in fact apprehended as Now is a piece of good fortune for which the theory does not account."[21] In phenomenology, however, nothing should be gratuitous. In fact, Husserl's own "law of modification" according to which every original impression *necessarily* becomes modified into a primary memory implies a necessary and specific temporal characterization of immanent content. As Husserl realizes: "The primary contents that spread out in the now *are not able to switch their temporal function*: the now cannot stand before me as not-now, the not-now cannot stand before me as now. Indeed, if it were otherwise, the whole continuum of contents could be viewed as now and consequently as coexistent, and then again as successive. That is evidently impossible" (Hua X, 322 [334–335]). Beneath the surface of immanent givenness, sensible content must already be *pre-given* in a temporal manner. With this glimpse into the realm of "pre-givenness" – more fully explored and exploited under the heading of genetic phenomenology (cf. chapter 7) – Husserl begins to overcome the last vestiges of a conception of subjectivity that purportedly first encounters the world at the level of constituted objectivities.

The temporal neutrality of immanent content generates, however, a second difficulty, and specifically for Husserl's construal of the consciousness of the immediate past, or just-now. Primary memory is meant to provide for a perceptual consciousness of the immediate past that is given along with the perceptual grasp of the now and the perceptual anticipation of the now yet to come. In effect, Husserl wants us to see that we *see* the running-off of the now through the eyes, so to speak, of primary memory. If immanent content is temporally neutral, any given sensible content could equally well serve as an immanent footing for either the apprehension of a now-phase as now or as just-now. As Husserl (self)-critically remarks: "We everywhere supposed here that what is in time is constituted by a content, really experienced in time-consciousness, that is animated by a temporal representation, by the time-apprehension. The *question* then is: Cannot the *same* content that is now the *presentant in a perception arbitrarily* function as the *representant in memory*?" (Hua X, 317 [329]).[22] In one regard, the issue raised here is comparable to the difficulty sketched

---

21 John Brough, "The Emergence of an Absolute Consciousness in Husserl's Early Writings on Time-Consciousness" in: *Husserl: Expositions and Appraisals*, ed. F. Elliston and P. McCormick (Notre Dame: Notre Dame University Press, 1977), 92.

22 Here is an example of how inattentive Husserl can be with his terms: by "memory" he means "primary memory."

above: is it viable to argue that *this* lived experience of perceiving could equally function as the experiencing of perceiving the now or as the experiencing of perceiving the just-now? Yet, this difficulty reveals a deeper problem with Husserl's conception of immanence. That Husserl could even consider immanent content as temporally neutral reflects his underlying conception of such content as immanent, and of immanence as intrinsically "self-given" or "self-present": sensations *and* phantasms are equally, and thus primordially, immanently "present" or "there" (*da*) *in* consciousness, as its real (*reell*) content. But if "phantasms" are contained immanently in consciousness, the apprehension of a now-phase *as past* would thus be based on an immanent content in consciousness that itself is "present" or "there" (*da*), or, in other words, a consciousness that is itself present, or now, for itself. But can the past be given as past on the basis of the present? Indeed, instead of delivering the perceptual givenness of the just-now, primary memory only provides a "ghostly presence" ("phantasm") of the now-phase. Although neither an image nor an imaginary semblance, the just-now of primary memory remains nonetheless the ghost – something present – of absence; not absence itself. Husserl has yet to exorcise the ghosts of the past from the consciousness of the past; indeed, this paradox of primary memory is a relic of Husserl's immediate past, since it repeats Husserl's original censure of Brentano's original association: "*The being-present of an A in consciousness* through the annexation of a new moment, even if we call that new moment the moment of the past, is incapable of explaining the transcending consciousness: A is past" (Hua X, 18 [19]; my emphasis). Much as with Brentano's original association, Husserl's primary memory lacks a radicalism of vision. In light of this impasse, Husserl, however, begins to formulate the genuine requirements for an intuitive consciousness of an immediate past, not only more clearly, but in terms more clearly phenomenological. As Husserl writes: "In short, there is a *radical alteration*, an alteration that *can never at any time be described in the way in which* we describe the *changes in sensation* that lead *again to sensations*" (Hua X, 324 [336]). This internal critique of the apprehension/content of apprehension schema points in the direction of a temporal transcendence that is originary, yet *non-objectifying*. When viewed from the theory of retention, which would replace primary memory (to which we turn in the subsequent chapter), primary memory appears in hindsight as still beholden to an "objective" character of apprehension, and thus structurally similar to the objectification of "secondary memory." The givenness of the immediate past, as a genuine instance of transcendence, cannot be based on something present – whether an image, a modified sensation, an immanent content, or even a ghostly consciousness that is itself present.

Yet a third difficulty further compounds the inadequacy in Husserl's proposed account. As noted, the non-temporal givenness of immanent contents is a function of their (self-)givenness as real (*reell*) content of immanent consciousness. Upon closer consideration, however, immanent contents cannot truly be characterized as temporally neutral since, insofar as "apprehension contents are *there* [*da*] *at each moment*: sensations for the now and phantasms for what is past, to the extent that the past was actually intuitive [*anschaulich*]," immanent contents are by default *now* or *present* (*da*) in or "as" immanent consciousness. Moreover, since immanent contents (as well as acts of apprehension) are given in the flow of immanent time-consciousness, their immanent temporality remains thus far *unquestioned*. This unquestioned default status of immanent contents further implies that each sensible moment within the three-fold declension of time-consciousness must be given simultaneously – precisely because "apprehension contents are *there* [*da*] *at each moment*," i.e., *there together*. The intentional declensions of original impression, primary memory and primary expectation are neither past, present, or future with regard to each other, nor with regard to the constituted phases of the time-object. Given that the three declensions of original time-consciousness are thus by default simultaneous with each other, Husserl unwittingly succumbs to the dogma of momentary consciousness *despite* his explicit appropriation, as discussed earlier, of Stern's "mental presence-time" and its inspired critique of Brentano; once again, Husserl repeats a critique, first raised against Brentano, verbatim against himself: "Can a series of coexistent primary contents ever bring a succession to intuition?" (Hua X, 323 [335]).

These three difficulties surrounding the function of primary memory within the apprehension of a time-object rest on an unquestioned assumption of the temporality of immanent consciousness. In addition to the "objective" orientation of phenomenlogial analysis towards the temporal givenness of time-object, Husserl also fashions, as discussed earlier, a second line of inquiry directed towards the temporality of a perceptual act. Indeed, the perceptual apprehension of a time-object implicates directly the temporality – or mental presence-time – of a perceptual act itself. And yet, the course of Husserl's analysis thus far has mainly followed the first line of inquiry without any extended pursuit of the second line, even though the argumentative burden of Husserl's account clearly falls on the implied, but as yet little examined, temporality of time-constituting consciousness. This double-line of inquiry reflects the structure of intentionality: "every experience is 'conscious' and consciousness is consciousness *of* ... but every experience is *itself experienced* [*erlebt*] and *to that extent* is also conscious" (Hua X, 291 [301]). Expressed in yet another manner: "Perceiving is the consciousness of an object. As consciousness, it is at the same time [*zugleich*] an impression, something immanently present [*ein immanent Gegenwärtiges*]" (Hua X, 89 [94]). If time-objects are constituted through the temporality of perceptual acts, how is the temporality of perceptual acts constituted, that is, how is immanent consciousness itself constituted in its own temporal self-givenness? How is the consciousness of time at the same time, the time of consciousness?

At this stage of thinking, Husserl relies on the "apprehension/ content of apprehension" schema of intentionality in order to illuminate *both* dimensions of intentionality of time-consciousness, despite the stated *difference* between the manner in which intentional objects appear to consciousness and the manner in which consciousness experiences itself, or is "self-given." As Husserl remarks: "we would prefer to avoid, then, the use of the word 'appearances' for the phenomena that constitute immanent time-objects; for these phenomena are themselves immanent objects and 'are' appearances in an entirely different sense" (Hua X, 27 [29]). On the one hand, immanent contents and acts of apprehension are "immanent time-objects" in a qualified and guarded sense since, as a lived experience necessarily possessing a temporal form, the immanent unity of act and content must *in some sense* have the form of an "object," albeit an "inner" object of consciousness: the immanent unity of act and content is "something immanently present." Yet, on the other hand, an immanent unity of act and content cannot possess the form of an object since, *as an immanent consciousness*, they are "lived through" (*erlebt*) or experienced in an intimate manner. Indeed, consciousness is "something immanently present" *for itself*, or, in other words, intrinsically self-conscious or self-aware.[23] Consciousness does not *appear*

---

23 Of the different expressions (inner consciousness, self-awareness, self-consciousness, absolute consciousness), I prefer "self-consciousness" and "absolute consciousness," with an occasional use of "inner consciousness."

for itself unless consciousness renders itself an object through an act of reflection; in the midst of lived experience, however, consciousness is conscious for itself – experiences itself – in a pre-objective and pre-reflexive manner. In his 1906/1907 lectures, Husserl fashions the concept of "absolute consciousness" to speak of the "time-stream" (*ein Zeitstrom*) in which acts of consciousness and their sensible underpinning are constituted. "Absolute consciousness" is the intrinsic self-experiencing (or self-consciousness) of consciousness that is always manifest in any consciousness of something, but which is itself neither a perception nor an objectifying consciousness (Hua XXXIV, 246).

With this conceptual precision in mind, Husserl proposes to distinguish *within* immanent consciousness between constituting "absolute" consciousness and constituted immanent consciousness. The reasoning here is that if consciousness is *itself* "an impression, something immanently present [*ein immanent Gegenwärtiges*]," then immanent consciousness must in turn be constituted: the distinct phases within the consciousness of time must also be constituted as phases belonging to consciousness. This distinction within immanence between a constituting absolute consciousness and a constituted consciousness mirrors structurally the distinction between constituting act and constituted (transcendent) time-object. In this manner, the temporality of immanent consciousness forces Husserl to recognize a transcendence within immanence, and so distinguish between two senses of immanence. As Husserl asks: "*What does immanence here mean? [Was besagt hier die Immanenz?]*" (Hua X, 279 [289]).

As Husserl observes: "*Immanent* can indicate the antithesis of *transcendent*, and then the temporal thing, the sound, is immanent; but it can also signify what exists in the sense of absolute consciousness, and then the sound is not immanent" (Hua X, 284 [294]). On the one hand, an intentional object is immanent in an intentional manner, that is, it is not a "real" (*reell*) content of consciousness; yet we can nonetheless speak of it as immanent in its transcendent character, since it is "contained" within the correlation of intentionality. On the other hand, as Husserl specifies, "Immanence of the identical temporal object, the tone, must surely be distinguished from the immanence of the tone-profiles and the apprehension of these profiles, which make up the consciousness of the givenness of the tone" (Hua X, 283 [293]). The point here is that the *consciousness* of a time-object is also immanent, but in a different sense than the intentional object, and precisely in the sense of belonging to consciousness in a real (*reell*) manner. Insofar as an act of consciousness is itself a temporal manifold, an immanent unity, it must, by implication, draw its temporal determinations from a constituting act, or what Husserl identifies as an "absolute time-constituting flow of consciousness," which, as indicated by the term "absolute" is in turn not itself constituted by another aspect of consciousness. Due to an unquestioned assumption of an isomorphic relationship between the constitution of a transcendent time-object *in and through* immanent consciousness and the constitution of immanent consciousness *in and through* absolute consciousness, Husserl is led to distinguish between "the flow of absolute consciousness," a "pre-empirical time" of immanent consciousness (act of apprehension along with its sensible content) and transcendent time-object in its temporal determination (Hua XI, 73 [77]). Expressed in terms of the double continuum of "transverse" and "lengthwise" segments, the distinction between constituted transcendent object and constituting time-consciousness designates the transverse intentionality (phase-continuum), whereas the distinction between constituted time-consciousness (immanent unity of act and its content) and constituting absolute time-consciousness designates the "lengthwise" intentionality (stretch-continuum). The note as a time-object is constituted in the perceptual act of hearing, which is itself, as interpreted by Husserl, an inner temporal "object" constituted in an "absolute" consciousness. When I hear a note, I am at the same time conscious of myself as hearing the note.

Although this tripartite distinction within time-consciousness as the temporalization of transcendent objects as well as the self-temporalization of consciousness reveals the *unified* temporal constitution of intentionality in its two-fold orientation of "other-directedness" and "self-directedness" (and thus delineates time-consciousness as the opening of transcendence and immanence as such), this early sketch of absolute time-constituting consciousness suffers from an intractable difficulty since it generates the threat of an infinite regress.[24]

Husserl's proposed solution for how consciousness is self-constituted in a temporal manner requires the formulation of a distinction within immanent consciousness between the act of consciousness as an

---

24 For a fuller treatment, Dan Zahavi, *Self-Awareness and Alterity: A Phenomenological Investigation* (Evanston: Northwestern, 1999), 69ff.

"immanent time-object" and its constituting absolute consciousness. Yet the characterization of this distance, as a form of intentionality, within consciousness on the model of an apprehension/content of apprehension introduces a division between "reflected" and "reflecting" consciousness; an act of consciousness would thus be an "immanent time-object" for an *inner consciousness*, removed as a spectator from the act of consciousness itself. Time-consciousness is both spectator and actor: the consciousness of time is also the time of consciousness, yet paradoxically, since the act of consciousness is separated from the spectator, even though both belong to one and same consciousness insofar as the latter consciousness *is* the consciousness of the former. Moreover, we would have to inquire further into the constitution of absolute time-consciousness itself, since, according to the principle that the temporality of an object is grounded in the temporality of an act, that absolute consciousness in which an immanent time-object is temporally constituted, must itself be temporal, and must therefore thus itself be constituted through another consciousness ... Or else absolute consciousness avoids an infinite regress if it were unconsciousness, and so, paradoxically loses its meaning as consciousness at precisely the moment where consciousness finds its own origin as consciousness.

Within this distance, it is not so much the distinction, or difference, between immanent and absolute consciousness as such that is problematic, as it is the way in which this difference is cashed out in terms of the apprehension/content of appehension schema of intentionality. Husserl is led to apply this schema of intentionality to both dimensions of intentional consciousness given the unquestioned *simultaneity* between the consciousness of time and the time of consciousness. As quoted above: "Perceiving is the consciousness of an object. As consciousness, it is at the same time [*zugleich*] an impression, something immanently present [*ein immanent Gegenwärtiges*]" (Hua X, 89 [94]). The expression "at the same time" (*zugleich*) in *this* particular context means both "simultaneously" (*Gleichzeitig*) *and* "of the same sense of temporal presence," i.e., as a form of *perceptual presence*, as captured with the term "object" (as *something* that is present). This unquestioned simultaneity further implicates a form of linearity to *both* the progression of objective time (the succession of now-phases of a time-object) and the sense in which consciousness "flows." Husserl remains unable to establish a structural difference between the stream of consciousness and the succession of time-objects that is *required* in order to avoid the problem of an infinite regress *but also* the collapse of temporality into the blink of an eye.[25]

Given these seemingly intractable difficulties, one may question whether Husserl's phenomenology of time-consciousness has at all progressed beyond Husserl's own critique of Brentano or whether, instead, the difficulties and entanglements of the problem of time are rehearsed once again, this time *within* the framework of Husserl's phenomenological thinking. Contrary to Husserl's announcement in the *Ideen* that "the efforts of the author concerning this enigma [*the enigma of time-consciousness*], and which were in vain for a long time, were brought to a conclusion in 1905 with respect to what is essential; the results were communicated in lectures at the University of Göttingen" (Hua III, 198 [194]), nothing in fact has been brought to conclusion. As Husserl expresses himself in a revealing passage in the ITC lectures (and omitted from the 1928 edition):

> Meanwhile the longed-for clarity calls us after long hours, we think that the most glorious results are so near at hand that we have only to stretch our hands forth to grasp them. All difficulties seem to dissolve, our critical sense mows down contradictions one by one, until only one final step remains. We make a sum of our results. We begin with a self-conscious "therefore," and then at once a point of difficulty raises its head that begins to become larger and larger. It spreads and spreads into a form of horror that devours all our arguments and reanimates contradictions we have just mown down. The corpses all revive and grin at us mockingly. Our struggle and effort have to begin all over again. (Hua X, 393)

Are these revived corpses the ghosts of Brentano's original association? Husserl must begin again, and yet not from where he first began.

---

25 Rudolf Bernet, "Einleitung" to Edmund Husserl, *Texte zur Phänomenologie des inneren Zeitbewußtseins (1893–1917)* (Hamburg: Meiner, 1985), xxx.

# 4

# THE RETENTION OF TIME PAST

*Ja am liebsten spreche ich selbst über Dinge, die noch nicht erledigt, vielmehr im Fluß begriffen sind.*

— Husserl

## An indirect approach

Husserl conceived of the ITC lectures as an initiation to the phenomenological problem of time-consciousness, not only for his students, but most of all for himself. As with many of his lecture courses, Husserl develops a path of thinking without any preconceived expectation of where it might lead. As Husserl remarks, "we want to follow [these problems] however far we can. Where we can proceed no further, we at least want to formulate clearly the difficulties and possibilities of interpretation; we want to make clear to ourselves where the genuine problems reside, and how to give them a conclusive formulation" (Hua X, xvii). Such an experimental approach allows Husserl to suspend the demand for conclusive results and polished arguments that a presentation in book form would necessarily have required. As Husserl candidly notes, "where I remain silent as an author I can therefore speak as a teacher [*Worüber ich mich als Autor ausschweige, darüber kann ich mich als Lehrer darum doch aussprechen*]" (Hua X, xvii). Subsequent writings on time-consciousness remained committed to exploring the "difficulties" and "possibilities of interpretation" that rapidly elevated the theme of time-consciousness to "the most difficult of all phenomenological problems" (Hua X, 276 [286]). Indeed, as discussed in chapter 3, despite significant progress, difficulties originally manifest in Brentano's original association still haunt Husserl's thinking, resurfacing in unexpected form within Husserl's own analysis.

The 1928 publication of the ITC lectures (but also in the *Husserliana* edition) misleadingly suggests that the theme of time-consciousness developed in isolation from other themes in the constellation of phenomenological philosophy. Such a view is in fact reinforced by Husserl's comment in the *Ideen* that "time is the name for a completely *delimited sphere of problems*" (Hua III, 198 [193]). The absence of an extended treatment of time-consciousness in Husserl's published writings and lecture-courses, with the exception of the ITC lectures, further reinforces the view that the problem of time-consciousness ran its course in isolation from the other developments of Husserl's phenomenological thinking. Yet, as Husserl himself stresses, "it is inherent to the inner entwinement as well as the uniqueness of phenomenological problems that they cannot be solved in isolation, and that first one part, and then another part is required, such that every progress in clarification shines back onto another part" (Hua X, xvi).

In fact, in the original format of the 1904/05 lecture-course from which the ITC lectures are drawn, the theme of time-consciousness is treated in connection with perception, the imagination and remembrance, each a type of sensible presentations (*sinnliche Vorstellungen*) that were excluded from the *Logical Investigations*. As Husserl remarks: "The entire sphere of remembrance and therefore also the entire problems of a phenomenology of originary time-perception are in the *Logical Investigations* so to speak 'left unspoken'" (Hua X, xvi). Aside from belonging to the class of sensible presentations, presupposed by acts of judgment discussed in the *Logical Investigations*, an analysis of time-consciousness and remembrance each respectively confronts the issue of how an object that is not-present is nonetheless given in the present, and in an intuition, as befitting the fundamental *sensible* character of both time-consciousness and remembrance. As examined in chapter 3, Husserl's analysis of time-consciousness remained haunted by the issue of how to account for the perceptual apprehension of the "just-past" or "immediate past," which Husserl struggles to describe in terms of primary memory. In the case of remembrance, a phenomenological investigation of its form of intentionality also addresses the issue of how an object of the past is apprehended in an intuitive manner without conflating, however, the distinction between primary and secondary memory. As we shall discover in this chapter, the themes of "de-presentification" and the "givenness of absence" addressed in the analysis of

time-consciousness resurface in different forms in Husserl's investigation of imagination, remembrance and image-consciousness: in
each of these cases, an object is apprehended as "absent," as either
"unreal," "having once been," or "present in image only."

As argued in this chapter, Husserl's analysis of the imagination and
remembrance, launched in a systematic manner in the 1904/05 lectures, provide the decisive clue for resolving the *aporia* of primary
memory encountered in chapter 3.[1] Over the course of his investigation of the imagination and memory, Husserl discovers what he calls
the "double-consciousness" or "double-intentionality" of "representi-
fication" (*Vergegenwärtigung*) – an original mode of consciousness in
which an object is given as "not-present." In light of the "double-
intentionality" (or "double-consciousness") of remembrance, Husserl
is guided towards the double-intentionality of what he now calls, after
shedding both the term and the concept of primary memory, "retentional consciousness."[2] This understanding of retentional consciousness in turn delivers a further clue for resolving the infinite regress of
absolute time-constituting consciousness.

## The decoupling of image-consciousness and imagination

Not surprisingly, Brentano's "psychology without a soul" provided the
catalyst for Husserl's phenomenological investigations of image-
consciousness (*Bildbewußtsein*), imagination (*Phantasie*) and remembrance (*Erinnerung*). In lectures on descriptive psychology during the
1880s, Brentano undertook a systematic examination of different types
of presentations (*Vorstellungen*), and centered his investigations on clarifying the difference between perception, or "authentic presentation"
(*Wahrnehmung* or *eigentliche Vorstellung*), and the imagination, or
"inauthentic presentation" (*Phantasie* or *uneigentliche Vorstellung*) – a

---

1 Robert Sokolowski, *Husserlian Meditations* (Evanston: Northwestern, 1974), 152:
  "Husserl's new analysis of remembering and imagining allowed him to resolve the
  vexing problem of how retentional consciousness works. But he did not immediately
  apply his new principles to the description of actual, non-presentational experiencing.
  There was a period of time in which he had established the two-track structure of
  memory, but still thought a kind of phantasm or present datum was needed for
  'primary memory' or retention to apprehend."
2 Husserl's discovery of double-consciousness resurfaces in his analysis of the constitution
  of "other-ego" in the Fifth Cartesian Meditation (see chapter 7).

term that broadly covers any consciousness of an "unreal object" (*Unwirklichkeitsbewußtsein*), including remembrance and expectation, yet excluding hallucinations, illusions and dreams. These latter instances are characterized as either a deformation of perceptual consciousness (i.e., I hallucinate seeing an oasis in the desert) or a "quasi-perceptual" experience for a consciousness that is asleep (i.e., dreaming). In contrast to hallucinations and illusions, the imagination does not deceive, for the givenness of an imaginary object is not directly comparable to a distortion of a perceptual given. If I visualize in my mind's eye – imagine – a unicorn, I am not deceived into thinking that this imaginary unicorn is a real object present in the room. My imaginary unicorn is not comparable to the pink elephant in the corner. *Phantasie* (translated here as either imagination or fantasy consciousness) covers any act of wakeful consciousness in which an object is apprehended as "unreal" or "irreal." Whereas a perceptual object is given to consciousness "as its own person" (*als eigener Person*) and, in this manner, itself appears in the present (*selbst gegenwärtig erscheint*), an imaginary object is given in an "inauthentic" presentation, in which it appears *as if* it were actually present and perceptually given.

In the *Logical Investigations*, Husserl followed in his mentor's footsteps by considering the inauthentic presentation of the imagination as a form of "image-consciousness" (*Bildbewußtsein*).[3] Ever since Plato, the imagination has been interpreted as the faculty for the production of mental images (*phantasmata*). In the *Philebus*, Plato describes the soul as inhabited by both a scribe and a painter, each diligently recording the soul's perceptions through inscriptions (or impressions) and images drawn on its surface. Imagination is "inauthentic" in the sense that its objects are not authentically present, even though they are presented to consciousness in the sensible form of an image. In the 1904/05 lectures "Imagination and Image-Consciousness" (*Phantasie und Bildbewusstsein*), Husserl submits this traditional conception of the imagination as an image-consciousness to closer phenomenological scrutiny in the wider context of examining different types of "inauthentic presentations." In the process, Husserl sets into motion a descriptive argument that decouples the classical union of image consciousness, the imagination and remembrance. This decoupling stands as one of the seminal achievements of Husserlian

---

3  LU V, § 14; Hua XXIII, 109 [117].

phenomenology, long over-shadowed by his analysis of time-consciousness, yet indispensable for grasping how Husserl moved beyond the "difficulties" and "problems of interpretation" discussed in chapter 3.

Although Husserl still considered the imagination to be a kind of image-consciousness in the *Logical Investigations,* and thus still remained beholden to Brentano's construal of the imagination as an inauthentic presentation, the seeds for its eventual decoupling from image-consciousness were first sown in Husserl's critique of the "image-theory" of perception. In an appendix to §§ 11 and 20 of the Fifth Investigation, Husserl takes issue with the classical image-theory of perception, according to which internal mental representations purportedly mediate between mind and world, by arguing that an image is constituted as its own form of intentionality that is distinct from, and not to be conflated with, the intentionality of perception in which an object is given in "flesh and blood," as itself actually present before my eyes. As Husserl argues: "[t]he constitution of the image as the image takes place in a peculiar intentional consciousness, whose *inner* character, whose *specifically* peculiar mode of apperception, not only constitutes what we call image-representation as such, but also, through its particular inner determinateness, constitutes the image-representation of this or that definite object."[4] The difference between the perception of a table and the perception of *the image* of a table is irreducible. In the former instance, it is the object itself – the table – that we perceive, not an image or other surrogate. In the latter instance, consciousness intends an object as the depiction of something; we perceive the *image itself* as the depiction of something. As Husserl observes: "In a representation by images the *presented* object (the original) is *meant,* and meant by way of its image as an apparent object."[5] An appeal, however, to an inherent quality of resemblance between the image and its depicted object still falls short of a satisfactory answer as to how an image refers to its object; that two objects resemble each other does not yet render one of these objects the image of the other. Husserl suggests that the relationship between an image and its depicted subject must itself possess the form of an intentional accomplishment on the basis of a specific function of meaning-bestowal. Any theory of consciousness that conflates the

4 Hua XIX.1, 437 [593].    5 Hua XIX.1, 437 [593].

perception of a table with the perception of its image remains blind to the distinct meaning-accomplishments of perception and image-consciousness. Yet, this phenomenological expulsion of the image from perception still leaves unexplored the type of intentionality responsible for the constitution of image-consciousness. Although the Fifth Investigation offers a few clues, it is not until the 1904/05 lectures "Imagination and Image-Consciousness" that Husserl undertakes a fuller, but by no means conclusive, phenomenological investigation of image-consciousness.

## Image-consciousness

As I admire Constable's *Wivenhoe Park*, the painting's depicted landscape is seen in a manner unlike how other perceptual things – this chair, this lamp – are seen or unlike how the trees, the mill, and the stream are seen outside my window. Although Constable's painting is a physical object hanging in a room surrounded by other objects, its depicted trees can neither be said nor seen to exist, nor not exist, in the same manner in which the trees outside my window can be said and seen to exist, or cease to exist, should I decide to cut them down. An image essentially consists in what Husserl identifies as the phenomenon of "seeing-in."[6] An image is a perceptual object *in* which something is depicted, which I perceive without perceiving in the manner in which I actually perceive this chair, this table, and the very canvas of the painting itself. As Husserl notes, "image presentations [are] remarkable presentations in which a perceived object is designed to present and is capable of presenting another object by means of resemblance; specifically, in the well-known way in which a physical image presents an original" (Hua XXXIII, 17 [19]). We do not look at an image without seeing something in an image. When I look at this table, even if this table is ornately decorated, I do not in truth see something *in* its givenness other than that which is given to me as the table. The appearance of the table does not contain an appearance of something other than that of the table. In contrast, an image is a perceptual appearance (a physical object that appears) *in* which something other, which itself does not *actually* appear, is given in light of a resemblance between an image and its depicted subject.

6 Husserl speaks of "hineinschauen" and "hineinblicken" (Hua XXXIII, 34; 31).

An image is therefore not simply the appearance of an appearance; it is the appearance of something other in an appearance, not as another appearance, but as *other* than the appearance in which it is given.

As with any objectifying form of consciousness, the intentional object – the image – transcends consciousness, but whereas in the case of perception, consciousness intends a single object (as given in a synthetic unity of changing adumbrations and situated within shifting perceptual horizons), in the case of an image, consciousness apprehends a stratified intentional object, composed of an "image-thing" (*Bildding*), an "image-object" (*Bildobjekt*) and an "image-subject" (*Bildsujet*) (Hua XXXIII, 19 [20]). Each of these objects is the intentional correlate of a respective form of apprehension; taken as an integrated whole, these three intentional acts, along with their intentional objects, constitute the consciousness of an image in its full significance. In looking at an image, I perceive a perceptual object (an image-thing) as an image (an image-object), and apprehend this image-object as the image *of* something (an image-subject). When looking at a photograph of my mother, for example, I see a physical photograph, an image, and my mother's face. These three modes of seeing are not constitutionally identical (as each gives its corresponding object in a distinctive manner), yet they are nested within each other on the basis of an underlying perception of the image as an image-thing. An image is a thing like other things – an image can be hung on a wall, placed next to a table, or hidden from view. But, an image is significantly unlike other things, for something is given in an image unlike how other things (this table, this pen) are perceptually given in this room. If I only perceived the image *as a thing* (if I only perceived the canvas as a canvas), I would not be able to perceive an image-object; I would only see a thing like other things. Yet, if I perceived only the image-object, entirely detached from its perceptual footing in an image-thing (i.e., the canvas), I would equally be unable to perceive an image, since some perceptual thing must at least be perceived in order for something to be seen as an image in which something is depicted. Image-consciousness requires that I apprehend this image-thing as an image (I recognize this painted surface as an image); on the basis of this apprehension of an image, I apprehend the depicted subject ("the trees") through a form of resemblance between the image-object and image-subject. Image-consciousness is a double-apprehension: seeing a depicted image-subject *in* the image-object is based on, but not the same as, the seeing of an image-thing *as* an image-object.[7]

Due to its hybrid form of appearance, the image-object plays a crucial role in the constitution of image-consciousness. The image-object is neither a straightforwardly perceptual object (i.e., an image-thing) nor entirely severed from a perceptual given, as the image-subject and its purely "imaginary" or "spiritual" (*geistig*) manner of givenness in those cases in which the depicted subject is itself not an actual object that I could perceive.[8] Neither entirely visible nor entirely invisible, the image-object straddles the visibility of the perceptual image-thing and the invisibility of the "spiritual" image-subject. The image-object is a "perceptual fiction" (*perzeptive Fikta*), "image-fiction" or "image-figment" (*Bildfiktum*). In Husserl's description: "The image-object is a *figment* [*ein Fiktum*], but not an *illusory* figment, since it is not – as in the case of an illusion – something harmonious in itself that is annulled by the surrounding reality (or, correlatively, in the positing in which something harmonious conflicts with something harmonious)" (Hua XXXIII, 490 [585]). The image-object is not posited as "something that actually exists." Its mode of givennness is neutralized by the imagination in so far as the image-object is given in the consciousness of a "fiction" or "figment," immune and indifferent to the positing of actual existence, through which an image-subject becomes depicted. And yet, this image-object nonetheless retains a *perceptual* character, albeit in a modified form. As Husserl argues, "the image-object is however given in a perceptual apprehension modified by the characteristic of imagination ... The appearance belonging to the image-object is distinguished in one point from the normal perceptual appearance. This is an essential point that makes it impossible for us to view the appearance belonging to the image object as a normal perception: it bears within itself

---

7 The stratified intentional objects of image-consciousness need not be each "awakened" in the constitution of any particular experience of an image, though some thing must be seen as an image at the very least; otherwise, we would simply perceive a thing, not an image.

8 Take for example a photograph of my mother: both the image-thing (the photograph as a physical object) and image-subject (my actual mother, to which corresponds the imaginary image-subject qua being depicted in the photograph's image) enjoy an independent perceptual existence; in the case of an image of a unicorn, the image-subject exists solely as an object of my imagination that I apprehend through the image-object (its image).

the characteristic of *unreality, of conflict with the actual present*"
(Hua XXXIII, 47 [51]). Perceptual consciousness has undergone a
modification through the protrusion of the imagination *into* perception; the imagination modifies the perceptual given in such a manner
that it enters into conflict with perception. In setting itself into conflict with the perceptual present (of the underlying perceptual apprehension of the image-thing), consciousness "de-temporalizes" the
image-object; it modifies the temporal form of the image-object's
manner of givenness. In this manner, image-consciousness is the
"appearance of the not-now *in the now* ... '*In the now*', in so far as the
image-object appears in the midst of perceptual reality and claims, as
it were, to have objective reality in its midst ... Yet, on the other hand,
a '*not now*' insofar as the conflict makes the image-object into a *nullity*
[*einem Nichtigen*] that does indeed appear but is nothing, and that may
serve only to *exhibit* something existing" (Hua XXXIII, 47 [51]). On
the basis of this conflict within the double-apprehension of image-
thing and image-object, the image-subject can saturate or penetrate
(*durchdringen*) the apprehension of the image object. Husserl
construes the dynamic relation between the image-object and the
image-subject in terms of gradients of intuitive fulfillment – the
apprehension of the image-subject *through* the image-object may
approach, but never completely attain, the ideality of the image-
subject. The image-subject fulfills the image-object as a function of
increased attainment of the ideality of a perfect resemblance between
the two; however, the image-object can never entirely fulfill the empty
intentionality of the image-object.

Image-consciousness is a consciousness of "otherness" (*Bewußtseins
des "Andersseins"*) based on a conflict (*Widerstreitsbewußtseins*) and
"doubling of consciousness" (*Verdoppelung des Bewußtseins*) through a
*temporal* modification of consciousness and its intentional object (Hua
XXIII, 32; 46–47). Consciousness is "doubled," in conflict with itself:
the image is more than its physical embodiment to the degree that
what I see in the image is more than just the physical object as an
image. I must perceive the lines, brush strokes, etc., as the resemblance of the depicted image-subject, for what constitutes the consciousness of an image is precisely the consciousness of the *difference*
between image-object and image-subject. This consciousness of difference constitutes a *temporal distance within appearance* in the specific
sense of "an appearance of a not-now in a now" (*eine Erscheinung eines
Nicht-Jetzt im Jetzt*) or "something that does not appear in something

appearing" (*eines Nicht-erscheinenden im Erscheinenden*) (Hua XXXIII, 29 [30]). As soon as the difference, or distance, between image-object and image-subject collapses, the god, so to speak, becomes the image – the otherness of the "something that does not appear" appears *as* the present image. As soon as the conflict between image-thing and image-object collapses, the image becomes mundane, a non-image, just another thing.

## The imaginary

Whereas image-consciousness is based on the conflicted double-apprehension of image-thing as image-object and image-object as image-subject, the imagination, Husserl argues, does not possess the *same* form of constitution (i.e., the same type of "double-consciousness"). In contrast to image-consciousness, the conscious-ness of an imaginary unicorn is *not* based on the perception of an image-thing and an image-object. To see an image of a unicorn is not the same as to imagine seeing a unicorn. In the case of the imagin-ation, the relation between the imaginary unicorn and any perceptual given is severed; the givenness of an imaginary unicorn is not based on "something that is present" (*Gegenwärtiges*). As Husserl notes, "in phantasy, we do not have anything 'present' [*Gegenwärtiges*], and in this sense, we do not have an image-object ... the relation to the present is lacking completely in the appearance [*of the imaginary*]" (Hua XXXIII, 79 [86]). As with image-consciousness, Husserl also speaks of the imagination as a consciousness of otherness, but the sense of otherness that enters into play in the imaginary is of a more radical form, given the absence of any basis on something present. Indeed, the imaginary is an intuitive consciousness of that which is not present, not, however, in the sense of something that once was present (as is the case with objects of remembrance), but in a stronger sense of something that could never be actually present. As a form of intentionality, imagination is an objectifying and self-transcending act of apprehension that intends an intentional object as irreal. As an irreal object, the imaginary unicorn transcends consciousness, yet its transcendence does not move in the same register as perceptual transcendence.

In acts of the imagination, an imaginary object is given to con-sciousness in and through a self-produced semblance of its own per-ceptual activity, or "quasi-perception." If I sing a melody "in my head," and, in this fashion, imagine hearing a melody, I do not actually hear the melody I nonetheless "hear." To imagine a melody is to imagine "hearing" a melody much as to imagine a unicorn is to imagine "seeing" a unicorn. In the imagination, the noematic *and* the noetic components of intentionality are appropriately modified; it is *as if* I *perceived* a unicorn. On the noetic side, consciousness has undergone what Husserl calls a "reproductive modification" in so far as consciousness reproduces, or replicates, its own perceptual activity in a modified form. On the noematic side, the intentional object is modified in terms of its givenness (*im Wie seiner Gegebenheit*). On this account, "pure" imagination (*reine Phantasie*) is distinct from the envisioning of a possibility or wishing a certain state of affairs, in which cases the imagined object is posited as a "possible actuality," and thus indexed with the temporal character of a possible now. When I purely imagine a unicorn, however, I do not imagine this unicorn as something that I might possibly perceive or would want to possibly perceive; the character of the "as if" should not be conflated with the character of the "might be."[9]

The upshot of Husserl's suggestive analysis is that the imagination is the appearance of appearance with a double meaning. As a "double-consciousness," consciousness induces within itself a semblance of its own perceptual activity in a modified form. I create a semblance of my own perceptual activity in imagining that I "see" a unicorn. This semblance of perception constitutes the "as if" (*als ob*) or "irreal" character of the imaginary object: the unicorn appears to me *as if* it actually appeared to me because in imagining the unicorn it is *as if* I actually perceived a unicorn. When I imagine a unicorn, I am not seeing an image of a unicorn in my mind; instead, I seem to see a unicorn, when in actual fact I am not actually seeing anything at all. And in resembling what it is like to see a perceptual object without actually believing that

---

9 For a discussion of the relation between the consciousness of possibility and the imagination, see Hua XXIII, 546–564 [659–677]. Under the heading of *Vergegenwärtigung*, Husserl distinguishes between "pure imagination," anticipation (*Vorerinnerung*) and "the entertaining of possibility" (*Gegenwartserinnerung*). Instances of *Vorerinnerung* (e.g., "I expect that so and so will happen") and "entertaining of possibilities" (e.g., "I wonder what's on the other side of this wall") entail the positing of an intentional object as possible, either as a possible future or as a possible present. Acts of the imagination, however, do not entertain their objects as possible objects in the temporality of inter-subjective transactions since the presence of the imagined object is characterized as irreal.

I am in fact looking at something real, consciousness comes to "neutralize" itself to the extent that its own simulated or "quasi" perception is experienced, yet not posited for itself as an actual experience.[10] Consciousness in this fashion becomes itself "irreal" and comes to experience itself "at a distance," in Rudolf Bernet's apt characterization.

Husserl thus convincingly uncouples the imagination from image-consciousness in distinguishing between the "consciousness of conflict" (*Widerstreitsbewußtseins*) of image-consciousness and the "consciousness of the not-present" (*Nichtgegenwärtigkeits-Bewußtseins*) of the imagination (Hua XXIII, 59 [64]). Both forms of double-consciousness, however, belong to the genus of "re-presentification" (*Vergegenwärtigung*), as defined in contrast to perception, or "presentification" (*Gegenwärtigung*). *Vergegenwärtigung* is a consciousness of alterity or "other-being" (*Andersseins*). But, whereas image conscious-ness is a mediated consciousness – something other is presented in something presently given – imagination is, in one sense, akin to perceptual consciousness since its object is directly given, without the mediation of an image or intermediary representation, internal or otherwise. Contrary to Brentano's assessment of the imagination as an "inauthentic presentation," a phenomenological examination reveals the imagination as an authentic presentation in its own right.[11]

The flowering of Husserl's central insight into the discontinuity between the imagination and perception (as well as the decoupling of the imagination from image-consciousness) hinges on a subtle attentiveness to the fluctuating character of an imaginary object's manner of givenness. When I imagine "seeing" a unicorn, this imagin-ary appearance, as an intentional object of consciousness, fluctuates before me, much as an object under the pulsing illumination of a strobe light. An imaginary object is Protean in its givenness; not only in the sense that an imaginary object can be transformed into another object (a unicorn into a griffin), but just as importantly, in the sense that an imaginary object is unstable in its givenness. Imaginary objects lack the stability of perceptual objects and, in this regard, possess a different form of objectivity by virtue of their modified form of tem-porality. Due to an imaginary object's discontinuity with any actual

---

10 This form of neutralization, under the heading of the "as if," must not, in turn, be conflated with the neutrality-modification of which Husserl speaks in *Ideen* (§§ 109–112).
11 Hua XXIII, 86 [93]: "Vergegenwärtigung ist ein letzter Modus intuitiver Vorstellung."

perceptual presence, the imaginary object cannot be constituted as an individual object in the same sense as a perceptual object. In fact, it is questionable whether we can even speak of an imaginary object as an individual object given its temporal form: in what sense is the imaginary unicorn, imagined today, the same as the imaginary unicorn, imagined yesterday? As Husserl claims: "An individual cannot be properly and fully imagined" (Hua XXIII, 552). An imaginary unicorn, as given in a determinate act of consciousness, does not occupy a fixed and determinate objective temporal position. When I imagine "seeing" myself on an Italian beach while at the same time half-listening to my colleague speak, the imaginary scene conjured in my mind's eye and the dull voice of my colleague cannot be placed in any direct temporal relation. Strictly speaking, the imaginary object and the perceptual object are not simultaneous with each other. As Husserl argues: "Aber eines fehlt notwendig in der bloßen Fiktion, das, was wirklich existierende Gegenstände auszeichnet die absolute Zeitlage, die 'wirkliche' Zeit, als absolute, ernstliche Einmaligkeit des in Zeitgestalt gegebenen individuellen Inhaltes. Deutlicher: Zeit ist zwar vorgestellt, sogar anschaulich vorgestellt, aber es ist eine Zeit ohne wirkliche und eigentliche Örtlichkeit der Lage – eben eine Quasi-Zeit."[12] This insight into the constitutive connection between different senses of objective being and different forms of temporality will gain in profile in the Bernau Manuscripts where Husserl more fully exploits an ontological correlation between regions of being and structures of temporality, and as grounded in the temporality (*Zeitlichkeit*) of original time-consciousness.

Even though an imaginary object is given in a "quasi-time," and, in this regard, cannot be placed in any direct relation, temporal or otherwise (i.e., causal, spatial), with perceptual objects, Husserl argues that imaginary objects can, in hindsight, be placed in such a relationship. An imaginary object can be "re-temporalized" in a secondary apprehension on the basis of an association, or analogy, with a perceptual object. In this manner, the imaginary object becomes objectified into an image. However, in such a secondary apprehension, we can only apprehend the act of imagination as "having been present," but not as having *once been actually present*, as would be the case in remembrance, when I recall an object as having once

---

12 Edmund Husserl, *Erfahrung und Urteil*, ed. Landgrebe (Hamburg: Felix Meiner, 1999), p. 197.

been actually perceived. In the case of the imagination, the imaginary object was never given as actually present, and thus, its secondary apprehension in the form of an image is (paradoxically) a reactivation of a past without an assignable present.[13] An intriguing problem, however, is that this implies a reactivation and reawakening of an experience that was never constituted as an actual experience in the first place – a remarkable consequence that we shall touch upon below.

This insight into the alternative temporality of the imagination as a double-consciousness forces Husserl to revise his apprehension/content of apprehension interpretation of intentionality. If interpreted along the lines of this "apprehension schema," the difference between perception and imagination would exclusively reside in different forms of apprehension. On this interpretation, an underpinning non-intentional sensation is itself present, as actually lived experience, on the basis of which an intentional act of apprehension intends its transcendent object. But in that case, the underlying sensible content would remain neutral. Even though Husserl speaks of "sensation" and "phantasm" as different kinds of immanent content correlative to perceptual acts or acts of the imagination, in both instances, the apprehension of an object, as either perceptual or as imaginary, is based on an content that itself is present in immanent consciousness. In fact, in the *Logical Investigations*, the imagination is considered to be an intentional consciousness with an imperfect manner of fulfillment in contrast to the perfect manner of fulfillment of perceptual experience in which the object is itself given "in flesh and blood." Yet, Husserl's analysis of the imagination has shown that the imaginary object is given as irreal in the temporal sense of "not actually something present." In perceptual consciousness, my consciousness relates to an object in an intentional manner while also "relating" to myself, that is: I experience myself as perceiving the tree. I experience myself as actually perceiving. In the case of the imagination, if such acts are based on "irreal" phantasma, in what sense does consciousness "actually experience" itself – how are irreal "phantasma" experienced? Immanence cannot be construed as an indifferent, monotone grey of "reell" content that is differentiated at the level of

---

13 For a useful discussion of this issue, see Samuel Dubosson, *L'imagination légitimée: La conscience imaginative dans la phénoménologie proto-transcendentale de Husserl* (Paris: l'Harmattan, 2004), 143ff.

acts of intentionality. Rather, consciousness becomes itself modified in modifying its own sense of experiencing. The imagination is a modification of consciousness *through and through*, in which its self-presence becomes modified; its own character *as consciousness*, as lived experience, becomes re-constituted in a different form of immanent temporality or self-temporalization. The act is experienced as not-actually-present; consciousness is given to itself as not-present in the specific mode of "irreal." Immanent sensible content must already be temporalized along with the specific temporal character of its act of intentionality; in other words, the imagination is constituted *as a form of inner time-consciousness, as a modification of time-consciousness itself.*

## Headless temporality

In this context of investigation, Husserl draws a fundamental distinction between "presentification" (*Gegenwärtigung*) and "re-presentification" (*Vergegenwärtigung*) as two irreducible forms of consciousness. Husserl no longer considers the difference between perception – as a "presentification" in which the intentional object is given as real – and the imagination – as a "re-presentification" (*Vergegenwärtigung*) in which the object is itself given as irreal – as constituted through different kinds of apprehension, as first advocated with an apprehension/content of apprehension schema of intentionality. As Husserl remarks: "the difference between perception and imagination does not reside on the side of their respective and shared objectification [*since, indeed, both are forms of transcending consciousness*], in which the 'appearance' of objects is accomplished, but rather in the characterization, which respectively constitutes the difference between 'presentifying' [*gegenwärtig*] or 're-presentifying' [*vergegenwärtigt*]" (Hua XXIII, 101). On the strength of a newly discovered phenomenological insight into the temporal modification of consciousness in its *character as consciousness*, the "double-consciousness" of the imagination is an "authentic" (*eigentlich*) mode of givenness in which an intentional object is *itself* apprehended as "not-present" or "absent" in the specific temporal sense of "irreal." In the term "re-presentification" (*Vergegenwärtigung*), the prefix "re-" (*Ver-*) does not signify the giving again of an original in the manner of a copy. "Re-presentification" is robust as an original mode of intentionality; the object of consciousness is "re-presentified" to the extent that an underlying imaginary consciousness of the object (i.e., semblance of perception or "quasi-perception") is "re-produced," that is, given for itself as other than itself, as itself not *actual*. Under the heading of "reproductive modification," Husserl designates how consciousness can modify itself in a temporal fashion such that it replicates itself, produces its own semblance, and in so doing gives itself an irreal (and modified) object of consciousness in and though an "irrealization" of consciousness itself.

In distinguishing between *Gegenwärtigung* and *Vergegenwärtigung*, Husserl finds a solution to one problem – the constitution of the imagination as a form of intentionality not based on something present – only to discover another: how is one and the same consciousness aware of both imagining and perceiving if imagination (*Vergegenwärtigung*) and perception (*Gegenwärtigung*) are beyond each other's reproach? When I imagine singing a tune "in my head" while listening to the radio, although these experiences are discontinuous in the sense argued above, I am nevertheless aware of both experiences as belonging to the same unity of my time-consciousness; the act of imagining as well as the act of hearing are constituted in immanent time-consciousness. Strictly speaking, an imagined melody and a melody actually heard are "contemporaries," since they are both given in immanent acts within one and the same – my – consciousness, yet they are not simultaneous with regard to each other. An imagined melody cannot be placed in any direct temporal relation with a melody actually heard. And even though music can only be imagined if it is "played" in the imagination, and thus, constituted as a time-object in a quasi-temporality, an imagined melody cannot be individuated in objective time. The imagined melody that I "hear" is not given as an actual experience; it is an irreal experience, with another form of temporality, and yet it is nonetheless still actually lived as an experience: I am aware of myself as imagining that I "hear" a melody. To imagine that I "hear" a melody *is* to actually experience the quasi-perception of hearing a melody without thereby actually hearing the melody. The consciousness *that* I am imagining is not an imagined consciousness. As Husserl insists: "the imagination is inconceivable without an actual ego that imagines [*Phantasie ist nicht denkbar ohne aktuelles Ich, das phantasiert*]" (Hua XIII, 303).

This difference between the consciousness *that* I am imagining and the imagined act of consciousness (the quasi-perception of hearing along with its intentional object) is internal to time-consciousness. As a "doubling of consciousness," consciousness reproduces itself, as we have argued, in a modified, temporal form, and in such a manner that consciousness provokes an internal "division" or "splitting" within its own ego (*Ich-Spaltung*). In Husserl's vocabulary, the imagined object (i.e., the imagined melody) is nested (*eingelegt*) within a quasi-perception of an "imaginary ego" (*phantasie-ego*), which, in turn, is nested in the "present ego" (*das jetztige Ich*) (Hua XXIII, 113; 115). The "imaginary-ego" is the non-actual and "irreal" consciousness to which an imaginary object is given; yet the "imaginary-ego" is itself "given" to an actual ego or actual consciousness *that* I am imagining that I hear a melody. As a modification of consciousness, the "imaginary-ego" is an irrealization of consciousness, and lacks any actual empirical determinateness. The double-consciousness of the imagination is a *(self-)transcendence within immanence*: consciousness transcends itself within its own immanence, and becomes other than itself in its own transcendence: in inducing a semblance of my own perceptual activity, I am for myself in a manner that is other than how I actually experience myself.

Whether I imagine "hearing" a melody or "seeing" a unicorn, my consciousness is primarily directed towards the imaginary object as such. It is the melody that I "hear"; it is the unicorn that I "see." As with any experience of intentionality, I am not explicitly directed towards the imagined act of hearing (or "quasi-perception"), even though I may, as with any kind of experience, attend directly in reflection to my imagined act of hearing, in which case the imagined act becomes the object of my consciousness. As a lived experience, self-consciousness – non-objective and pre-reflexive – intrinsically belongs to the imagination. As I sit during a department meeting and imagine myself on an Italian beach, this flight of fancy takes the significance of a wishful evasion from my current boredom. Although this act of imagination is discontinuous with perceptual experience (I can pretend to pay attention to the professor while imagining my day at the beach), it is nevertheless nested within the broader horizon of other possible experiences, perceptual or otherwise. And yet, in spite of the discontinuity between imaginary consciousness and perceptual consciousness, a fundamental continuity remains untouched, namely, that in both instances, looking at my colleague and giving myself to acts of my imagination are lived experiences, of which I am aware, and as such are constituted in and through inner time-consciousness. I can always be called back to a perceptual awareness of my surroundings and become explicitly aware that I was daydreaming: the chairperson

calls on me to answer a question I failed to hear; in heeding her
call, I am caught unaware that I had been elsewhere in my flight
of fancy.

Acts of the imagination in and through which imaginary objects are
given to consciousness are constituted immanent time-objects, yet
Husserl does not regard the embedded character of imagined acts, or
quasi-perceptions, within immanent time-consciousness on the model
of an inner perception. The imagined act of "hearing" a melody is not
an inner "object" of consciousness. In fact, if an imagined act was based
on an inner perception, it would forfeit its character of "re-presentification" (*Vergegenwärtigung*) by virtue of having a basis on an inner *perceptual* act, and thus being based on "something that is present." Even
though the imaginary act is a self-affected modification of my own
consciousness, I am conscious of myself as imagining in a pre-reflexive
and non-objective manner. Both dimensions of consciousness – as
imagined consciousness and as imagining consciousness – within the
"two-sidedness of consciousness" (*Zweispaltigkeit des Bewußtseins*) are
given in different, yet inseparable ways, as incompatible with each
other.[14] The character of consciousness as experiencing itself as
*actually* (absolute time-constituting consciousness) imagining is *incompatible* with the irreal, yet lived, character of "hearing" (i.e., the quasi-
perception as constituted immanent time-object). In this manner,
double-consciousness is contemporaneous, without being simultaneous, with itself. As Husserl remarks: "the imagined ego is *at the same
time* and one with the absolute subject, and yet different from it [*Das
imaginierte Ich ist zugleich und in eins mit dem absoluten Subjekt, und doch von
ihm verschieden*]" (Hua XIII, 314; my emphasis). Consciousness "exists"
on two planes without coinciding, and it is this "non-coincidence" that
structures the unity of consciousness as a distance within itself, as its
own distant contemporary. The freedom of the imagination is the
freedom of a strangeness I become for myself.[15]

This constitutive incompatibility of consciousness with itself
leads to a remarkable implication for Husserl's conception of time-
consciousness, the full significance of which eluded Husserl's thinking,

---

14 Rudolf Bernet, *Conscience et existence. Perspectives phénoménologiques* (Paris: PUF, 2004),
   112.
15 Bernet (117) suggests the illuminating idea that the imagination provides the
   phenomenological "matrix" (*la matrice*) for every other form of alterity and, indeed,
   for the exteriority of subjectivity.

even if an occasional sign attesting to the possibility of something "really strange" can be found in his writings. As Husserl once remarks: "what about a phantasy understood as the re-presentification (of a present) [*Vergegenwärtigung (eine Gegenwärtigung)*]? It grants a possible present but not an actual present, and accordingly not an individual present. But that is really strange" (Hua XXIII, 552 [623]).

In a perceptual consciousness, or "presentification" (*eine Gegenwärtigung*), consciousness actually experiences itself and its intentional object. As detailed in the previous chapter, a perceptual act is temporally constituted in terms of the three-fold declension of an original time-consciousness. Time-consciousness, in other words, is not a specific form of intentionality; it is the foundation for intentionality as such, as is in evidence with Husserl's analysis of different forms of intentional consciousness as different forms of temporalizations of consciousness, or what Husserl captures under the heading of different "characterizations" of consciousness. In our discussion of the intrinsic temporality of perceptual acts, we already noted that original time-consciousness is an "operative intentionality" in terms of which intentionality is at all possible. The actual lived experience of perceptual consciousness is constituted temporally in and through an original time-consciousness; its three-fold structure of temporalization springs from the axis of an original impression in terms of which the intuitivity of experience – the "force," "vivacity" and "immediacy" of experience – is originally constituted. By contrast, an imaginary consciousness is not itself actually experienced, even though it is lived, but from a distance. Indeed, if it were the case that my imagined "seeing" of a unicorn was actually experienced as a perceptual act of seeing, I would, strictly speaking, no longer imagine that I "see" a unicorn, but instead be under the spell of a hallucination or illusion. But if an imaginary act of consciousness, along with its corresponding "irreal" objects, is not constituted as an actual experience of consciousness, it means, by implication, that the temporality of such imaginary acts cannot be based on the wellspring of an original impression. The source-point of temporality becomes suspended or neutralized, modified in its temporality, and, in this manner, deprived of its originary meaning of actual givenness. The temporalization of the imaginary is, in other words, *headless*, bereft of an ever-renewing stabilizing axis of temporalization. As Husserl expresses the matter: "While at the head, the living process receives new, original life, at the feet, everything that is, as it were, in the final acquisition of the retentional synthesis, becomes steadily sedimented."[16] In the case of an imaginary consciousness, we are faced with a "really strange" situation of a temporality bereft of an actual present. It is in terms of a headless temporalization that the fluctuating givenness of the imaginary object (and therefore in truth: a quasi-object) is constituted. An imaginary object is *originally* constituted in retentional modification and protentional indeterminateness without passing through or emerging from the wellspring of an original impression. The reactivation of an imaginary object in its re-constitution as an image, in which we have an equally remarkable re-doubling of consciousness, produces an "image phantasis" (as opposed to the "pure phantasis") (Hua XXIII, 84). But in such an instance, we have the paradoxical feature of a reactivation of a past without an assignable present. The imagination differs from remembrance, in which case, we have the reactivation of a past with an assignable present.[17] It is in this emphatic sense that the intentionality of the imagination gives to consciousness not only an imaginary object as "not-now," but also its own quasi-perception, or imaginary consciousness, as "not-now." As Husserl repeatedly asserts: "in the imagination we do not have anything present [*Gegenwärtiges*]" (Hua XXIII, 79) – indeed: not even something present of consciousness itself.

### The decoupling of imagination and remembrance

Ever since Plato and Aristotle, the theory that memory requires a "memory-image" has rendered the imagination indispensable for the possibility of memory.[18] The problem of memory finds its classical formulation in Aristotle's short treatise *De memoria et reminiscentia* where memory is explicitly defined with a reference to time: "memory is of something that has passed or elapsed the past" (449b15). With this definition in hand, Aristotle rejects the possibility of a memory of the immediate past since memory presupposes temporal distance, that is, an *elapsed* – as opposed to an *elapsing* – duration, or interval (451a29–31). As noted in chapter 2, Aristotle also distinguishes

---

16 Hua X, 178 [227].
17 See the excellent discussions in Dubosson (149ff.) and in Marc Richir, *Phénoménologie en esquisses: Nouvelles fondations* (Bruxelles: Jérôme Millon, 2000).
18 In this discussion, I shall not distinguish terminologically or conceptually between memory and remembrance.

between memory and perception with reference to time. To perceive something is to perceive an object as actually present: perception reveals the present. By contrast, to remember an object is to remember an object that was once experienced as present. Based on an "inactual sensation," memory possesses an indirect relation to perceptual presence since memory refers to a present that once was, and which, by definition, can no longer be given. According to Aristotle, all thinking is based on images, and memory is no exception; memory thus belongs to the imagination (449b30; 450a22), yet the images of memory have the specific function of establishing an indirect relation to perceptual presence. These images (*phantasmata*) are produced in the soul by the "sense-images" (*aisthemata*) of perception. Memory-images are thus not the same as the sense-images created in the soul, but are subsequently produced, once a perception has expired.

The argument that memory is based on images has yet to demonstrate how these memory-images refer to the past. As Aristotle remarks, "one might be puzzled how, when the affection is present but the thing is absent, what is not present is ever remembered" (450a25). An image is something present in the soul ("the present state or affection"), yet if memory reveals the past, how can something present reveal something absent? Faced with this difficulty, Aristotle offers a solution that has long prevailed in the history of philosophy by proposing that a memory-image is a "copy" or "exhibits a likeness." A memory-image is "like a sort of picture or painting" of what is absent (450a29). Much as a painting, a memory-image (*phantasma*) represents what is "other" (*heteron*) than itself by virtue of being a copy of what it is not. Accordingly, a memory-image can be taken as either an image or as an image of something, as the copy of an original. As Aristotle states, "the figure is either a figure or a copy, and while being one and the same, it is both, even though the being of the two is not the same" (450b21).

In his fragmentary reflections on memory leading up to his 1904/05 "Imagination and Image-Consciousness" lectures, Husserl subscribed to Aristotle's view of memory as belonging to the imagination. In these early reflections, Husserl often speaks of "memory-images" (*Erinnerungsbild*), and describes memory as picturing (*abbilden*) its objects. In light of Husserl's decoupling of image consciousness from the imagination, it is evident that Aristotle's definition of memory as "a kind of painting" is phenomenologically untenable. Aristotle's "image-theory" of memory presupposes the consciousness of the past it is meant to explain: to recognize an image as a copy of an object that is past presupposes that I can already identify that object as past in order to recognize this image as an image of that object as past. Nothing inherent in the consciousness of an image signifies something as past. To express the matter differently, the Aristotelian distinction between figure (*zoon*) and "copied likeness" (*eikon*) matches Husserl's distinction between "image-object" (*Bild-Objekt*) and "image-subject" (*Bild-Sujet*), and must therefore be classified as an example of image-consciousness, but not as a convincing account of memory.[19] As Husserl acknowledges, "memory is not image-consciousness but something entirely different." Yet, the argument that memory does not involve memory-images still leaves intact the connection between memory and the imagination. In fact, in the beginning sections of the 1904/05 lectures, Husserl retains Brentano's view that memory and the imagination belong to the same act type of "re-presentification" (*Vergegenwärtigung*). Memory is based on a *phantasm*, even if this *phantasm* is no longer interpreted as an image.

Arguably, the persistence of the imagination's perceived indispensability for memory is a function of the sharp distinction made between memory and perception – a distinction that is reinforced with Husserl's distinction between secondary memory and primary memory, and the inclusion of the latter within the temporal arc of perception, the living present. Memory and the imagination are mutually defined, and thus bound, by a shared contrast with perception; memory and the imagination are both forms of givenness of "something that does not appear" (*Nicht-Erscheinende*). This shared boundary with perception obscures, however, any further distinction between the imagination and memory. Even if memory is "entirely different" from image-consciousness, it does not therefore automatically follow that memory is entirely different from the imagination, once the imagination has itself been shown to be entirely different from image-consciousness. Husserl's decoupling of image-consciousness and the imagination, however, places the relation between the imagination and memory on a new footing *while at the same time* reinforcing a shared contrast with perception. The imagination is an intuitive givenness of the alterity of "not-now" (*Nicht-Jetzt*) in the form of a "double-consciousness." With this basic feature of "re-presentification" (*Vergegenwärtigung*) in

19 Cf. Paul Ricoeur, *La mémoire, l'histoire, l'oubli* (Paris: Éditions du Seuil, 2000), 56.

hand, Husserl explores the double-structure of consciousness in the specific case of memory. Is there an essential – eidetic – difference between the double-consciousness of memory and the imagination?

There is an apartment in Paris where I once lived as a child that I remember fondly and clearly. I remember this apartment's individual rooms, the smell of the wooden floors, the color of its walls, the plaster ceiling molds, and countless other details, large and small. In addition, I also recall specific events – reading a particular book in a green upholstered chair, watching a certain film on television, various discussions with family members. These experiences are given to me directly in remembrance. Expressed in Husserl's terminology, each of these intentional objects (the green chair in the dining room, the book shelf in the main hallway, etc.) is given to me as being-past; indeed, the givenness of these objects in remembrance is indistinguishable from their givenness as objects *of the past*. By definition, these objects (as Aristotle already stressed) are "of the past," no longer to be regained. If I thought that an object of the past was given to me again as actually present, I would be in the grips of a hallucination. If I believed that these footsteps were actually present again – footsteps ostensibly remembered from my past – I would no longer recall them as past, but hear them as present, and would thus be hallucinating that my brother was creeping up along the same hallway as he once did in the past. In remembrance, I can hear again the footsteps on the wooden floor, yet I do not actually hear these steps as I would hear the steps of a person actually entering my room. But I also do not hear them in the manner as I would hear a melody that I imagine singing in my head. When I remember a melody from the past, the object is given as an object that was once actually given to me in a now that is no longer. The remembered past is always relative to the present in which it is remembered since the past, as having once been present, is understood as no longer present only in relation to the present in which I remember. The now of remembrance is never simultaneous with the now that I remember. I remember something *from* my past and "relive" the past from the point of the view of the present. I see the distance between the present and the past in recognizing the past as past. When I remember something from the past, I apprehend the remembered object in its "distance from the present of the actual now" (Hua X, 182). Past experiences are "lived again" or "lived afterwards" (*Nachleben*); they enjoy an "after-life" in the living present of my consciousness.

As I remember my childhood apartment, different objects can be grasped with a relative degree of "emptiness" and "fulfillment." I may recall an object only vaguely, having only the dimmest of memories; I may remember incorrectly the placement of a table in a particular room and stand corrected when another person remembers more clearly. I may search in my memory for the details of an experience that seems too distant to recall. In each of these examples, the object of memory is given to me in a sliding scale of possible fulfillment. Remembered objects can be intended in an "empty" fashion, as when I recall having met a person without recalling that person's name or face. They can also be recalled in a "fulfilled" manner, as when I finally remember where I left my keys. Cases of mistaken memories can be explained with reference to the play of "empty" and "fulfilled" intentions, but only an act of memory can fulfill or "cancel" ("strike-out"), validate or invalidate, another act of memory.

This dynamic of empty and fulfilled intentions also characterizes another aspect of remembrance. As I remember my childhood apartment, I can move among my memories, "walk again" in different rooms and inspect its various items. I can recollect a past conversation or the unfolding of a past experience in detail. In short, I can narrate my past and follow the chains of memories from one event to another, either by reconstructing an original chronological sequence or meandering by way of association, random or deliberate. A memory is never given in isolation; memories always arrive in the *implicit* plural. Every act of remembrance casts a net of implications around itself that calls upon the original "context" or "environment" (*Umgebung*) of its past occurrence. Of course, we need not pursue the many paths of memory. We may remain content to entertain a single memory. Yet the possibility of expanding the orbit of light around any given memory is in principle always available to us. In Husserl's terminology, an act of remembrance is inseparable from "intentions of together-ness" (*Zusammenhangsintentionen*) (Hua XXIII, 262). Much as an act of perceptual consciousness is situated within a perceptual nexus of determined and undetermined intentions, as well as inner and outer horizons, an act of remembrance is likewise determined by inner and outer horizons that reach beyond it along tracks first constituted in the original perceptual experience. These surrounding intentions refer either "backwards" to earlier experiences or "forwards" to later experiences. Ideally, the entire course of my life between any given memory of the past and the present in which it is remembered could be recalled, and made explicit step by step in sequences of memories. I can never become stranded in the past, since from any given memory of the past, I can always find my way back to the actual present in which I remember.

As a double-consciousness, remembrance does not possess the same eidetic structure as the imagination, even though both belong to the broad category of "re-presentification" (*Vergegenwärtigung*). In both cases, the intended object is given in a dimension of experiencing that is discontinuous with perceptual experience. Neither the rooms I remember nor my imagined unicorn can be placed alongside, in objective space-time, objects in my perceptual field. Nothing speaks directly for or against my memories *as memories* in the perceptual world. Even if another person remembers correctly what I have incorrectly remembered, I must still acknowledge this correction by either correctly remembering myself ("yes, I do remember it that way") or suspend the reliability of my memory in accepting the truth of another's testimony ("it *must* be as you remember it"). In the case of the imagination, as discussed above, the imagined object is given as "not-now" (as "what does not appear as present"), or "irreal," *in and through* a "quasi-perception" that is itself not present, or now, in immanent consciousness. Consciousness has "doubled" and transcended itself by producing a semblance of its own perceptual activity. In the case of memory, double-consciousness also has a specific temporal character. The intentional object of memory is grasped as having once been actually present. As Husserl remarks: "Pure imagination is a 'semblance' of perception; it is as if I perceived an object in a here and now, however 'seeing' is not 'seeing-again' and 'having already seen,' and the object is not 'past' and posited as past here and now" (Hua XXIII, 287).

Remembrance emphatically relives the past because in recalling an event or object of the past, consciousness replays itself by "implicating" its (own) past consciousness. As Husserl notes, "Every remembrance of A is at the same time a remembrance of an earlier perception of A" (Hua X, 197). To take an example from *À la recherche du temps perdu*: in remembering his childhood in Combray, Marcel perceives his past experience from the perspective of his present self. Marcel relives the consciousness of his childhood: he experiences again waiting for his mother's kiss, the dread of his father's anger, his fascination with Swann's ways. An object of past experience (the imposing shadow of Marcel's father lumbering up the stairs) is given *as past* in reliving the original consciousness of the object (Marcel's dread of his father's anticipated anger). In other words, the remembered object is given to me only in so far as it is given in a "reliving" of the past intentionality of its experience. The remembered object is not made into a real (*reell*) content of my present consciousness; as a remembered object, it transcends my present consciousness much as my own past perceptual experience is given again, but as transcending the now in which I remember. My own past consciousness is given to me again in a reproduced manner, but given to me as transcending the immanence of my (now) consciousness. Marcel's illumination of his childhood experiences, that is, the understanding of his past self that only a perspective removed from the past is able to generate, is a function of the overlapping of "time lost" and "time regained" in a "doubling" of consciousness.

In remembering a past experience, I must recognize that experience as having once been actually present and experienced by me. I must be implicitly aware of having once perceived this table as present in the past. Since perception is the act that constitutes the present, to be given again as having been present means to be given as having been perceived, or what Husserl calls a consciousness of "having-been-perceived" (*Wahrgenommengewesen-sein*) (Hua X, 57). I must be aware of the "otherness" (*Anderseins*) of the reproduced consciousness in contrast to my present act of remembering. I confront myself as other and this otherness that I am for myself temporarily inhabits the actual consciousness of the now. The flow of consciousness transcends and suspends itself by reproducing its past self. In this regard, memory has a double intentionality: the past object is intended as well as the past perceptual consciousness in which that object had originally been given (i.e., constituted) as present. Yet these two branches of intentionality do not refer in the same manner. The intentionality directed towards my past perception is not of the same character as the intentionality directed towards the past object. The intentionality directed towards my past consciousness should not be conflated with an act of reflection, in which an act of consciousness becomes "thematized" and "objectified."[20] Whereas the

---

20 However, I can make this original act thematic in remembrance: I can reflect upon a past experience without having originally reflected upon my perceptual consciousness. In these instances, Husserl speaks of a "double-reflection" (*doppelten Reflexion*) (Hua X, 208 [215]): a reflection directed towards the "actual" consciousness in which the "not-presented" is given and a reflection directed towards the imagined experience of imaginative consciousness.

remembered object is meant directly as an object, my past perception is meant in an implied and non-objective manner. I implicate myself as a witness in giving testimony about the past. To remember something of the past is automatically to betray my own presence as having once been there in the past. In recalling a past perception, memory must also reproduce the elapsed portion of time-consciousness in which the perception was originally constituted. Husserl thus considers that, "the whole [entire consciousness] is reproduced, not only the then conscious present with its flow, but rather 'implicit' the whole stream of consciousness up to the living present" (Hua X, 54).

Remembrance is an "again consciousness" (*Wiederbewußtsein*). We neither return to the past nor experience what we once experienced again as if for the first time. Remembrance is the consciousness of return as much as the return of consciousness (*Wiederbewußtsein*). The past returns to us as a ghost; it continues as an "after-life" (*Nachleben*) that inserts its own time within my actual presence-time. As Proust understood, time is regained, not in the sense that I come to possess those objects once experienced; rather, I come to repossess myself in light of those objects once experienced. And yet, since acts of remembrance are rooted in the living present, remembrance regains time only to relinquish it again.

### Double intentionality of retentional consciousness

Husserl's investigations of different types of "re-presentification" (*Vergegenwärtigung*) provoked a substantial and far-reaching reformulation of intentionality, not only in terms of uncovering different senses of transcendence, but also, and of equal significance, in terms of refashioning the underlying conception of consciousness as immanence, or "self-givenness," through the discovery of "transcendence within immanence." The double intentionality of remembrance in particular gave renewed impetus to the problem of time-consciousness; as detailed in the previous chapter, Husserl's analysis of time-consciousness remained haunted by the ghosts of Brentano's original association in its failure to adequately describe the perceptual grasp of the immediate past in tandem with the perception of the present, as well as the temporality of absolute time-constituting consciousness, on the basis of which the apprehension of temporal succession is possible. These difficulties stem from an underdeveloped phenomenological description, or "seeing," of the constitutive performance of original time-consciousness and the apprehension/content of apprehension schema of intentionality, both of which are directly challenged by Husserl's phenomenology of "re-presentification" (which could also be called a phenomenology of alterity).

As Husserl's investigations reveal, in an act of remembrance, the tree that I perceived yesterday is not given to me on the basis of something present *in* consciousness, in which case I would apprehend the tree as past on the basis of something present, and thus not grasp the tree as past, but as re-presented as past. Remembrance, however, is a transcendence in which the memorial object is itself, and intuitively (*anschaulich*), given as past, and with a double significance. In an act of remembrance, the memorial object is given to me on the basis of an underlying, and implicit, reproduction of the original act of perception; this act of consciousness from the past is "re-lived" or "given again" in the present, but not as actually experienced once again, but as having once been actually experienced. In intending the memorial object, consciousness intends itself as past in an implicit manner. This reproductive modification of consciousness is not a thematic object of consciousness; on the contrary, my past consciousness is reproduced as a lived experience, and thus does not appear as an object. In remembering the past, it is as if I were seeing again through the eyes of the past, yet in the doubled-vision of the present, in the consciousness of the *difference* between the present in which I remember and the past that is remembered. The character of consciousness *as consciousness*, in its own "self-givenness" or immanence, is temporal (and temporary). As a double-consciousness, remembrance opens a (self-) transcendence within immanence in which a reproduced (transcendent) consciousness of the past is given in an immanent consciousness of the present. Moreover, in remembering the tree that I perceived yesterday, I implicitly remember *myself* as having perceived the tree yesterday, and thus as having originally constituted that experience in its temporality. Although the memorial object itself is directly intended, the act of remembrance entails an intentional implicature of my own (past) consciousness. I am both the subject and the object of remembrance, but not an object in a thematic and objectified manner (even though, of course, I can make my past act of consciousness into a thematic object through an act of reflection within the arc of remembrance). Remembrance in this fashion presupposes the unity of absolute time-consciousness. I must have originally been self-conscious *when* I originally perceived the tree in order to remember myself as having once perceived this tree; remembrance entails that I remember myself as having perceived the object, and this in turn entails that I was aware of myself as perceiving the object. In this light, the double-consciousness of remembrance must imply a comparable, yet *constituting*, double intentionality within original time-consciousness, and, specifically, within "primary memory," or "retentional consciousness," as Husserl comes to designate his refashioned conception of the consciousness of the immediate past. If remembrance entails an intentional reference to the memorial object as well as an implicit intentional relation to myself, as having once perceived, this implies that the object *and* my act of perceiving were originally constituted in a temporal manner, and that each is retained in its temporal constitution. Moreover, it further suggests that in retaining myself, I am aware of myself as (self-)retaining. With this set of implications in mind, let us return to the analysis of time-consciousness.

In the years (1906–1908) following the ITC lectures, Husserl continued to struggle with the problem of time-consciousness in the new light, however, of his parallel investigations into the imagination and remembrance. If we return with fresh eyes, informed by the object-lesson of remembrance, to the quintessential phenomenon of temporal succession in the perceptual act of hearing a melody, elapsed notes are still retained within the arc of perceptual consciousness, and to the extent that these notes remain apprehended by consciousness as "just-past," fall within the three-fold intentional declension of original time-consciousness. On Husserl's account in the ITC lectures, the *phantasma*, on the basis of which primary memory apprehends a note as just-past, is an immanent component (or sensible content) of consciousness, since it is the form of apprehension (primary memory) that confers onto the *phantasma* the significance of exhibiting the intended note as "just-past." The apprehension of the note as just-past has its footing in a consciousness that is immanent for itself in the lived immediacy of sensing itself as now. Yet, this means that the apprehension of the note as just-past is based on something-present, namely, on consciousness as itself sensed in a real (*reell*) manner. And yet, a note that is just-past cannot be considered as still sensed in a "real" (*reell*) manner, that is, in the same manner as the actual lived immediacy or sensing underpinning the apprehension of the note as now. Moreover, if the *phantasm* of the just-past and the sensation of the now are each immanently contained in consciousness, both are simultaneous with each other. But if the consciousness of the just-past is simultaneous with the consciousness of the now, Husserl unwittingly succumbs to the dogma of momentary consciousness. But as Husserl critically remarks: "[t]he just-past tone, as far as it falls into the present-time [*die Präsenzzeit*], is still conscious, but not in the sense as if it were actually 'sensed' and 'reell,' not in the sense of a now-tone ... in this consciousness [*of the just-past tone*], no actual tone can be found, only a tone that has been. In short, there is a radical alteration, an alteration that can never at any time be described in the way in which we describe the changes in sensation that lead again to sensations" (Hua X, 324 [336]). With this observation in mind, Husserl is able to challenge the previously unsuspected assumption that a consciousness of the past is itself present for itself. In light, however, of the double-consciousness of remembrance, the consciousness of the past is itself an "again-consciousness" (*Wiederbewußtsein*); the givenness of the memorial object as past is not based in a consciousness that is actually, and immanently, present, but in a "re-lived" and reproduced past-consciousness, a transcendence within immanence. In this manner, the consciousness of the past entails a radical alteration of consciousness in the form of a temporal modification of immanent time-consciousness. This insight, in an appropriately different form, resurfaces within retentional consciousness. As Husserl now recognizes: "the still living tone, the tone 'still' standing in the view of temporal intuition, no longer exists; and what pertains to its appearance is not a 'tone-sensation' (an actually present now) but an 'echo' [*"Nachklang"*] of the sensation, a modification that is no longer a primary content in the sense of something actually present (not an immanent tone-now). On the contrary, it is something modified: a consciousness of past sensation. In this consciousness, however, no actual tone can be found, only a tone that has been" (Hua X, 324 [336]). The expression "echo" (*Nachklang*) serves in this instance to describe the consciousness of the tone as just-past, or "retentional consciousness." As a term of art, "echo" is deliberately marked off in quotes (*"Nachklang" der Empfindung*) since the note that is no longer is not really contained or present (*nicht reell vorhanden*) in retentional consciousness. As Husserl rhetorically asks: "But does the given not transcend retention since retention posits the given as "this is just past" rather than as "this is now"? (Hua X, 350). Retentional consciousness, as a unique form to temporal transcendence within immanence, provides for an "intuition of the past" (*die Vergangenheitsanschauung*) as an "original consciousness" (*ein originäres Bewusstsein*): as an original consciousness, retentional consciousness is neither an image nor other kind of representation, nor based on an immanent content, nor a perception in the strict sense of "making present" (Hua X, 311 [323]).

Retentional consciousness is a transcendence of "de-presentification" (*Entgegenwärtigung*) in the sense of counter-intentionality.[21] In this, its primary significance, retention is not a "holding back," as if I were holding back something running away from me, but a withholding or not making present. Retentional intentionality runs counter to the givenness of an original impression. Whereas an original impression is the wellspring of the visibility and affective force of lived experience, retentional consciousness "de-presentifies" in the sense of "emptying," or reversing, the intuitive fullness of the now. Importantly, retentional modification does not "negate" or "infect" an original impression with a difference, or differentiation, of alterity "from the outside," so to speak, since an original impression *necessarily* succumbs to, or becomes, its own retentional modification (in keeping with Husserl's focus, we suspend here the function of protentional consciousness); indeed, as Husserl routinely emphasizes, an original impression is itself an abstraction, as it always irrupts in an interplay of retentional consciousness. The now is no longer than the retention of what it no longer is. In retentional modification, an original impression is thrown "outside of itself." Time-consciousness is, in other words, "ecstatic." Expressed in Husserl's native conceptuality: time-consciousness is transcendence within immanence. In its necessary retentional modification, an original impression is defined by the law of its modification of presence; but by the same logic, the absence of presence inscribed in the essence of the now allows for a renewal of its presence. The entwinement of original impression and retentional modification reflects the entwinement of absence and presence, running-off and renewal, at the origin of time-consciousness. As Rudolf Bernet remarks: "The 'Ineinander' of original impression and retention is in truth an original 'Auseinander,' in which both poles presuppose the other and in which neither can be derived from the other."[22] If we recall Husserl's neologism of *Entgegenleben*,

---

21 *Entgegenwärtigung* is a terminological innovation from Husserl's later writings in the C-manuscripts to describe the emptying function of original time-consciousness, and specifically in its retentional component; the term was adopted in turn by both Fink and Heidegger.

22 Bernet, "Einleitung," lv.

introduced in chapter 3, we are now in a position to better fathom its meaning: the consciousness of the now is both one and two, a consciousness of an in-betweenness that is itself caught in-between, and thus, situated, within the intersection of the absence and presence of time-consciousness itself. Time-consciousness is an original dispersal; the origin is always a diaspora, the loss of a renewed presence, an invitation for the retrieval of an origin already past in the opening of a new beginning.

As a transcendence within immanence, retentional consciousness of the just-past is not simultaneous with the original impression, nor does retentional consciousness arrive "after" an original impression. If, as Husserl first conceived, retentional consciousness and original impression are simultaneous, the consciousness of temporal succession would be based on the dogma of momentary consciousness. On the other hand, if the relationship between retentional consciousness and original impression were one of succession, Husserl would conflate the succession of consciousness with the consciousness of succession. In addition, a successive relationship between retentional modification and original impression would provoke an infinite regress, since this temporal form of succession would in turn require a further act of consciousness for its constitution. As encountered in chapter 3, the original impression and its retentional modification are given together in an originary manner, and at the same time (*zugleich*), but they are not, therefore, simultaneous (*gleichzeitig*) with each other. As Husserl remarks: "the flow of the modes of consciousness is not a process; the consciousness of the now is not itself now. The retention that exists together with the consciousness of the now is not 'now,' it is not simultaneous with the now, and it would make no sense to say that it is" (Hua X, 333 [345]).

The sense in which original impression and retentional consciousness are neither simultaneous nor successive resides in the double intentionality of retentional consciousness. In the act of hearing a melody, the retention of each note as just-past entails an implicit retention of my own act of having just heard the note. Looking once again to the double continuum of time-consciousness, Husserl designates each continuum (i.e., phase and stretch continuums) as an intentionality of retention within a unified double intentionality: a "cross-intentionality" (*Querintentionalität*) is reponsible for the retention of the now-phase of the time-object while a "lengthwise intentionality" (*Längstintentionalität*) is responsible for the retention of elapsed phases within the consciousness of hearing in which the now-phase of a time-object is given. As Husserl specifies, retention is "an intentional relation of phase of consciousness to phase of consciousness" (Hua X, 333 [345]). This retention of phases of consciousness is pre-reflective and non-objectifying: consciousness implicitly retains itself as having just heard the note while retaining the note as just-past. Indeed, if consciousness did not retain itself as having just heard the note, it could not retain the note it just heard as just-past, since a consciousness of the note as just-past is impossible without an implicit self-consciousness that I have just now heard this note. This double intentionality of retentional consciousness is embedded in absolute time-consciousness since in retaining the act of just having heard a note, I am conscious of *myself* as having just heard the note as just-past. In this regard, the transcendence of "de-presentification" carries a dual significance. On the one hand, as Husserl remarks: "Certainly retention, which is an act now living and an act that can be made to be given itself, transcends itself and posits something as being – namely, as being past – that does not 'reell' inhere in it" (Hua X, 344 [356]). But, on the other hand, as Husserl equally observes, "Immediate retention, which still retains the just elapsed experience as it recedes – but retains it only in the mode of what is just past – already seems to be afflicted with the problem of transcendence. The retention certainly no longer possesses the *cogitatio* itself that existed" (Hua X, 350 [361]). In other words, the note as just-past as well as the act of hearing as just-having-been experienced are held onto (*noch im Griff*) as *past*, and thus, as transcending the now-phases of both consciousness and its time-object. Not even consciousness "really inheres" within its own retentional consciousness.

The double intentionality of retentional consciousness constitutes the unity of the time-object (retention of now-phases belonging to a time-object in the "cross-section" or "phase-continuum") as well as the unity of constituting time-consciousness itself, as manifest along the length-wise intentionality of (self-)retention (in so far as consciousness constitutes itself as retained through its own self-retention). Within this "stretch-continuum" of consciousness, each retentional modification, as discussed in chapter 3, re-retains, and, in this manner, re-capitulates, the entire continuum of earlier retentional phases, which are, in this manner, sedimented within the continuity of the retentional past as a whole. The flowing of retentional consciousness is continually re-iterated and displaced relative the incessant resurgence of an original impression. The irruption of every original impression in its retentional modification is the renewal of time-consciousness in its constitutive accomplishment. On the one hand, this iteration of overlapping retentions (where each retention retains, or better, re-retains, earlier retentions) along the lengthwise continuum provides for a continuous act of hearing a melody; if this were not the case, Husserl's argument could not account for the temporal continuity on the noetic side of consciousness, but could only account for succession on the noematic side of a time-object. But since the succession of a time-object is constituted in the temporality of time-consciousness, such a one-sided account is in fact no account at all. On the other hand, if the continuous retention of overlapping phases of consciousness is imbued with a temporality, as flowing or "running-off," this should not, in turn, be considered as a form of succession. The construal of the continuous flowing-away of conscious-ness itself, in its continuous retentional (self-)modification, as a succes-sion would provoke the specter of an infinite regress since a further act of time-constituting consciousness would be required in order to constitute the immanent "succession" of retentional consciousness.

Indeed, this was precisely the problem that faced Husserl's attempt to describe the temporality of perceptual acts of consciousness through the schema of apprehension/content of apprehension. Armed with the double intentionality of retentional consciousness and its unique form of self-transcendence, Husserl is now in a position to recognize that "the phases of consciousness and continuities of consciousness must *not* be regarded as time-objects themselves" (Hua X, 333 [345]; my emphasis). The flow (or stream) of original time-consciousness along the lengthwise continuum of self-retention does not have the form of a duration or succession; the manner in which consciousness retains, and re-retains, itself, relative to the renewal of an original impression is not comparable to a rectilinear movement or flow in the metaphorical sense of a river. This iterative structure of retentional implication, as Rudolf Bernet notes, "shatters" the linearity of time-consciousness, metaphorical or otherwise.[23] My sedimented past is not "behind" me as something that I have left behind. In an emphatic sense, *I am* the accomplishment of my past.

23 Bernet, "Einleitung," li.

As built of sedimentations, as we explore in chapter 5, experiences of the past are compacted into a depth or reservoir of meaning that can be recollected in the present as well as provide, in the form of what Husserl will identify as "far retentions," a horizon of possibilities and implicit meaning through which I already navigate the world yet to come. As a further consequence of this shattering of linearity, the phases of time-constituting consciousness must therefore *not* be considered as time-objects since this would require (as Husserl considered in the ITC lectures) the formulation of a distinction between "act of apprehension" and "object of apprehension" *within* immanent time-consciousness, and thus the creation of an inner distance, or duality, between reflected ("acts and content of immanent time-consciousness") and reflecting ("absolute time-consciousness") consciousness. The threat of an infinite regress, as provoked by a "reflection theory" of absolute time-consciousness, rests on the unexamined assumption of the *linearity* or succession of time-consciousness. The double intentionality of retentional consciousness is an essentially *constitutive* accomplishment, in constituting *both*, but each in a *different* manner, the temporal givenness of a time-object as well as the temporality of perceptual consciousness. Indeed, retentional consciousness, as a continuous of self-differentiation or flow, constitutes itself as this *difference* itself. The characterization of time-constituting consciousness as flowing or streaming underlines the "lived" and "concrete" weave of original time-consciousness as a *condition of possibility*. Original time-consciousness, in its three-fold temporal declension, is not an *act of consciousness*, but its inner condition of possibility, or "operative intentionality" on the basis of which the "thematic intentionality" described in the *Ideen* (noetic–noematic correlation) is grounded. The meaning behind our earlier introduction of original time-consciousness as an operative intentionality can now be grasped in its full *transcendental* significance. Through the lengthwise intentionality of retentional consciousness, consciousness "bears within itself the heritage of the past" (Hua X, 327 [339]), but since this branch of intentionality is unified with, and unifies, the "cross-intentionality" directed towards time-objects, consciousness bears within itself the heritage of the world as its own constitutional accomplishment. In this manner, Husserl's seminal insight into the double intentionality of original time-consciousness frames his later investigations of the retrieval and reactivation of the origin of constitutional accomplishments, as, for example, in "The Origin of Geometry." But since, as we have discussed, it is the retentional modification of an original impression that retains the origin as origin, and thus *opens* the origin as a beginning, it is only because of the retentional fissure from within that the retrieval of an origin is at all possible.

The "transcendental absolute" that the *Ideen* prepared, but did not reach, is finally uncovered. Absolute time-consciousness retains itself – in both senses: it retains *itself* and is *itself* retaining.[24] In this self-temporalization, absolute time-consciousness *differentiates itself* in a two-fold manner along the lines of the double intentionality of retentional consciousness: as a differentiation *from* itself in terms of the transcendence of constituted time-objects ("cross-intentionality") vis-à-vis constituting immanent consciousness ("length-intentionality"); as a differentiation *of* itself in terms of the transcendence of absolute time-constituting consciousness vis-à-vis constituted immanent consciousness. As Husserl expressed in *Idea of Phenomenology*: "the *cogitationes*, which we regard as simple givens and in no way mysterious, hide all sorts of transcendencies" (Hua II, 11 [8]). As we shall discover in the next chapter, absolute time-consciousness, as an impossible puzzle, hides itself within its own self-transcendence.

24 Following Bernet, "Einleitung," li ff.

5

# THE IMPOSSIBLE PUZZLE

*Wir bohren und sprengen, wie in diesen Abhandlung überhaupt,
allseitig Minengänge nach all möglichen Seiten.*

— Husserl

## The Bernau Manuscripts

Husserl's penchant for self-critique is nowhere more exemplified than in research manuscripts dedicated to the analysis of time-consciousness written during two vacations in the Black Forest town of Bernau in the fall of 1917 and spring of 1918. In the summer of 1917, Edith Stein began preparing the ITC lectures for publication, and eventually joined Husserl in Bernau to inform him of her progress and continue her work in consultation with the "master."[1] Stein's visit provided a catalyst for Husserl to return to his investigations of time-consciousness, suspended since 1911, in light of the transcendental expansion of phenomenology since the *Ideen*. In reading these manuscripts, perhaps the most obstinate of Husserl's corpus, one cannot help but think of the circumstances in which they were written, in the catastrophic twilight of a war to which Husserl had personally sacrificed, but which he also, as with other German philosophers, patriotically supported. One is also reminded of the reflections on time and narration, within a sweeping examination of European culture and the sources of its

---

1 As Stein remarks in a letter of September 8, 1917 to Roman Ingarden: "Ich bin drei Tage hier beim Meister, es wird eifrig Zeit gearbeitet" (Roman Ingarden, "Edith Stein on her Activity as an Assistant of Edmund Husserl (Extracts from Letters of Edith Stein with a Commentary and Introductory Remarks)," *Philosophy and Phenomenological Research*, 23 (1962), 173.

undoing, in the reclusive setting of a Swiss sanatorium in Thomas Mann's *Magic Mountain.* In his retreat in the Black Forest, Husserl returns to the theme of time-consciousness just as the wider promise of transcendental phenomenology for the crisis of European rationality began to take shape in his mind, and which would mature into the unfinished *Crisis of the European Sciences,* published in piecemeal at the dawn of the next world war. Spurred to tackle once again "the most difficult of all phenomenological problems," Husserl considered these writings as feeding into a larger projected work entitled "Time and Temporalization" (*Zeit und Zeitigung*). Sustained reflection on time-consciousness ended, however, in 1918 without yielding a completed manuscript, even though Husserl continued to speak of his Bernau writings as his *Hauptwerk,* and optimistically announced their imminent publication in letters to both Roman Ingarden and Alexandre Koyré as late as 1931.[2] As with other literary ambitions, Husserl's envisioned *magnum opus* on time-consciousness never advanced past the stage of incubation in the remoteness of the Black Forest.

Far from constituting a cohesive body of work moving in a single direction of argument, the Bernau Manuscripts present a written labyrinth. On the one hand, they share with other manuscripts the function of serving as a conceptual laboratory for Husserl's solitary reflections. On the other hand, when compared with earlier manu-scripts on time-consciousness, the Bernau Manuscripts achieve a daunting level of technical difficulty, even by Husserl's own demanding standards. Newly minted terms are unceremoniously introduced, yet seldom defined; established concepts, forged in earlier stages of analysis, are displaced in one manuscript, rehabilitated in another; and discarded terms from Brentano's descriptive psychology (for example: primary and secondary objects) surprisingly resurface on more than one occasion. To make matters more complicated, Husserl circles back on the progress of his own thinking, and revisits the abandoned apprehension/content of apprehension schema of inten-tionality in his continuing struggle with the self-constitution of abso-lute time-consciousness – even the expression "inner perception" still lingers in his thinking. Yet, the principal difficulty of the Bernau Manuscripts resides in the challenge to us readers, in our eavesdrop-ping on Husserl's inner monologues, to attain phenomenological

---

2 Roman Ingarden, *Briefe an Roman Ingarden. Mit Erläuterungen und Erinnerungen an Husserl* (Den Haag, 1968), 154.

fluency in a discourse that remained perpetually in motion, and yet
which clearly has something firmly in its sights, drawing ever closer
to the desired ideal of clarity that defined the ethos of Husserl's
phenomenological endeavor. Appearances to the contrary, the Bernau
Manuscripts are not a technical apparatus of specially minted phenomenological concepts, or "jargon," in which concepts are tailored
to fit for the sake of fitting together; in fitting, we are invited to see.
We must learn how to see the "things themselves" under their phenomenological description. And yet, repeated and emphatic gestures "to
see" on the part of Husserl offer no guarantee for showing us how
to perceive, let alone how to think – how does one enter a space of
thinking that, by Husserl's own reckoning, runs against our ordinary
habits of mind, including his own?

Looking back to our point of departure in the ITC lectures, the
trajectory of Husserl's thinking – from the critique of Brentano to the
broadening of time-consciousness in the torso of the ITC lectures, and
subsequent manuscripts, to the wider, and diverse, landscape of the
Bernau Manuscripts – contravenes Husserl's pronouncement in the
*Ideen* that the problem of "time is a name for a completely *delimited
sphere of problems* and one of exceptional difficulty. It will be shown that
in order to avoid confusion our previous presentation has remained
silent to a certain extent, and must of necessity remain silent about
what first of all is alone visible in the phenomenological attitude and
which, disregarding the new dimension, makes up a closed domain of
investigation" (Hua III, 198 [193]). In the *Bernau Manuscripts*, the task
of describing the complex accomplishment of time-consciousness
along the two lines of inquiry first established in the ITC lectures
continues to preoccupy Husserl's interest. The three-fold temporal
declension of time-consciousness, the constitution of time-objects,
the fixed order of objective temporal positions, and the vexing issue
of absolute time-consciousness are submitted once again, and repeatedly, to meticulous phenomenological scrutiny. Yet, Husserl also
begins to explore the further significance of these temporal analyses
for the broader sweep of transcendental phenomenology. The one-
sided analysis of noetic temporalization of earlier investigations is
redressed with a balanced reflection on both the noetic and noematic
poles of time-consciousness. Along with this noematic expansion, the
ontological import of time-consciousness is brought to transcendental
fruition: the connection between different forms of objectivity and
different forms of temporality is fashioned with an emphasis on the

contribution of temporality for the problem of individuation. Different regions of beings – real, irreal, ideal – are correlated to different forms of temporality, with each transcendental schema of being and time constituted in original time-consciousness. And lastly, the Bernau Manuscripts' sophisticated revision of the three-fold temporal declension of retention, original impression and protention delineates the contours of genetic phenomenology: the themes of affectivity, motivation and genesis – key elements in analysis of passive synthesis developed during the 1920s (cf. chapter 7) – are compellingly introduced within the spectrum of the temporalization of consciousness.

## Transcendental temporality

As Husserl stresses again in the Bernau Manuscripts, time is not only in the "knowing subject" (*Erkenntnissubjekt*), but also the necessary form of individual entities (*Form individuellen Daseins*) within the form of an individual world (Hua XXXIII, 90). Although the scope of the Bernau Manuscripts does not extend as wide as Husserl's later thinking in the 1930s, with its fashioning of the concept of life-world, the temporal analysis of the Bernau nonetheless broaches the transcendental solidarity between temporality and the horizon of the world (or "world-at-large" as we discussed in chapter 1) that circumscribes and situates the universality of possible experience. The apriori correlation between noetic forms of consciousness and noematic forms of objects is constituted in a transcendental temporality explored in the Bernau Manuscripts. The two lines of inquiry that shaped a phenomenology of time-consciousness in the ITC lectures and subsequent manuscripts during the years 1906–1911 are reformulated into a fully transcendental conception of temporality, as constituted along the parallel lines of "noetic" and "noematic temporalization."[3] The formal structure of Husserl's argument as well as its strategy of investigation is thus retained in the transcendental reflections of the Bernau Manuscripts; the noematic temporalization of objectivity is founded in the noetic temporalization of consciousness, both of which are constituted in the double-consciousness of "crosswise" and "lengthwise" intentionality of absolute time-consciousness. As Husserl emphasizes, "it belongs to the essence of time-consciousness that it is

3 Klaus Held, *Lebendige Gegenwart* (Den Hague: Martinus Nijhoff, 1966), 48.

at the same time the unity-consciousness of its own immanence and the unity-consciousness of a transcendent object" (Hua XXXIII, 164).

In the Bernau Manuscripts, Husserl broadens this investigation of the constitutive accomplishment of time-consciousness in two mutually supporting ways, and with a marked *ontological* inflection, as evidenced with terms such as "existence" (*Dasein*), "being" (*Seiende*) and "reality" (*Wirklichkeit*). The now-phases of a time-object are described as the "modes of being" (*Seinsmodus*) of the noematic object. Every time-object "fills" a determinate duration and occupies a determinate time-position, on the basis of which measurement and chronological ordering is possible. This "rigid form" (*starre Form*) of time stands in contrast to the flow or stream of original time-consciousness, or the "manners of givenness" of time (*Gegebenheitsweisen der Zeit*). Husserl reiterates that objective time (or "clock-time") is a "rigid time-order" (*die starre Zeitordnung*) that is itself neither "present, past, nor future" (Hua XXXIII, 181). These temporal determinations, as should be familiar to us from earlier discussion, are acquired on the basis of time-constituting subjectivity. The form of an object's temporality is consequently inseparable from possible modes of temporal givenness to consciousness, and these modes of givenness are "modes of orientation" of subjectivity itself. Only a subjectivity that is itself in time can encounter events in time.

This noetic–noematic formulation of the constitutive accomplishment of time-consciousness is inseparable from an expansion of the range of objectivities that fall under the purview of transcendental temporality. As explored in chapter 4, a phenomenological investigation of the imagination, image-consciousness and remembrance reveals a constitutive correlation between the sense of an object's givenness and its form of temporality. Different ontological regions are constituted as different forms of temporality. The many *senses* of being are constituted as different forms of temporal givenness, or temporalization. Husserl in particular is keen to cast the problem of individuation in the mold of time, and considers temporality, and not spatiality, as the origin of individuation, where individuation is understood as an object's "actual identity with itself and as differentiated from other objects" (Hua XXXIII, 330). In essence, Husserl argues that the temporal determination of an object depends on an identification of an identity, or "substratum," and that such an identical substratum individuates an object. A substratum is the objective unity of a noematic kernel of sense and the noematic pole of "object X" that endures in time, in the specific sense of its own constituted temporal form. Whereas logical predication and the articulation of judgments presuppose an object's substratum as an individual, the analysis of time-consciousness uncovers the origin of individuation, and thus the foundation of "substrates" and the structure of predication. In this manner, a phenomenology of individuation delineates a focal center for a transcendental genealogy of logic, as Husserl developed in *Experience and Judgment*. In the Bernau Manuscripts, we find detailed reflections on the different types of temporal individuation for imaginary objects, perceptual objects and ideal objects. Husserl continues to probe the quasi-temporality of the imagination but also expands the scope of his reflection to include the constitution of the "omni-temporality" of ideal objects (*Allzeitlichkeit*). The difference between these three basic forms of temporality – real, irreal and ideal – are constituted in original time-consciousness. In this manner, Husserl tacitly distinguishes between an ontological and transcendental significance of temporality. The different senses of being (perceptual things, imaginary objects and ideal objects), as different senses of temporal transcendence, are constituted in the transcendental temporality of absolute time-consciousness. The transcendental temporality of absolute time-consciousness – the "true absolute" excluded from the *Ideen* and other writings – constitutes the difference between mind and world, between possible modes of my own consciousness and objects other than me (*egofremde Gegenstände*) (Hua XXXIII, 91). As Husserl writes,

> Behind the subjectivity of this sphere of time [*the temporal horizon for constitution of the many senses of objective-being*] resides a further transcendental-subjective sphere, the sphere of "experiences" (and with a new level and new meaning), in which this temporality is constituted, that is, experiences that bring to appearance temporal objects with a form of time, but which is itself not temporal, either in an objective-temporal sense or in the sense of an event in the transcendental time in the first sense. (Hua XXXIII, 184)

### Near and far retention

Husserl's entry into the problem of time-consciousness passed through the gateway of the ubiquitous phenomenon of temporal passage that stood at the focal center of Brentano's theory of original association. From this inconspicuous beginning, the plot of Husserl's unfolding investigation of time-consciousness centered on describing the constitution of time-objects, and with this passage of objects in time, how consciousness itself, on the basis of which the constitution of time-objects is possible, itself passes, and comes to past. As Rudolf Bernet notes, "the crucial question is that of the phenomenology of absence. It is in the presence of the past that Husserl sees the fundamental mode of this presence of absence."[4]

In the life of consciousness, the past, however, is present in many ways. As we have seen, Husserl characterizes the temporal modification of retentional consciousness with different registers of meaning – as a compression of perspective; as the opening of distance; as the production of depth; as a gradual loss of "intuitiveness" (*Anschaulichkeit*); as the diminishment of affectivity. The richness of these descriptions is testimony to the diverse meanings of retentional temporalization. In its primary function, retentional consciousness constitutes the now-phase as just-past that is given along with, but not in the same manner as, an original impression in which a now-phase is constituted as now. Through the retentional modification of an original impression (along with, of course, a protentional consciousness) the living present is constituted in its temporal passage. Retentional consciousness, we recall, is not a reproductive mode of consciousness in which an object, already past, is brought back into the present; on the contrary, as the condition for the reproductive consciousness of remembrance, retentional consciousness is an original accomplishment in which the just-past is constituted as just-past. Retentional consciousness "de-presentifies" in the sense of withholding or running counter to an original presentation. It does not remove from the table, so to speak, what has already been given; rather, it already withholds or inhibits from within the constitution of a presence unpunctuated by absence. Retentional modification reverses in mid-stream an original impression without entirely effacing it. Since retentional modification and original impression are given pairwise "at the same time," but not as the same time, the event of time-consciousness is constituted as an original difference or pure "in-betweenness." The retentional modification of an original impression is an original transcendence within immanence with a dual significance, as prescribed by the double

---

4 R. Bernet, *La vie du sujet: Recherches sur l'interprétation de Husserl dans la phenomenologie* (Paris: PUF, 1994), 216.

intentionality of retentional consciousness. Time-consciousness opens a transcendence towards objects other than itself along the axis of cross-intentionality while at the same time, but not in the same manner, opening itself for itself along the axis of lengthwise intentionality; along the lengthwise intentionality, absolute time-consciousness constitutes itself as "self-transcendence."

Another way to formulate this constitutive significance of retentional consciousness is to underline its function of reversal. Retentional modification reverses the givenness of original presentation in mid-stream, in such a manner that an original impression becomes folded, not back onto itself, but outside of itself. The sense or direction of constitution is reversed in retentional consciousness in so far as it robustly constitutes the givenness of absence through an internal tension with the givenness of presence. In fact, there are different senses in which the absent, or the not-now, is constituted in retentional consciousness. As discussed in chapter 4, the headless temporalization of the imaginary operated in the anarchic interplay of retentions and protentions in the absence of any mooring in an original impression; as we shall further suggest in chapter 6, retentional consciousness resurfaces, in a different form, to play a significant function in the constitution of the Other.

In the Bernau Manuscripts, Husserl returns to the analysis of retentional consciousness by submitting its array of descriptive characterizations to further phenomenological scrutiny. A change in Husserl's terminology announces this renewed concern with describing how retentional modification is responsible for a contraction of perspective, an opening of distance, a loss of intuitive fullness, a diminishment of affectivity, and the creation of depth within consciousness in becoming its own past. Husserl replaces his earlier use of the terms "run-off" (*Ablauf*) and "running-off" (*Ablaufen*) with the newly promoted term *Abklang* – "subside," "decay," "fading" – arguably because "subsiding" (*Abklang*) evokes more sharply the aural experience of hearing the decay of a note. As a note "subsides" or "fades" (*abklingt*), it progressively loses its intuitive presence as a function of its increased "fading" or "sinking away" from the actual now (Hua XXXIII, 65). This renewed attention to the "fading away" (*Abklang*) of the immediate past is connected with Husserl's replacement of the term "original impression" with "original presentation" (*Urpräsentation*) or "originarity" (*Originarität*), as well as other terms, which we shall take up below. As with the choice of *Abklang*, the abandonment of "original impression" is a terminological acknowledgment of his rejection of the "apprehension/content of apprehension" interpretation of retentional intentionality. Any lingering suggestion of an initially temporally neutral sensible content is removed by adopting the terms "original presentation" and "originality," both of which emphasize the "originality" and "presentational" character of the now-phase of time-consciousness.

As should be familiar from earlier discussion, retentional consciousness is a continually re-iterating modification of an original presentation; each phase of retention re-retains by way of an intentional implication the stretch of earlier retentional phases, each of which, in turn, re-retains the "stretch continuum" of the immediate past. Within the arc of the living present, Husserl insists that we perceptually grasp not only the temporal distance between an elapsed phase and the actual now-phase, but also, the relative lengths (or intervals) within the continuum of elapsed phases. Elapsed phases of time-consciousness, along with their retained phases of intentional objects, become sedimented and compacted into the past. As Proust describes the inattentive hearing of church bells while reading in the garden of his Aunt Leonie's house in Combray: "when an hour rang in the bell tower of Saint-Hilaire, of seeing fall piece by piece what was already consumed of the afternoon, until I heard the last stroke, which allowed me to add up the total . . . at each hour it would seem to me only a few moments since the preceding hour had rung; the most recent would come and inscribe itself close to the other in the sky, and I would not be able to believe that sixty minutes were held in that little blue arc comprised between their two marks of gold." Although we do not attend directly to the discrete and successive modifications of each elapsing now-phase ("each stroke of the bell"), retentional consciousness passively constitutes the crystalline sequence of elapsed phases. In hindsight, we may tally exactly the number of clock-strokes just heard, but we may equally become aware that *a* number of strokes was just heard without being able to give their number exactly. A silent assembly of distances is retained in retentional consciousness that exceeds what consciousness can at any moment grasp thematically.

The tolling of the bells comes to an end; the bells are heard no longer. Yet, from within the perspective of time-consciousness, retentional modification is a passive and overlapping synthesis that continues *in infinitum*. As Husserl writes, "The past is an unceasing transforming itself; the transformation goes on ideally *in infinitum* [*Die Vergangenheit ist eine unaufhörlich sich wandelnde; der Wandel geht ideell in infinitum*]"

(Hua XXXIII, 293). The retentional modification of an original presentation cannot abruptly come to a halt nor could consciousness, by way of an intervention on the part of the ego, bring to an end the unfolding destiny of the past's fading away. The unending iteration of retentional modification is tethered to the ever-renewing actual now-phase. This iteration of retentional modification *in infinitum* underscores the eternal recurrence of the now as always different, as a principle of temporal differentiation. Retention is unending because the now is always returning. The temporal distance between any given elapsed now-phase and the actual now is never finally settled to the degree that every actual now is never actually the last. This restlessness of the past runs in tandem with the restlessness of original presentation. One must not, however, be seduced by the *image* of the stream lodged in Husserl's construal of the "infinite" re-iteration of retentional consciousness. When Husserl speaks of retentional modification as an "unceasing transformation," we must take the expressions "stream," "flow" and "transformation" in their transcendental significance, that is, as indicating the unsettled and unfinished accomplishment of constitution. As a transcendental description of the "movement" of constitution, the past is never completely settled as long as the future is not yet completely past.

And yet, even though retentional modification is infinite in the constitutional significance just indicated, Husserl also recognizes that retentional consciousness is finite in character in so far as the fading-away of the bells as just-past inevitably reaches a vanishing point or "null-point." At this vanishing point, the tolling of the bells has slipped beyond the arc of my living present, as if my own consciousness of time has slipped beyond the orbit of its own living present. Indeed, if Husserl did not insist on the finite breadth of the living present, that is, if retentional modification did not reach a horizon of emptiness in which the intuitive visibility and charge of affectivity had entirely dissipated, time-consciousness would resemble the madness of Ryunosuke Akutagawa's "Spinning Gears" – I would succumb to the deafening madness of hearing without end the resonances of every experience ever had. *Nothing* would be past for me, and my life would be strangled in the amassing clutter of my living present as it expands unendingly.

The issue of how to juxtapose the infinite and finite aspects of retentional consciousness partly resides with the noetic–noematic correlation of time-consciousness. The phenomenon of "fading off" (*Abklang*) refers to the continual and incremental impoverishment of an elapsing now-phase's "fullness" and "visibility," or "intuitivity" (*Anschaulichkeit*). As noted, the continuum of retentional modification (or stretch continuum) eventually reaches a "vanishing" or "zero point" (*Nullpunkt*); the tolling of the bells is over and done with; an intentional unity of consciousness has entered the realm of "sleep," as Husserl describes the continued retentional consciousness within the deep reaches of my past. It is important to bear in mind an observation earlier raised (cf. chapter 3), the significance of which can now be fully grasped: as I hear a musical note fading away, the noematic sense of this time-object does not in any meaningful manner "decrease" or "fade." The *sense* of a note, for example, as a middle C, does not "fade." It is no less a middle C in its givenness as just-past as it is now. The retentional modification of "de-presentification" describes the "fullness" of a noematic object's givenness in consciousness, its "realization" as an experience. In other words, the gradual fading of a time-object's fullness over the life of its duration describes the noematic object's "modality of concretion" as an event in the life of consciousness. Once the time-object has exited the arc of the living present, its manner of givenness continues as an empty consciousness (*leeres Bewusstseins*) (Hua XXXIII, 67). What we experience never leaves us, but remains "dormant" within our past, open to reactivation in acts of reawaking called remembrance or reflection.

The primary function of retentional consciousness can be further specified in terms of its generation of "emptiness" or "nothingness" in the sense of "empty consciousness." Within the sedimentations of retentional consciousness, objects of the past are still given in the manner of an "empty consciousness," as covered over in layers of sedimentation, not, however, as devoid of noematic sense, which remains intended in the form of retentional consciousness; retentional consciousness is empty in the sense of lacking intuitive fullness (*Anschaulichkeit*). As Husserl remarks: "Every accomplishment of the living present, that is, every accomplishment of sense or of the object becomes sedimented in the realm of the dead, or rather, sleeping horizonal sphere, precisely in the manner of a steady order of sedimentation: While at the head, the living process receives new, original life, at the feet, everything that is, as it were, in the final acquisition of the retentional synthesis, becomes steadily sedimented" (Hua XI, 178 [227]). This characterization of the depths of sedimentation within retentional consciousness as a "sleeping horizonal sphere" invites a number of intriguing questions regarding the metaphorical meaning of sleep within phenomenological description, as well as the phenomenological constitution of the distinction between waking and sleeping consciousness. These issues, however, lead beyond the scope of our study since Husserl does not address the problem of sleep in the Bernau Manuscripts.[5] However, the metaphor of sleep in these manuscripts implicitly introduces the distinction between "near" and "far" retentions. The gradual slackening of visibility and affective charge in the fading away of retentional modification is metaphorically compared to "falling asleep" until that vanishing horizon at which retentional consciousness has passed into the inner night of sleep.[6]

The distinction between near and far retention is already implicit in the presentation of retentional consciousness in chapters 3 and 4. The retentional modification of an original presentation within the arc of the living present is "near" in the sense of maintaining a proximity to the renewing axis of the original presentation. In the experience of hearing a melody, near retentions are integral to the constitution of the living present. By contrast, once the melody as a whole has elapsed and sinks into the darkness of the remote past, it nonetheless remains – is continually retained – in consciousness in the form of *far* retentions. If, after a prolonged silence after the hearing of a melody, I hear a second melody that immediately brings to mind the melody previously heard, this first melody remained within my consciousness in the form of far retentions – held at a distance beyond the arc of the living present. As Husserl remarks, "The present turns into the past as the past that is constituted for the ego through the lawful regularity of retentions; and finally, everything that is retentional turns into the undifferentiated unity of far retention of the one distant horizon, which extinguishes all differentiations" (Hua XI, 288 [422–423]).

This transformation of near retentions into far retentions represents a transformation in the *constitutive* function of retention. As with our earlier caveat regarding the transcendental significance of Husserl's metaphors of "movement," "transformation" and "stream,"

---

5 I address these issues in a forthcoming paper, "The Inner Night: A Phenomenology of (Dreamless) Sleep."

6 This distinction is implicitly drawn in the Bernau Manuscripts; it is only in passive synthesis lectures, as well as in unpublished manuscripts from the 1920s, that Husserl uses the terms "near" and "far" retentions.

the characterization of the flipping of near retention into far retention as a movement into the realm of sleep characterizes a transformation of the sense in which retention continues in its constitutional
accomplishment. The sense in which the past is given as no longer
present undergoes a significant transformation in far retentions.
In fact, Husserl subsumes under the capacious heading of far retention *different* senses in which the past is present without being present.
Husserl tends to bundle together these different senses without
explicitly and carefully distinguishing among them. Here, however,
is not the occasion to undertake this task, which would, in addition,
call upon the resources of genetic analysis. However, we can make
passing mention of three functions of far retentions. Far retentions
constitute the tacit dimension of knowledge in two related senses. On
the one hand, the "extinguishing" of differentiation noted by Husserl
in the quote cited above indicates the manner in which far retentions
generate generalized schemas and typifications of objects. As Lanei
Rodemeyer argues, "differences between specific experiences disappear in far retention as similar experiences become grouped
together into general 'memories.'"[7] In a manner recalling Hume's
account of the formation of general terms, repeated experiences are
grouped together through associations into a generalized type. These
"memories" qua "far retentions" can be awakened within the arc of
the living present through what Husserl calls an "original association"
(Hua XI, 286 [420–421]). An original association (not to be confused, of course, with Brentano's "original association") in which
I recall something of the past is not an act of remembrance; remembrance is a reproductive modification of consciousness in which
I remember myself as having once perceived this pen before. When
a pen strikes me as familiar in a manner that does not recall a
particular memory, yet nonetheless invokes the consciousness of
having seen this type of pen before, this experience is constituted in
an original association between the perception of this pen in the living
present and the far retention of this type of pen.

The third function of far retention carries the most significance
within the reflection of the Bernau Manuscripts. We have already noted
that the primary function of retentional consciousness, as both near
and far retention, is the constitution of empty consciousness in the

7 Lanei Rodemeyer, *Intersubjective Temporality: It's About Time* (Dordrecht: Springer Verlag, 2006), 94.

dual significance of empty consciousness of an object and as emptied consciousness of, or from, itself. The "emptying" of consciousness of its lived experience represents the constitution of potentiality within the life of consciousness. Consciousness retains itself as the potentiality for its own remembrance. As Husserl writes, "The given time-field, which is always something filled, is surrounded by an empty time-horizon, which belongs to the essence of time-consciousness as a potentiality of an empty component of consciousness, which can become fulfilled in chains of remembrances" (Hua XXXIII, 71). As discussed earlier, remembrance is an awakening of consciousness within consciousness itself. When I remember an experience from my past, a past act of consciousness is "reproduced" in the consciousness of the present. Remembrance is, in this sense, an "again consciousness" (*Wiederbewußtsein*) and, in this manner, relives the past because in recalling an experience of the past, consciousness reawakens itself by "implicating" its (own) past consciousness. Within original time-consciousness, retentional consciousness is both a retention of an elapsed now-phase as well as a self-retention, that is, a retention of my own perceptual act of having just heard the note. Indeed, it is only on the constitutive basis of the double intentionality of retentional consciousness that remembrance is itself possible: if I did not retain myself as having just heard the note, I could not remember that note once experienced by me. In remembrance, it is as if that particular act of consciousness has regained consciousness, not in its original form, but in a reproduced form, as given again in a now that it no longer calls its own.

As the condition of possibility for remembrance, retentional consciousness, in both of its functions as near and far retention, constitutes the consciousness of possibility along with the possibility of consciousness itself. Objects of experience are intended in an empty manner; they are present in their absence. But what does this mean that consciousness intends its object in an empty manner and that the object is given precisely to the degree that it is not present?

In an embarrassing moment of absent-mindedness, the name of my wife escapes me; her name is on "the tip of my tongue," as such a condition is commonly expressed.[8] In one sense, her name is given to

---

8 An example examined by Husserl (Hua XXVI, 30–34). Or take Proust's description of involuntary memory: "I feel something quiver in me, shift, try to rise, something that seems to have been unanchored at a great depth; I do not know what it is, but it comes up slowly; I feel resistance and I hear the murmur of the distances traversed" (46).

me to the extent that I am certain of the name that I have forgotten
and thus know the name I nonetheless cannot produce for myself.
Indeed, when I am presented with a list of names by a group of
concerned friends, I am able to reject each possible name presented
to me without thereby producing her still elusive name. In other
words, my wife's name is given to me as an empty consciousness.
Although I intend her name in an empty manner, I cannot bring to
fulfillment the name itself: her name itself is not forthcoming in an act
of remembrance. Having a name on the tip of my tongue is a condition
of inner restlessness; the empty consciousness of the forgotten name
"begs," so to speak, to be remembered. I become single-mindedly
preoccupied, indeed, obsessed, with remembering my wife's name.
In Husserl's thinking, the relationship between empty and fulfilled
consciousness is teleological: an empty intention motivates its fulfillment without being able on its own to render present its intended
object. Only the name itself can fulfill its empty intention. This tendency to produce my wife's name is not essentially motivated by the
practical worry of avoiding her ire. The consciousness of possibility is
intrinsically a movement towards its fulfillment in actual self-givenness.
The consciousness of possibility is a movement towards actuality in a
dual significance: the consciousness of possibility is also the possibility
of consciousness. As exemplified in having a word on the tip of
my tongue, my wife's name is given to me without actually being
genuinely given. Her name does not appear in my mind and yet it is
nonetheless given as this non-appearance, as this determinate forgetting, namely, as an "empty consciousness." This empty consciousness
is produced by consciousness itself, since I must have at one point in
my past learned my wife's name in order to stand here and no longer
remember the name I know to have forgotten. The opacity of an
empty consciousness defines its mode of givenness as well as the style
of its motivation.[9] In Husserl's thinking, this empty consciousness –
forgetting – of my wife's name has the form of a far retention; when
and if I finally remember her name, in a moment of relief (the tension
of an empty intentionality becomes relaxed and "satisfied," as Husserl
would say), far retention, as an empty consciousness, becomes fulfilled

9 Cf. James Dodd, "Expression, Ideality and the Ego: Some Remarks on the 1913
Revisions of Husserl's *Logical Investigations*," in *Husserl's* Logical Investigations, ed.
D. Dahlstrom (Dordrecht: Kluwer, 2003), 167–187; esp. 173–174.

in an act of remembrance in which her proper name is actually given to me. I now remember my wife's name.

When first asked the question of my wife's name, and which I could not initially recall, a far retention containing her name was awoken through an original association in the living present, without, thereby giving rise to a reproductive association of remembrance. Finally, I successfully recall my wife's name, yet this success does not need to take the form of a reproduction of the consciousness in which I first learned her name, even though, on Husserl's reckoning, it remains an ideal possibility. Husserl's metaphorical comparison of the emptiness of far retentions to a sleeping consciousness can, in light of our discussion of having a word on the tip of my tongue, be seen as the characterization of far retention as an *original forgetting*. In a striking parallelism between Husserl and Augustine, Augustine argues in his discussion of forgetting in the *Confessions* that "memory retains forgetting."[10] In order to understand how we know what it is we have forgotten, without thereby actually remembering it, Augustine proposes the suggestive simile that in forgetting, memory is divided from itself, working with parts from an encompassing whole that thus remains, as a whole, divided from itself. When I have a word on the tip of my tongue, as Augustine continues, it is as if memory was "limping" or operating with a missing element, and yet demanding its return. Expressed in a phenomenological framework, forgetting is an empty consciousness that awaits fulfillment; forgetting "resides in memory" or, in other words, in far retention. Far retention is, in this sense, the constitution of forgetting. If retentional consciousness opens the possibility of remembrance, the possibility of remembrance *presupposes an original forgetting*. The temporal constitution of experience in the living present is already the constitution of its withdrawal or forgetting – its "self-emptying": the sense in which consciousness becomes buried over, or sedimented, in the very movement of its originary givenness. Transcendentally speaking, an origin is constituted as an origin in its being sedimented or forgotten. The origin covers over its own accomplishment as an origin.

And yet, this phenomenological implication of an original forgetting does not contradict Husserl's argument for the eidetic necessity that every experience is temporally constituted as the possibility

10 Augustine, *Confessions*, 223.

of remembrance and reactivation. The possibility of remembrance intrinsically belongs to the temporal constitution of the living present. In this regard, Husserl contends that a total remembrance of my life is *ideally* possible; indeed, as discussed in chapter 4, individual memories are implicitly connected, and thus implicate in an intentional manner, the total chain of all possible memories. In the Bernau Manuscripts, Husserl entertains the thought of "an all-knowing 'divine' consciousness" (*ein allwissendes "göttliches' Bewusstsein"*): ideally speaking, the empirical limitations that prevent me from "total recall" could be removed such that the circle of remembrance could be expanded *in infinitum* to encompass the totality of my life. In such an ideal scenario, I would become the god of my own consciousness. The point of entertaining this idea of total recall is, however, to accentuate the relationship between forgetting and remembrance. As Husserl expresses himself, consciousness "swims" within the empty horizons of the past (retentional consciousness) and the future (protentional consciousness). In so far as the horizon of retentional conciousness, and its pluralization of pasts (i.e., "comet tails"), delineates the condition of possibility for both remembrance and reflection, in other words, for different ways in which consciousness either gives itself again in a modified way or objectifies itself in reflection, retentional consciousness delineates the horizon of my own self-understanding. But since this retentional horizon, as just argued, represents an original self-forgetting, any understanding of myself in either remembrance or reflection operates against the background of forgetting. When measured against the ideal possibility of total self-understanding, any actual self-understanding remains finite in character. Indeed, as we shall discuss in chapter 7, consciousness remains marked by the finitude of its own self-understanding or self-manifestation. Whatever I grasp of my own lived experience remains surrounded by the inner night of my own consciousness.

## The future is never now

One of the more central transformations in the analysis of time-consciousness in the Bernau Manuscripts is signaled with the replacement of the term "original impression" with a string of interchangeable terms: "original datum" (*Urdatum*), "original core" (*Urkern*), "original phase" (*Urphase*)," "original presentation" (*Urpräsentation*) and "originarity" (*Originarität*). This terminological change marks Husserl's recasting of the constitutive function assigned to what had earlier been designated as "original impression" within a sweeping recallibration of the three-fold declension of original time-consciousness as a whole. Original presentation still retains its stabilizing function as the wellspring of "intuitive fullness" (*Anschaulichkeit*) and the unmodified "zero-degree" of temporal orientation. Original presentation is not an "aspect" of time *towards which* we have an orientation: it is that *from which* we have any temporal orientation towards the past or the future, understood here in terms of their respective foundation within the three-fold declension of original time-consciousness. In this manner, an original presentation is akin to the absolute here of the lived-body. An original presentation is not "there," i.e., "now," in time, but rather, as embedded in the weave of retention and protention, provides the condition of possibility for temporal distances and positions. When Husserl speaks of an original presentation as the "zero-degree" and "maximal point of visibility," he seeks to characterize the sense(s) in which the stream of time-consciousness is always *situated* in a continually refreshed original presentation. Retention and protention are responsible for the shifting temporal distances – for the sense of time as a distance – within a temporal landscape that is incessantly recallibrated with reference to a renewed original presentation. Husserl therefore still adheres to his fundamental claim that the "basic fact" (*Grundtatsache*) of time-consciousness is its original consciousness of the present. Yet, this "basic fact" that defines the life of consciousness becomes reformulated around the conception of an original presentation as a mode of "fulfillment" and maximal intuitiveness within the three-fold intentionality of time-consciousness. This characterization of original presentation as the site of temporalization gives a sharper profile to the basic figure of "Entgegenleben," but by the same token, to the definition of the now-phase of consciousness as the pure difference of an "edge-consciousness" (*Kantenbewusstsein*), as Husserl introduces in the Bernau Manuscripts.

In the Bernau Manuscripts, the function of original presentation is minimized, yet not entirely suppressed, due to a sophisticated conception of original time-consciousness as an "intertwining" or "weave" (*Verflechtung*) of retentions and protentions. Within this weave of temporality, the now-phase is the point of fulfillment in which a protention becomes fulfilled with a maximal degree of intuitive fullness while at the same time becoming emptied, through retentional modification, of its newly acquired plenitude. Every now-phase of time-consciousness contains a stretch of retention, a point of original presentation as fulfilled protention, and a stretch of unfulfilled protention (Hua XXXIII, 14). On this account, every retention implies a protention, whereas every protention implies a retention: protention emerges from an "earlier" retention much as a protention intends the now-yet-to-come in its temporal constitution of running-off. The original presentation, as the phase of fulfillment, defines a new and "differential" consciousness, or "edge-consciousness," lodged within the entwinement of retention and protention. This increased attention to the protential dimension of time-consciousness does not simply add a missing piece to an otherwise settled framework; on the contrary, Husserl rearticulates the meaning of his earlier insights into the three-fold declension of original time-consciousness in light of this newly discovered complexity on the side of protention and its entwinement with retention. As Husserl notes, Hume had already emphasized that every "impression" necessarily becomes modified in a retentional manner, that is, that every (simple) impression loses its original force and vivacity in becoming a (simple) idea. Hume, however, remained blind to the protential modification of impressional consciousness and the manner in which retention (speaking here in Husserlian terms) generates a protentional consciousness (Hua XXXIII, 13).

Husserl's increased attention to the protential dimension of time-consciousness brings with it an added degree of descriptive complexity. Contrary to the narrow view of protention in Husserl's earlier descriptions, a protention is not simply a next now-yet-to-come. In addition to this protention of the next now-phase, Husserl stresses that protential consciousness intends the "flowing event-horizon," by which he means, that every protention carries (*trägt*) in an intentional manner – in an intentionality of implication – all future protentions, much as every retention implies the continuum of all earlier retentions (Hua XXXIII, 8; 10). In the case of protention, however, its structure of "double intentionality" adds a further degree of complexity.

On the one hand, a distinction must be drawn – following Dieter Lohmar's illuminating proposal – between "near" and "far" protentions, much as Husserl explicitly distinguishes between "near" and "far" retentions.[11] Within the three-fold declension of original

11 Dieter Lohmar, "What does Protention Protend? Remarks on Husserl's Analysis of Protention in the *Bernau Manuscripts* on Time-Consciousness," *Philosophy Today*, 46, 5 (SPEP Supplement), 154–167.

time-consciousness, near protentions contribute directly to the overall constitution of the living present. Far protentions, by contrast, exceed the arc of the living present and, in this sense, are "layered-over" the near protentions. Perhaps we could also speak of far protentions as "asleep" in the sense of "protentional expectations" projected into the remote future, which are "pre-given" as horizons on the outer limit of the living present, lurking and waiting to be discovered.

I am impatiently waiting at a traffic light for the signal to turn from red to green.[12] The continued presentation of the red light is constituted as an unfolding temporal experience – or continuous time-object – on the basis of the three-fold temporal declension of time-consciousness in which its protentional component is motivated to expect the continuation of "red" on the basis of retentional consciousness of earlier phases of red as just-past. With each passing phase, the near protention constitutes the consciousness of the now-yet-to-come, yet this protention has the determinate shape of intending the *continued givenness of the signal as red*. Protentions are thus generated, i.e, motivated, in retentions (Hua XXXIII, 13). A retentional phase retains and thus "empties" a fulfilled protention – fulfilled in an original presentation – while at the same time motivating the continued protentional intention of the now-yet-to-come. Yet, as I perceive the red light, I also expect that this light will change to green and, in this sense, already project in a protentional manner ahead of the living present. In my impatience, I can already "see" the light turn green and, in this sense, intend through a far protention, itself motivated by a far retention, that this light will turn green; with every moment in which the light remains red, my far protention is dissapointed, and yet renewed; and with every moment of waiting, my impatience is the consciousness of diminishing the distance between near and far protentions.

On the other hand, in addition to this distinction between near and far protention, a further distinction must be formulated between the determinate protentional consciousness (in both senses of near and far protentions) and the indeterminate protentional consciousness in light of the double intentionality of protention. In addition to the protentional intention of the future in a determinate manner, as illustrated with the example of waiting at a traffic light, consciousness

---

12 I owe this example to Dieter Lohmar, and use it here with fond memories of our instructive conversations in Cologne.

also intends the protentional future in an *indeterminate manner*. Over and above the protention towards a particular future, there is the protention towards the future as such, towards the open-ended character of the future. With the fulfillment of each protentional consciousness in an original presentation, the horizon of the future as such remains as remote or distant from the now as it was just a moment ago. This indeterminate protention of the entire protentional horizon is akin to the retention of retentions, namely, that every retentional modification entails a (re-)retention of the entire continuum of retentions, such that, the "heritage" of the past is "contained" within every retentional phase. In the case of protentions, every protention already implicates the entire continuum of protentions yet to come. Original time-consciousness is "infinite" in its protentional direction, yet infinite in a specific manner, as reflecting its double intentionality. Protention and retention are each a "mediated intentionality" (*mittelbare Intentionalität*), which Husserl (surprisingly!) still characterizes with Brentano's terminology of "primary and secondary objects." As Husserl notes, "On both sides [*retention and protention*], we have a mediated intentionality, and to every mediated intentionality belongs the double 'direction' of intentionality, towards the primary object and towards the secondary object, that is, towards 'acts' and the primary object in the how of its manner of givenness" (Hua XXXIII, 10). Although Husserl never explicitly addresses the self-manifestation of consciousness along the lengthwise intentionality of protentional consciousness, he clearly implies the protention of my own consciousness, and thus, the pre-reflective consciousness of myself as "anticipating" myself in the future, with this offhand remark that an infinite regress is avoided "on both sides" of retention and protention (Hua XXXIII, 10). The thought must be – never explicitly stated by Husserl – that much as retention retains the act of perceiving as just having been experienced, protention projects my own act of perceiving as yet-to-come. But if I implicitly project myself in the future, original time-consciousness is infinite in the sense that *my* consciousness can never imagine its own death or absence since any projection into the future implicates the survival, i.e., self-projection, of my own consciousness as a spectator. Consciousness is conscious of itself as immortal: the consciousness of any possible future involves an implicit consciousness of myself as "there" in the future, as both spectator and actor, even of the event of my death that I secretly witness. In this manner, I implicitly anticipate myself, that is, my consciousness anticipates its own temporal self-constitution and anticipates *itself* as the source of any possible future constitution of experience, including its own. As Husserl writes, protention is directed towards "future constitution as well as the future event" (Hua XXXIII, 16). The death of my consciousness is, in other words, not an experience that I myself constitute; it takes me by total surprise, but always in the implicit hope that this moment is not my last.

### "Kantenbewusstsein"

In order to further detail this complex rearticulation of the three-fold declension of original time-consciousness, Husserl proposes a sweeping and striking reconstruction of his basic time-diagram (Figure 3). Husserl's revised diagram is based on its earlier versions in which a horizontal axis represents the successive now-phases of a time-object. Along this succesion of now-phases, if we focus on the segment $E_1$-$E_2$, for example, the three-fold temporal declension of original time-consciousness is represented through the "cross-section" of $E'_3$-$E_2$-$E_1^2$ and the "length-section" in *both* of its retentional and protentional dimensions. Within the cross-section (or phase-continuum) in which the objective succession $E_1$-$E_2$ is constituted, the retentional modification $E_1^2$ implicates protentional modification $E'_3$ in the sense that protentional consciousness is motivated or generated through a retention. In this manner, as Husserl remarks, "'the style of the past is projected into the future' ['*der Stil der Vergangenheit wird in die Zukunft projiziert*']" (Hua XXXIII, 38). The running-off of retentional consciousness affects (*wirkt auf*) protentions in a determinate manner by prescribing the content of what is to come (i.e., the continued "red traffic light"). A protention can only intend the future in an empty manner, as motivated by the sedimentation of the past. If protentional consciousness could originate its own content, this would amount to prophecy, in which case the future would be "seen" with the intuitiveness and visibility of an original presentation. Although the future, here construed in its primordial sense as protention, is determined and motivated by the heritage of the past – through a combination of far and near retentions – protentional consciousness also intends the future as such, as an indeterminate and open horizon of what is to come. With the "running-off" of each original presentation, a new protentional consciousness "emerges" that "meets halfway" (*entgegenkommt*) the new original presentation. Expressed in Husserl's visual idiom, the protentional consciousness as both determinate and indeterminate is awkwardly represented by the segment $E_3' E_3$ (determinate protention intending the now-yet-to-come) and the vertical segment shooting into "infinity" through $E_3'$. With the arrival of $E_3$, the empty protention $E_3'$ is fulfilled through the eruption of an original presentation while at the same emptied of its intuitiveness through a retentional modification. Moreover, the protentional consciousness $E_3'$ intends in an empty manner its own retentional modification, that is, it intends the now-yet-to-come in its temporal constitution as running-off. In this manner, Husserl argues that protention implicates a (future) retentional modification much as a retention implicates a (past) protentional consciousness. The emergence of each now takes the form of a primordial hope – an indeterminate expectation that rises above the determinate projections emerging from the sedimented past – waiting for another now yet to come. The now arrives within a continuous consciousness of the possibility of what is to come (*Möglichkeiten des Kommenden*) that is perpetually renewed (Hua XXXIII, 24). And yet, the novelty of each now is absolute, and, in this sense, never entirely expected since each now-phase is not identical with the now-phase just past. We repeatedly expect another now, but its arrival always surpasses our expectation, taking us by surprise in catching us, so to speak, from behind.

In an evocative expression, Husserl writes that "protention is a retention inside out [*Protention ist umgestülpte Retention*]" (Hua XXXIII, 17). Husserl's choice of the verb "umstülpen" is indeed suggestive as the term can mean "to turn inside out" or "to turn upside down." In this phenomenological context, Husserl seeks to describe how retentional consciousness is thrown ahead of itself as a protention as well as how protentional consciousness is behind itself in its emergence in a retention. Protentional consciousness is *ahead* of the now to the extent that it emerges *behind* the now, in the wake of a retentional consciousness. Retentional consciousness is already *behind* the now in so far as it is projected *ahead* of the now. As we have seen, protentional consciousness is the modification of retentional consciousness much as retentional consciousness is a modification of protentional consciousness. The intentionality of the now-phase, in its three-fold temporal declension, involves a double-modification of "fulfillment" and "emptying." The tension between the poles of retentional modification and original presentation in Husserl's earlier analysis is thus displaced onto the poles of retention and protention; their point of intersection is the fulfillment of a protention in an original presentation. Significantly, Husserl argues that we should not think that we *first* have an original presentation, which is *then* retentionally modified, which *then* motivates a protentional consciousness directed towards the now-yet-to-come. Such a construal tacitly assumes a linear progression from original presentation into its retentional modification into its protentional projection. But, this would be to presuppose constituted time, not to provide its phenomenological provenance, and to conflate the self-constituting flow of consciousness with the constituted succession of events in time. This interplay of "self-fulfillment" (*Sich-Erfüllen*) and "self-emptying" (*Sich-Entleeren*), of presentification and "de-presentification," passes through the original presentation, which, for its part, is caught in-between the pull of two opposing intentional tendencies.

Husserl characterizes retention and protention as different tendencies. A protention is a tendency towards "maximal fulfillment" in an original presentation that "climbs" towards the horizontal line of constituted succession whereas retention is a tendency towards "maximal emptiness" that "sinks away" towards the horizon of far retentions. With these characterizations, Husserl underlines that retention and protention must be seen as a "double-branch" (*Doppelzweig*) of intentionalities whereby one should not read the diagram as indicating that the "above section is simply protention and the bottom section is simply retention" (Hua XXXIII, 28). This manner of characterizing protentions and retentions as tendencies is further developed by speaking of their respective "charge" (*Belegung*) of affectivity. This language of each stream of retention or protention as "charged," as visually indicated by Husserl's darkening of the retentional and protentional sections in his diagram (not visually represented in our rendition), is a direct ramification of the "shattering of linearity." It also shows how Husserl has silently overlaid his descriptions of originary time-consciousness with *affectivity*, and thus, already interjected the problem of genesis. The "future" and "past" are neither "in front" nor "behind" us. That is, the past can be "in front of us" in the sense of an empty intentionality that motivates a protention with charged affectivity to be fulfilled in the future. But the future can also be behind us, in the sense of a minimal affectivity towards which we do not take heed and respond – a warning, for example, that we do not heed. Nevertheless, even though the linearity of time is shattered and, indeed, reconfigured into this weave of "streams" of retentions and protentions, the primary directedness of the life of consciousness remains towards the future. The "tendency towards" is the basic form of intentional consciousness. Significantly, tendency does not mean desire (*Streben*) or counter-desire (*Widerstreben*), but rather "graduality towards" proximity and distance. As a tendency, consciousness has both a *terminus ad quem* and a *terminus a quo*, and every momentary phase of consciousness is "both in one" (Hua XXXIII, 39). A momentary phase is both a being-directed towards and a being-directed away; it is caught in-between two opposing tendencies of presence and absence, and, in this sense, escapes a characterization as simply either present or absent.

In this fashion, Husserl's intention is to reject the suggestion that retentional consciousness and protentional consciousness are *symmetrical* and that the transition from protention to retention is a "constant and uniform transition" (*stetiger gleichartiger Übergang*) from "positive" to "negative" (Hua XXXIII, 34). In order to illustrate this argument, Husserl ingeniously suggests that his diagram must become transformed from a two-dimensional to a three-dimensional diagram; he folds the paper in half along the crease or edge (*Kante*) designated by the horizontal axis E-E. Husserl proposes to think of the horizontal axis E1-En as an "edge-consciousness" (*Kantenbewusstsein*) that is produced as a tension between two opposing and colliding planes, each with a different "charge," and which is indicated in the diagram with "+" and "−"signs. The positive process of "increase" (+) (*Steigerung*) streams *towards* the edge (*Kante*) and attains its maximal point of fulfillment in this crease of original presentations while retentional modifications cascade away from the crease in their progressive decreasing of intuitivity. The consciousness of the now is an "edge-consciousness" (*Kantenbewusstsein*) that is produced in the *collision* (*stoßen*) of protentional and retentional streams. The consciousness of the now is an edge-consciousness in two senses: as the consciousness on the edge of past and future; as the edge of consciousness, as the sharpness of the limit or cut called the now, but also, as the moment just before the marked event of the renewal of the now. It is in fact structurally indecidable whether consciousness is ever on the mark – in the now – or on the edge of the now. If we return to the argument just advanced regarding the novelty of the original presentation as surpassing its necessary protentional expectation – as taking us from behind, so to speak – we can further specify the sense in which consciousness *misses itself* in the now in terms of this indecision of the now as both "on time" or on the edge of time. It is in light of this conception of original presentation as the eruption of the new and as an edge-consciousness that Husserl reaffirms the pre-reflective self-consciousness, and hence, self-temporalization, within this unfolding of original time-consciousness. As Husserl remarks: "in so far as the consciousness of the past and the consciousness of the future is continually transformed, and, in this manner, transformed in so far as consciousness streams, and thus continually new, there is also a consciousness of this consciousness [*indem es immerfort neu ist, indem es strömt, sich wandelt und sich so das Bewusstsein von Vergangenen und Künftigen wandelt, ist auch Bewusstein davon da*]" (Hua XXXIII, 47). As Husserl asks himself: "And is this not entirely comprehensible? [*Und ist das nicht voll verständlich?*]" (Hua XXXIII, 48).

## An impossible puzzle

But what is here entirely comprehensible? How is consciousness "also there" (*auch da*), or for itself, as an intrinsic self-consciousness caught and constituted in the stream of its own consciousness? As covered in chapter 3, Husserl's argument that the consciousness of a time-object was based on an inner temporality of consciousness (the perception of duration presupposes the duration of perception) generated the problem of an infinite regress that proved insurmountable in Husserl's earlier analysis. To recall: guided by the schema of appre-hension/content of apprehension, Husserl distinguishes within immanent consciousness between constituted acts of consciousness (and their sensible content) and the constituting "absolute" con-sciousness. This distinction within immanent consciousness mirrors the distinction between constituting act of consciousness and its con-stituted, transcendent perceptual object. Although the formulation of this schema of intentionality within immanent consciousness, as a means to explain the sense in which consciousness experiences itself as its own intentional activity, recognizes as a problem the issue of how the temporality of consciousness is itself constituted, the diffi-culty with this proposed account, as already noted, stems from the introduction of a distinction between "constituting act" and "consti-tuted object" *within consciousness*. Consciousness becomes divided from itself in such a way that begs the question of how to account for its intrinsic self-consciousness. If an act of consciousness is consti-tuted as an "inner time-object" for an absolute consciousness, this would imply that the constituting act of absolute consciousness must itself be constituted by another, "deeper" and separate act of consciousness, thus leading to an infinite regress – the consciousness of any time-object – including "inner time-objects" – would depend on a temporality of consciousness that in turn would have to be constituted in a further consciousness.

In the years (1906–1908) subsequent to the formulation of this dilemma, as charted in chapter 4, Husserl discovered the double intentionality of retentional consciousness in light of which the status of absolute time-constituting consciousness could be revisited and more sharply delineated. As presented earlier, absolute time-consciousness retains *itself* and is *itself* retaining. In this self-temporalization, absolute time-consciousness *differentiates itself* in a two-fold manner along the lines of the double intentionality of retentional consciousness: as a differentiation *from* itself in terms of the transcendence of constituted time-objects ("cross-intentionality") vis-à-vis constituting immanent consciousness ("length-intentionality"); as a differentiation *of* itself in terms of the transcendence of absolute time-constituting consciousness vis-à-vis constituted immanent consciousness. At the point in which we left this development, we did not, however, address specifically how the revision of retentional consciousness contributes to resolving the dilemma of how consciousness constitutes itself without succumbing to an infinite regress.

In the context of the Bernau reflections and their ontological inflection of temporality, the two lines of Husserl's investigations represent the parallel structures of noetic and noematic temporalization, as reflected in the "cross" and "length" intentionalities of retentional consciousness, but also, of protentional consciousness. In this manner, temporalization is a process of "ontification." The sense of being is constituted as a form of temporalization. As Husserl remarks: "Reality is the realization of an anticipating consciousness [*die Wirklichkeit ist Verwirklichung eines antizipierenden Bewusstseins*]" (Hua XXXIII, 46). But this also means that self-temporalization is a "realization" or "ontification" of consciousness itself in so far as consciousness is itself temporally constituted. Thus, the problem of absolute time-consciousness encountered at the end of the previous chapter must be seen with this added significance, in terms of a *transcendental problem of the becoming or self-constitution* ("ontification") of consciousness, and not merely, in terms of the issue of self-consciousness. We shall return once again to this central problem of Husserl's thinking, with which he would continue to struggle in the C-manuscripts during the 1930s, in the concluding chapter of our book.

In order to avoid an infinite regress, Husserl must demonstrate how the absolute flow of time-constituting consciousness is structurally distinct from constituted time-objects, including the constituted temporality of immanent consciousness (acts of consciousness and their sensible underpinning), without, however, severing the unity of both, or succumbing to a reflection theory and internal separation between subject and object. Husserl's early view rested on the unquestioned assumption of the simultaneity of the flow of time-constituting consciousness and the constituted immanent time-objects along with their intentional correlation (the melody as a time-object). This presumed simultaneity implies that the consciousness of time coincides with the time of consciousness; in other words, absolute consciousness is, on this view, temporal in the same sense as the time-objects constituted on its basis.

In other words, demonstrating this structural distinctness must involve showing how the temporality of absolute consciousness is other-temporality, a different "kind" of temporality than that of constituted time-objects, which Husserl designates as "flow" or "stream" – a pair of extremely elusive terms in Husserl's thinking. Before we ask why "absolute time-constituting consciousness" is characterized as a "flow," and in what sense it is "in flux," let us first delineate how absolute time-constituting consciousness must be structurally distinct from time-objects, immanent and transcendent. Transcendent and immanent time-objects endure over time and have duration (*Dauer*), and this means that as time-objects, they have a beginning and an end separated by an interval. As something enduring over time, a time-object has the form of an identity: *it is something*. In this sense, an act of consciousness is an immanent unity since it also is constituted as having a beginning and an end: an act of hearing begins when I hear a tone and ends with the final phase of the tone. Even if an act of consciousness is only rendered into a time-object in an act of reflection, as an immanent unity within consciousness, an act of consciousness must have a beginning and end. Even if I am not attentively directed towards my lived experience of hearing a sound, I experience in its immediacy that act of hearing as beginning and ending. Otherwise, Husserl would have to claim that acts of consciousness do not have a beginning or end until we recognize them as objects in reflection. If we think back to Proust's description of the tolling of bells noted above, although Marcel was not explicitly attentive to how many times the bells tolled, his ability to recall in hindsight the number of rings, even if without precision, testifies to the pre-reflexive and non-objective lived experience of discerning the beginning and endings of acts of consciousness and their intentional time-objects. Indeed, reflection discloses its object; it does not give it attributes that it did not possess prior to reflection. Moreover, on Husserl's account of the retention of retention, the beginning phase of a time-object is continually retained and retained again, and this means that I retain implicitly the first phase of having heard the tone.

By contrast, the intrinsic consciousness *that* I am hearing a note does not have a beginning or end in the sense that each phase in the "row of adumbrations" of the constituting time-consciousness does not contribute to the formation of an enduring identity over time. The stream of absolute time-consciousness is not an "object" or "substrate" that changes or is changed over time. As Husserl insists: the flow of absolute consciousness is not an "object" or an "event in which something happens" (*Vorgang*) (Hua X, 370; 74). Husserl thus wants to argue that time-constituting consciousness is absolute, yet it is neither a "substrate" nor a timeless form, but a constant flux or continuous self-differentiation. Each phase or "flux" is not an identity (we are reminded of Brentano's waves within a wave) since each now-phase of time-constituting consciousness is a self-differentiating original presentation. An original presentation is the renewed principle of self-differentiation that becomes retained in its difference from other, already retained original presentations. The term "flow" means, therefore, not duration – something enduring or given over or in time – but the form of always becoming other than myself. And yet, if absolute time-constituting consciousness is neither an object nor an event, this does not mean – as Sartre draws the conclusion – that absolute time-consciousness is "nothing," in other words, that it is nothing other than the consciousness of not-being the object of its consciousness.

Husserl argues, however, that the flow of absolute time-consciousness "appears" in so far as "the phase of time-constituting flow *belongs* to a now, that is, belongs to a constituted now" (Hua X, 75). In this manner of speaking, the flow of absolute time-consciousness "belongs" to the now it constitutes without itself occupying, or being, in the now of its constitution. In this sense, the flow of absolute time-consciousness is itself not now – not in the now that it constitutes.

And yet, both must belong together – constituted temporality and temporalizing consciousness – otherwise we would revert to an infinite regress or some variant of a reflection theory. As Husserl argues: "There is one, unique flow of consciousness in which both the unity of the tone in immanent time and the unity of the flow of consciousness itself become constituted at once. As shocking (when not initially even absurd) as it may seem to say that the flow of consciousness constitutes its own unity, it is nonetheless the case that it does" (Hua X, 80/434).

Husserl looks to the double intentionality of retentional consciousness as providing a clue to this difficulty. As he writes: "The doubleness of intentionality in retention gives us the clue for resolving the difficulty of how it is possible to have knowledge of the unity of the absolute constituting flow of consciousness" (Hua X, 80). As presented earlier, Husserl distinguishes between the "cross intentionality" through which the original presentation of a tone is retained and the "length-intentionality" along which consciousness retains itself as having just heard the note; along this axis of intentionality, consciousness manifests itself in a pre-objective and pre-reflective manner. Retentional consciousness thus performs a double function: retention of the note as well as the retention of absolute consciousness (retention of myself as having just heard the note; but also retention of the act of perception that just occurred – in this sense, Husserl employs the same language to describe *both* constituted immanent time-objects (act of perception) and the absolute consciousness). In this regard, the self-manifestation of the flow is not temporal in the manner of a time-object since the retention of retention is not the same, nor analogous to, the retention of phase of the object. The cross-intentionality and length-intentionality are contemporaneous (*zugleich*) but not simultaneous (*gleichzeitig*). Husserl thus argues that, "the consciousness of now is itself not now. That which is 'together' with the consciousness of now, the retention, is not 'now,' is not simultaneous with the now." Absolute consciousness is not "in" time in the manner in which constituted objects are "in time." Instead, absolute consciousness is the "self-temporalizing" of consciousness *itself*, as the difference between "objective" (constituted time-objects) *and* "subjective" (constituted immanent time-objects).

As Rudolf Bernet suggests, absolute consciousness is a "pure difference" in the two senses of differentiation from self and differentiation of self.[13] In the first instance, consciousness differentiates itself from its constituted time-objects; in the second sense, consciousness differentiates itself from itself in the form of retentional consciousness in terms of which consciousness manifests itself for itself. At the heart of this constitutive accomplishment is an "unceasing self-loss of presence" that characterizes the "absolute" of consciousness.[14] The past that I have become for myself can only be recovered from a distance in which that lost presence is given again, but never from when the past first began. And yet, this loss of presence is part and parcel of the renewal of presence, not as the eternal recurrence of the *same*, but as the perpetual return of the difference between past and future – the "in-betweenness" of original presentation and retentional modification that impels, and so divides, the passage of time.

Despite the advantages of this newly articulated conception of absolute time-consciousness, Husserl's phenomenology still remains plagued with difficulties with which he would continue to struggle. Most strikingly, as Husserl remarks, "this flow is something we speak of *in conformity with what is constituted,* but it is not 'something in objective time'" (Hua X, 371 [382]). At this point, Husserl is quite frank about the difficulties that he is faced with. If we can only speak of the flow of absolute time-consciousness in light of constituted temporality, its attributes are negative: not an object, not an appearance, not a time-object. Any description of the stream or flow of absolute time-constituting consciousness must thus fall under the suspicion of missing the phenomenon in question – including the terms "consciousness," "stream" and "time." The absolute of absolute time-consciousness would be reduced to the "there is" of time-constituting consciousness: *it* happens. As Husserl writes: "It is *absolute subjectivity* and has the absolute properties to be designated *metaphorically* as 'flow'; the absolute properties of a point of actuality, of the original source-point 'now,' etc. In the actuality-experience we have the original source-point and a continuity of moments of reverberation. For all of this, we have no names" (Hua X, 371 [382]). The self-temporalization

13 Bernet, "Einleitung," lvi. For a more detailed study of Rudolf Bernet's reading, see Nicolas de Warren, "Time and the Double Life of Subjectivity: On Rudolf Bernet's 'Introduction to Husserl's Phenomenology of Time Consciousness,'" *Journal for the British Society of Phenomenology* (forthcoming, 2009). Parts of the discussion in this chapter overlap with this engagement with Bernet's thinking.
14 Bernet, "Einleitung," liv.

of absolute time-consciousness in which the differentiation between "subjectivity" and "objectivity" is constituted in a primordial form of temporal self-transcendence presents an impossible puzzle in which subjectivity's own transcendental accomplishment cannot be pieced together from the ontological pieces it has left behind, and in which, we catch but an image, or metaphor, of ourselves.

# 6

## THE LIVES OF OTHERS

*The mystery of the Other is nothing but the mystery of myself.*
— Merleau-Ponty

### The specter of solipsism

A specter of solipsism haunts transcendental phenomenology. From its methodological inception with the suspension of the natural attitude and the ensuing reduction to the field of pure consciousness, the descriptive science of transcendental phenomenology invokes a first-person point of view that might easily be seen as inviting the objection of solipsism. This vantage-point of transcendental *self*-explication is more than simply a stylistic device or a matter of descriptive convenience; it reflects the essential egological form of subjectivity in its singular accomplishment of transcendental constitution. As Husserl states, the *epoché* "affects the intramundane existence of all other egos, such that we should no longer rightly speak in the communicative plural" (Hua I, 58 [19]). Such a methodological commitment to the suspension of the natural attitude and the disclosure of transcendental experience do not look promising for avoiding what seems inescapable: the solipsism of transcendental subjectivity. If the transcendental reduction is *ipso facto* a reduction to the *solus ipse* of pure consciousness, how can such solitude render intelligible the transcendence of a world open to any possible experience, including the lives of Others? Must we conclude that "le solipsisme n'est ni une aberration, ni un sophisme: c'est la structure même de la raison"?[1]

---

1 Emmanuel Lévinas, *Le temps et l'autre* (Paris: PUF, 1983), 48.

Ironically, this fate appears to be confirmed by Husserl's account of the Other in the context of presenting phenomenology as the proof of transcendental idealism. In Derrida's assessment, the Fifth Cartesian Meditation contains the "very profound lesson [that] I have no originary access to the alter-ego . . ."[2] In the *Cartesian Meditations* and other writings, Husserl repeatedly defines the Other as an alter-ego who cannot be made present within the sphere of my own experience as she is present, or given, to herself within her own. To experience the Other is to experience her absence. Taken at face value, the consequence of this "lesson" is, however, not at all evident. If I am bereft of any original access to the Other, how can I truly experience the Other as Other, let alone pursue the question of how the Other is at all given to me? Faced with this mystery, any experience of the Other would by default either substitute something else – a machine, a zombie or other surrogate – for the Other or project onto the Other a surreptitious image of myself. Is this insistence of the Other's alterity just another name for solipsism, an acknowledgment that I am indeed self-enclosed, accessible in an original manner only to myself? "Were I perchance to look out of my window and observe men crossing the square . . . what do I see aside from hats and clothes, which could conceal automata?"[3]

## Time and the Other

In the Fifth Cartesian Meditation, Husserl confronts the objection of solipsism that shadows transcendental phenomenology from its inception with the suspension of the "universe of all being" and the disclosure of the transcendental field of experience. Lest we think that Husserl's concern with solipsism is here restricted to an account of the Other, and thus to a refutation of solipsism *stricto sensu*, we must note that this objection challenges "nothing less than the claim of transcendental phenomenology to be itself transcendental philosophy and therefore its claim that, in the form of a constitutional problematic and theory moving within the transcendentally reduced ego, it

2 Jacques Derrida, "Hospitality, Justice and Responsibility: A Dialogue with Jacques Derrida," in *Questioning Ethics*, ed. R. Kearney and M. Dooley (London: Routledge, 1999), 71.

3 René Descartes, *Meditations on First Philosophy*, trans. R. Ariew and D. Cress (Indianapolis: Hackett, 2006), 17.

can solve the transcendental problems pertaining to the objective world" (Hua I, 91 [89]). The philosophical claim of transcendental phenomenology remains in suspense as long as the objection of solipsism remains unanswered. At stake is nothing less than the promise of phenomenological philosophy itself.

And yet, of the different literary forms chosen for the presentation of phenomenological philosophy, it might still seem puzzling that Husserl confronts the specter of solipsism in a format most unashamedly Cartesian in orientation, and thus explicitly aligned to a conceptual framework most infamously associated with the problem of solipsism. This choice is especially bold since phenomenological reflection can neither appeal to an idea of God nor rest content with the ego as *substantia cogitans* in order to parry the threat of solipsism and reclaim the transcendence of the world in the manner attempted by Descartes in the *Meditations on First Philosophy*. Moreover, within the ambit of phenomenological reflection, Husserl clearly recognizes the insufficiency of the Cartesian way of the reduction with regard to inter-subjectivity well before the writing of the *Cartesian Meditations* (Hua VIII, 129; 432ff.). As Husserl remarks in the early 1920s, "for many years I did not see any possibility of how to shape the phenomenological reduction into an inter-subjective reduction. A way was finally opened that carried a decisive significance for enabling a complete transcendental phenomenology as well as for a transcendental philosophy of a higher order" (Hua VIII, 174). This extension of the reduction to inter-subjectivity takes its initial form in the 1910/11 lecture-course *Basic Problems of Phenomenology* (*Grundprobleme der Phänomenologie*); yet, it is not until the lecture-course *First Philosophy* (*Erste Philosophie*) (1923/24) that Husserl develops a fuller conception of an inter-subjective reduction in tandem with the opening of a "second way" to the reduction through intentional psychology. As Husserl notes, "The advantage of this second way to transcendental subjectivity vis-à-vis the Cartesian way is that this second way immediately opens the possibility of connecting inter-subjectivity into the reduction" (Hua VII, 313). And yet, the fount of Descartes's way to transcendental philosophy still remains a viable point of departure for a phenomenology of the Other that becomes deliberately exploited in the *Cartesian Meditations*.

In the context of Husserl's development, the treatment of solipsism in the *Cartesian Meditations* represents substantial progress with regard to the *Basic Problems of Phenomenology* and *First Philosophy*, as well as the

*Ideen,* which, in Husserl's own judgment, suffers from a "lack of completeness" due to its failure to offer "a proper consideration of transcendental solipsism or transcendental inter-subjectivity."[4] Whereas the former lecture-courses address the problem of inter-subjectivity in the absence of a robust formulation of its contribution to transcendental phenomenology, the *Ideen* presents the framework of transcendental phenomenology in the absence of a robust formulation of inter-subjectivity. And yet, while it is true that the *Ideen* remains silent on the problematic status of its methodological solipsism, the importance of inter-subjectivity nevertheless lurks beneath its surface. Indeed, the *Ideen* acknowledges the Other and the constitutive experience of empathy without yet submitting the Other to constitutional scrutiny.[5] Although the significance of inter-subjectivity for transcendental phenomenology cannot yet be discerned in the manner in which it is developed in the *Cartesian Meditations* and other subsequent writings, inter-subjectivity is minimally acknowledged to the degree that the inter-subjective identity of the thing (as a "spiritual" [*geistig*] and cultural object) refers to a higher-order constitution based on the primary constitutional accomplishment of perceptual experience.

As discussed in chapter 1, the constitution of a perceptual object is the accomplishment of an unfolding synthetic unity of actual and possible experiences. I perceive this table, for example, against a horizon of possible experiences – moving around the table to discover its other sides, etc. – the intentional lines of which extend not only across the breadth of *my* possible life, but beyond to implicate experiences other than my own. As the regulated system of actual and possible manners of givenness, every perceptual object is situated in a nexus of horizons that delineate, and so motivate, other possible experiences (turning the thing on its side, disappointing or fulfilling my expectations, etc.). The necessary inadequateness of the thing's objectivity constitutes its transcendence; yet, as the object of any *possible* experience, the table is not simply there to be seen by me, it

---

4  1931 Preface.

5  "Der Andere und sein Seelenleben ist zwar bewußt als 'selbst da' un in eins mit seinem Leibe da, aber nicht wie dieser bewußt als originär gegeben" (Hua III, 11). *Ideen II* does, however, provide an analysis of the constitution of the Other but not in terms of transcendental significance of the Other nor in terms of the significance of inter-subjectivity for the constitution of objectivity.

is there to be seen for any possible *other* consciousness. As Husserl argues, "anything objective that stands before my eyes in an experience, and first of all in a perception, has an apperceptive horizon – that of all possible experience, my own and foreign" (Hua XIV, 289).[6] Given this surplus of transcendence vis-à-vis my own consciousness, the transcendence of my consciousness towards the world is inextricable from the transcendence of my consciousness towards Others, for whom this *same* world is equally a horizon of actual and possible meaningful experience. As Husserl writes, "The sense of being of the world [*Seinssinn der Welt*], and in particular of nature, as objective, includes after all. ... thereness-for-everyone, which is always co-intended, when we speak of objective actuality" (Hua I, 94 [92]).

Seen in this fashion, the problem of solipsism complicates the guiding phenomenological thread into the transcendence of the world: if the Other is necessarily implied in the transcendence of the world, transcendental reflection must make room for an immanent temporality other than mine in and through which the world is *also* constituted, or "co-constituted," in an accomplishment not of my own. Such an account must furthermore clarify how a *plurality* of other subjectivities – as opposed merely to the dyad of myself and *one* Other – is both a constituted and constituting community, as what Husserl calls the "monadic community" of transcendental inter-subjectivity. The origin of the world coincides with the origin of inter-subjectivity; in this fashion, "transcendental subjectivity is broadened to inter-subjectivity – or rather, properly speaking, it is not broadened; rather, transcendental subjectivity only understands itself better" (Hua XV, 17).

The transcendence of the world thus implies the transcendence of the Other in two distinct, yet related, senses: as the Other who transcends me and as the Other for whom the world is also a transcendence. Although the Other is given to me as herself-there, as a life other than mine, the life of the Other is not given to me in an original manner, that is, she is not given to me in the manner in which I am given to myself nor indeed in the manner in which she is given to herself. The transcendence of the Other is in this sense primordial given its irreducibility to my own immanence. As Husserl writes, "we have the only transcendence that is genuinely worthy of its name, and

---

6 Cf. Dan Zahavi, *Husserl and Transcendental Intersubjectivity*, trans. E. Behnke (Athens, OH: Ohio University Press, 2001), 26: "Horizon intentionality is intersubjective in its very essence."

everything else that is also called transcendent, such as the objective world, rests upon the transcendence of foreign subjectivity" (Hua VIII, 495). Without this possibility *for me* of an ego other than my own, the world as there-for-everyone, as other than me, could not be given to me. As Husserl remarks, "the world is continually there for us; but in the first place it is there for *me*. This fact too is there for me; otherwise there could be no sense for me in which the world is there for *us*, there as one and the same, and as a world having a particular sense" (Hua XVII, 214 [242]). This recognition of the primordial transcendence of inter-subjectivity is *occasioned* by the problem of solipsism since the sense of the world as there-for-everyone must primarily be constituted as a world there for me. I must constitute the world *in me* as there-for-everyone by way of constituting *in me* an ego other than my own, as not (in) me.[7] We shall return to this seemingly impossible puzzle that defines the challenge of the Other for transcendental phenomenology.

The problem of the Other therefore completes the form of Husserl's transcendental idealism in two significant senses: (a) transcendental phenomenology would fall short of grounding the transcendence of the world without an account of the (primordial) transcendence of the Other; (b) transcendental phenomenology would fall short of a consistent transcendental idealism by virtue of having to adopt a position of transcendental realism (for which the Other remains a hypothesis, beyond the reach of transcendental constitution), without an account of how the transcendence of the Other is constituted in the unfolding of *my* experience.

In the Fifth Meditation, the aim is not only to reclaim the lives of Others as a theme of phenomenological inquiry but also, in so doing, to secure the philosophical claim of transcendental phenomenology. Much as the problem of time-consciousness is of fundamental significance for Husserl's phenomenological project, the problem of the

---

7  Zahavi's (57) claim is thus too strong: "It is not our thematic experience of the Other that makes our experience of the intersubjective world possible; rather, the horizon structure of our relatedness to the world points to open intersubjectivity, and it is the latter that makes our thematic and concrete experience of the Other possible." It is too strong because what is at issue is how the Other is constituted *for me* and how the Other-monads are carried *within me*. In the passage cited above, where Husserl says that subjectivity understands itself as inter-subjectivity, Husserl continues: "It [*transcendental subjectivity*] understands itself as the primordial monad that bears other monads intentionally within itself" (Hua XV, 17).

Other is likewise called upon to shoulder Husserl's transcendental ambition. In the Fifth Meditation, however, this intersection between time-consciousness and the Other for the argument of transcendental phenomenology is, on the one hand, recognized, yet, on the other hand, largely unexamined. In its explicit manifestation, the intersection between time-consciousness and the Other reveals an underlying affinity between how the Other, as irreducible to my self-presence, can nonetheless be given, and how the past, as irreducible to my self-presence, can nonetheless be given. In both instances, the challenge is to understand the givenness of absence without undermining the phenomenological adherence to the original givenness of presence, or evidence, as the foundation for all constitution. Yet even if time-consciousness and the Other find such a point of intersection, the confrontation between time-consciousness and the Other remains unexplored in the *Cartesian Meditations* (and, indeed, in other Husserlian writings within the chronological parameters of this study), in so far as both problems provoke a fundamental questioning of the "manner of being" (*Seinsweise*) of transcendental subjectivity from different angles that converge on its transcendental prerogative as the origin of the world.[8] Does the transcendence of the Other ultimately rest on the transcendence of absolute time-consciousness or does the temporalization of absolute consciousness rest on the alterity of the Other?

## In the name of solipsism

In an instructive essay on the Fifth Meditation, Ricoeur argues that "the problem of the Other is presented in a very abrupt manner by way of an objection *interrupting* the course of a meditation directed by the ego into itself." This "well-known" problem is "the common-sense objection to idealistic philosophies" that calls into question the reduction of the Other to my own consciousness. Even though common sense was explicitly suspended by the *epoché*, the banality of the Other still persists as an interruption of phenomenological reflection

---

8 Although Husserl rightly considered the treatment of inter-subjectivity in the *Cartesian Meditations* as a correction to the incompleteness of the *Ideen*, the *Ideen* remains in another sense incomplete; the *Cartesian Meditations* corrects the insufficiency of the *Ideen* vis-à-vis inter-subjectivity, yet nonetheless *repeats* the insufficiency of the *Ideen* vis-à-vis inner time-consciousness.

("received from without"); this intrusion, however, is "accepted" "entirely from within" by phenomenology as "a challenge."[9] On Ricoeur's view, the interplay between the externality of common sense's protest and its internal acceptance as a challenge mirrors a defining tension within the Fifth Meditation between following the reduction to its transcendental conclusion and accepting the sense of the Other as an ego radically other than mine. Given its significance for the argument of phenomenology, understanding the Fifth Meditation critically depends on whether the objection of solipsism issues from an external or internal point of view. Can one pose to phenomenology a question that phenomenology cannot pose for itself, especially if this question confronts transcendental phenomenology with its own promise?

Rather than immediately accept Ricoeur's thesis, let us insist on the prerogative of suspending any formulation of a question that phenomenology cannot *first* formulate for itself. If the objection of solipsism is to be accepted "*entirely* from within" (emphasis mine), are we not equally free to refuse entirely *from within* the intrusion of a naïve determination of what it means to confront, and to be confronted by, the problem of solipsism? In other words, let us imagine that Husserl considered the *Cartesian Meditations* completed with the constitutive problem of the ego presented in the Fourth Meditation. Can we nonetheless identify the problem of solipsism in the unfolding of transcendental self-explication *prior* to the conclusion of the Fourth Meditation, and thus identify a motivation for the *necessary* extension of the *Cartesian Meditations* to the Fifth Meditation without an appeal to a perspective outside the course of the meditations?

In deepening the reach of the reduction to the dimension of passive genesis and the self-constitution of the ego, the Fourth Meditation crystallizes around the claim that the transcendental ego constitutes the universe of all possible forms of experience. Within the movement of phenomenological reflection, the Fourth Meditation brings to light the full structure of *ego-cogito-cogitatum* that had only partially received explication in the Second and Third Meditations along the axis of the noetic–noematic correlation (*cogito-cogitatum*).

---

9 Paul Ricoeur, "Edmund Husserl. La cinquième *Méditation Cartésienne*," in *A l'école de la phénoménologie* (Vrin: Paris, 2004), 234. ("Husserl's Fifth Cartesian Meditation," in *Husserl: An Analysis of his Phenomenology*, trans. E. Ballard and L. Embree [Evanston: Northwestern, 1967], 116).

This introduction of genetic phenomenology further solidifies the transcendental significance of temporality as the foundation for the constitution of the world *as well as* the constitution, or becoming, of subjectivity. The temporal self-constitution of the ego contains "all possible problems of constitution." The progression of the *Cartesian Meditations* from the First to the Fourth Meditations encompasses the concreteness of transcendental subjectivity as "transcendence" and "immanence." In this regard, the Fourth Meditation brings the *Cartesian Meditations* to a conclusion with the insight: "Transcendence in any form is a constituted sense of being with the ego. Every imaginable sense, every imaginable being, whether immanent or transcendent, falls within the domain of transcendental subjectivity that constitutes sense and being." As Husserl further remarks, "The attempt to conceive the universe of true being as something lying outside the universe of possible consciousness … is nonsensical" (Hua I, 117 [84]). Having arrived at the Fourth Meditation, nothing remains beyond the sphere of egological subjectivity and its work of constitution; there appears to be no opening – textually and transcendentally – for a subsequent Meditation.

This central claim that "nothing is for me otherwise than by virtue of the *actual and potential accomplishment of my own consciousness*" must be grasped in its proper transcendental significance (Hua XVII, 241). As developed in earlier chapters, Husserl rejects a falsified notion of transcendence whereby objects of experience are conceived as things-in-themselves beyond the arc of meaning-bestowing consciousness. As Husserl writes, "experience is not an opening through which a world, existing prior to all experience, shines into a room of consciousness; it is not a mere taking of something alien into consciousness" (Hua XVII, 239). This lesson of transcendental phenomenology is grounded in the primary lesson of transcendental temporality. Time-consciousness is an original self-differentiation that opens the transcendence of the world in the midst of my own immanence. The "self-alienation" or "alterity" that I become for myself within the primordial constitution of time-consciousness (due to the "de-presentification" of consciousness itself through retentional modification) inaugurates the possibility of any form of givenness; on the basis of this "primary alterity" that I become for myself, other forms of alterity are rendered possible.[10]

10 Natalie Depraz, *Transcendance et Incarnation* (Paris: Vrin, 1995), 244ff.

As the crease of such temporalization, an original presentation is neither straightforwardly *received* nor *created* by consciousness. Nor is an original presentation a "sensation" or an "impression," the blind impact of exteriority. The retentional modification of an original presentation (along with its protentional dimension) constitutes the openness *to* givenness as such. Neither active nor passive, neither received nor created, time-consciousness is a medial form of self-constitution that cuts across – by undercutting and rendering at all possible – the distinction between "receptivity" and "spontaneity." In Husserl's Bernau writings, the temporalization of original presentation and retentional modification undergoes a substantial revision through an increased attention to the protentional dimension of time-consciousness. In these writings, an original presentation is recast as the unexpected irruption of *novelty*. Time-consciousness always anticipates the next original presentation, and, in this sense, anticipates itself, yet the arrival of the actual now is an irruption of the new, a beginning anew of the world and of consciousness itself in its constitutional significance. The alterity of the new does not interrupt the course of time-consciousness from the outside; it interrupts from within, by ejecting consciousness from itself, throwing consciousness ahead of itself in such a manner that consciousness cannot recuperate itself entirely despite its own accomplishment in the folding and unfolding of temporality. Consciousness illuminates the medium of its own opacity; from the beginning, I am already other than myself.

The Fourth Meditation thoroughly presupposes this transcendental significance of temporality as the universal form of egological genesis, especially in its concluding presentation of phenomenology as the proof of transcendental idealism. As Husserl writes, "now that the problems of phenomenology have been reduced to the unitary comprehensive title, 'the (static and genetic) constitution of objectivities of possible consciousness,' phenomenology seems to be rightly characterized also as a *transcendental theory of knowledge*" (Hua I, 114–115 [81]). Whereas traditional notions of knowledge operate on the distinction between appearance and the thing in itself, and thus, for Husserl, are committed to *some* form of representationalism, a phenomenological clarification of intentionality liberates us from the view of consciousness as a room in which the blind impacts of the given somehow come to be known. From Husserl's critical point of view, traditional philosophy is caught in the spell of an absurd problem: "can I ask seriously ... how I get outside my island of consciousness and how what presents itself in my consciousness as a subjective evidence-process can acquire objective significance?" (Hua I, 116 [83]). In the Fourth Meditation, the phenomenological proof of transcendental idealism reveals itself as the final overcoming of Descartes' legacy, baptized by Husserl as the "father of the absurd position of transcendental realism, an absurd position" (Hua I, 63 [24]). As the only meaningful theory of knowledge, transcendental phenomenology is uniquely capable of achieving a "kind of transcendental understandableness [*transzendentale Verständlichkeit*] [that] is the highest imaginable form of rationality" (Hua I, 118 [85]).

On Ricoeur's reading, we would not be able to identify an *internal* motivation for the problem of solipsism over the course of the meditations, had the *Cartesian Meditations* ended with the Fourth Meditation. Should "common sense" be the veritable source for the objection of solipsism raised at the beginning of the Fifth Meditation, the extension of the *Cartesian Meditations* would thus depend on a motivation external to the logic of transcendental reflection. The task of the Fifth Meditation would, on this view, consist in refuting the "well-known" problem of solipsism, which traditionally takes one of two forms: as a metaphysical question concerning the existence of the Other or as an epistemological worry concerning the certitude of other minds compared with the claimed indubitability of my own. Should Husserl find himself face-to-face with either of these "well-known" problems, or both, it would further imply that phenomenology is itself a "well-known" form of idealism, to the extent that it is at all susceptible to such a "well-known" objection. The game itself would be well known; everything would already be decided.

## Into a dark corner

Looking back to the beginning of the *Cartesian Meditations*, we find, however, that Husserl has already signaled the concern of solipsism – significantly – in the context of distinguishing between the pre-critical and critical stages of the transcendental reduction. These two stages refine the sense in which the reduction exploits the initial break with the naïve reliance on the world instituted by the *epoché*. In the first stage of the reduction (i.e., the reduction of the *Ideen*), phenomenological reflection discovers the noetic–noematic structures of intentionality, as given in immanent consciousness, yet as bounded to actual (pure) consciousness. Revealingly, Husserl remarks that phenomenological research operates at this stage much as an investigator of nature who naïvely relies on the evidence of natural experience. This is not to repudiate the break with the natural attitude instituted by the *epoché* or to conflate the field of transcendental experience with psychological experience; it is, however, to recognize a *transcendental naïveté* within the reduction in its initial stage of performance. Within this field of transcendental experience, actual consciousness is reflected upon without the benefit of a further questioning of its own temporal givenness. However, as Husserl learns in his analyses of time-consciousness, it is necessary to distinguish between two complimentary directions of reflection *within* immanent consciousness. On the one hand, we can direct our interest towards the actual objects of immanent consciousness (e.g., an actual act of perception as a constituted time-object within my stream of consciousness), in which case, we can descriptively track the streaming of consciousness along the spine of its recurring actual now-phases. But, as Husserl notes, "Even when I make what is immanent into a theme of reflection, and follow purely my flowing appearances of something objective, there appears something that I therefore do not make into a theme, namely, the appearances themselves in their flow are constituted as unities, and precisely as events of the flow itself" (Hua VIII, 467). Thus, on the other hand, we can direct our interest towards the implicitly accepted givenness of the stream of consciousness, in which case, we can descriptively track the temporal constitution of the immanent time-objects as such, with their retentional and protentional horizons. When operating within the narrow bounds of the "first" reduction to the actual givenness of consciousness, this depth and weave of immanent time-consciousness – encompassing the full concreteness of consciousness – remains naïvely given, that is, unquestioned (Hua VIII, 434).

The temporal constitution of consciousness, however, is not exhausted with this questioning of its temporal givenness. In addition, the transcendental field of experience entails horizons of *my own* consciousness, including other possible acts of consciousness, not only in the form of continued acts of perceptual experience (e.g., continued perceptual experience of this table), but also in the form of other possible kinds of consciousness, for example, remembrance and imagination. As with the deepening of the transcendental field to the full temporality of consciousness, the diverse temporal forms of consciousness, understood as different kinds of intentional acts, cannot be accepted naïvely as merely given, but must instead undergo a supplementary transcendental critique (Hua VIII, 159). Within the ambit of the first reduction, however, phenomenological reflection remains restricted to actual perceptual acts of consciousness, that is, acts of consciousness in which an intentional object is given as present (e.g., this table). In contrast to perceptual experience (*Gegenwärtigung*), acts of "re-presentification" (*Vergegenwärtigung*) possess a distinctive temporal form of intentionality in which intentional objects are given as "not-present." Acts of *Vergegenwärtigung* thus require what Husserl dubs a "double-reduction" since the reduction must target not only the actual act of *Vergegenwärtigung* as a lived experience (i.e., the double intentionality of remembrance), but also the object of remembrance entailed *within* the act of *Vergegenwärtigung*, namely, my past act of perception and *its* intentional object (cf. chapter 4). When described within the ambit of the first reduction, the remembered perception and *its* intentional object remain by default the experience of an "empirical ego" (i.e., a worldly ego) since its intentional object – the remembered perception – has yet to undergo a reduction. The double-reduction deepens the reach of transcendental reflection *within* consciousness while simultaneously extending its reach to all possible acts of consciousness (Hua VIII, 437). As Husserl remarks, "Does it not suffice to begin without the tiresome critique of world-experience and the bringing to evidence of the possibility of the non-existence of the world, and instead, to direct the *epoché* of disinterested self-reflection directly to individual acts? Does it not suffice, that I perform all of my acts within the brackets of the one *epoché*, whereby I must therefore attain my pure subjectivity?" (Hua VIII, 127). In the second stage of the reduction, phenomenological reflection matures into a *critique* of its own (pure) experience through which we can arrive at the sought-after "eigenartige Wissenschaft" of transcendental knowledge (Hua I, 70 [31]).

The implications of the double-reduction are far-reaching as it renders to phenomenological reflection the plenitude and concreteness of the *life* of transcendental subjectivity, "the entire empire of experiencing, intentional life as such" (Hua VIII, 434). Whereas the *Ideen* considered that the "residuum" of pure consciousness, as an immanent stream of consciousness, *unquestionably* delivered a field of "absolute evidence," Husserl discovers, through his reflections on inner time-consciousness and programmatically in *Erste Philosophie*, that the *Ideen* in fact naïvely presupposes the temporality of consciousness

(Hua VIII, 433). The recovery of the *entire* temporality of transcendental subjectivity overcomes a methodological solipsism *within* my own consciousness; in the first reduction, phenomenological reflection remains captive to the narrow band of my actual consciousness at the expense of my consciousness of the past and of the future, as if I could only occupy myself with an island of my consciousness within the stream of my own consciousness as a whole. This self-recovery of my own life of subjectivity also (as Husserl pursues in the Fifth Meditation) opens the way for the recovery of lives other than mine within the ambit of phenomenological reflection. As Husserl notes, "Not only do I, the ego who is the subject of the phenom-enological reduction, attain in this way myself as a transcendental ego – I also attain foreign subjectivity, as connected in this method" (Hua VIII, 129).[11]

This *dual* recovery of my own consciousness and of the Other delineates the world-horizon *within* transcendental subjectivity. As noted, the double-reduction extends the reach of phenomenological reflection to the full temporality of my own subjectivity by bringing into the fold of reflection the dimensions of subjective life that are not given in actual consciousness, but which consciousness nonetheless has access to in the form of non-originary manners of givenness, such as remembrance: when I recall an episode from my past, the past experience is given again to me in the present as having once been, yet not in the same manner as I experience the present, in which I remember the past. In exposing the temporal reach of remembrance to transcendental scrutiny, the double-reduction implicitly discloses the horizon of the world in its temporal form. As Husserl remarks, "Every living-present [*Lebensgegenwart*] contains in its concrete inten-tionality an entire life 'itself,' and within this perceptual present of consciousness, there is also contained, as one, objectivity with its universe of all objectivity, as a horizon, that could have validity for me and in a certain manner also that objectivity that will still have validity for me in the future" (Hua VII, 161).[12]

---

11  The double-reduction *ipso facto* extends the reach of transcendental reflection to inter-subjectivity, since the Other is constituted in a special kind of *Vergegenwärtigung* (i.e., empathy).

12  "Jeder Lebensgegenwart hat in ihrer konkreten Intentionalität das ganze Leben 'in sich', und in eins mit der in dieser Gegenwart wahrnehmungsmäßig bewußten Gegenständlichkeit trägt sich horizontmäß in sich das Universum aller

In an earlier chapter, we discussed Husserl's complex descriptions of how my entire past is entailed within time-consciousness and its interlaced structure of "distances" and "orientation," which we interpreted as a minimal transcendental architecture for the temporality of experience. At this point, we can draw a further implication from Husserl's reflections on the constitution of my life as the having of a past and its connection to the double-reduction. The expansion of the reduction to the entire temporality of transcendental life can be seen as the *functional* Husserlian equivalent to Kant's refutation of idealism. However, whereas Kant's argument hinged on demonstrating how determinations of inner sense depended on determinations of external sense, Husserl's refutation of idealism, in its methodological format as the double-reduction, turns on demonstrating how the temporal determination of *actual* consciousness requires the temporal determination of *possible* consciousness, in its horizons of past and future, which, in turn, given the intentionality of time-consciousness, entails the temporal form of the world. As noted at the end of chapter 4, "the pastness of my life" (*Lebensvergangenheit*) entails (in the sense of intentionality) the world of objects (*Gegenstandswelt*) and, in this sense, the pastness of my life implicitly contains the world as constituted within itself.[13] In other words, the Kantian notion of time as the form of inner sense should not be conflated with Husserl's conception of inner time-consciousness since the latter is not a medium in which *representations* are determined; time-consciousness is already a transcendence of the world in the midst of my own self-givenness.[14] The double-reduction performs a task comparable to Kant's refutation of idealism by recovering the entire temporality of transcendental subjectivity, in both its "objective" and "subjective" correlates. But we have also presented how the double-reduction recovers the Other within the method of

---

Gegenständlichkeit, die je für mich galten und in gewisser Weise sogar <die, die> noch zukünftig für mich gelten werden."

13  I follow here Iso Kern's excellent discussion in "The Three Ways to the Transcendental Reduction in the Philosophy of Edmund Husserl," in *Husserl: Expositions and Appraisals*, ed. P. McCormick and F. Elliston (Notre Dame: University of Notre Dame, 1977), 132: "in order to secure my past and future transcendental life as something valid, the past and future world as well must have validity for me. The objective time determination is thus the condition of possibility of the subjective time determination." Kern proposes the illuminating thought that Husserl's expansion of the reduction through the double-reduction effectively formulates a Husserlian version of Kant's refutation of idealism.

14  Cf. Hua XVII, 210.

reduction. How then does the problem of the Other find a place within Husserl's own refutation of idealism?

In the *Cartesian Meditations*, Husserl first formulates the problem of solipsism in the context of distinguishing between the pre-critical and critical stages of the reduction. This acknowledgment of solipsism does not anticipate an objection external to the trajectory of phenomenological reflection. On the contrary, as Husserl writes, "Without a doubt it [*transcendental phenomenology*] begins accordingly as a pure egology and as a science that *apparently* condemns us to a solipsism, *albeit a transcendental* solipsism" (Hua I, 69 [30]). These two phrases – *apparently* and *albeit transcendental* – define phenomenology's *internal* articulation of the problem of solipsism. Indeed, of the many ways in which Husserlian phenomenology presents a striking profile in the history of philosophy, this initial requirement of the posture of solipsism places Husserl in a singular category. Whereas philosophers commonly seek to avoid the consequence of solipsism, Husserl argues *for* solipsism as an initial methodological requirement. As Husserl remarks, "As beginning philosophers we must not let ourselves be frightened by such considerations" (Hua I, 69 [30]). In this temporary form, solipsism emphatically does *not* possess a common-sense meaning: it is not "well known" to us, since the well-known metaphysical and epistemological forms of solipsism each invoke the *existence* of the Other that the formulation of a *transcendental* problem of solipsism prohibits us from invoking as the focal point of questioning. As Husserl explicitly notes, "as a phenomenologist, I am necessarily a solipsist, however not in the usual and ludicrous sense, that has its roots in the natural attitude, but in the transcendental sense."[15] And yet, this preliminary acknowledgment of the transcendental problem of solipsism delineates a question – the question of the Other – that we have yet to learn how, within the field of transcendental experience, to pose, let alone to understand the reach of its transcendental implications. Phenomenology must necessarily begin with a methodological solipsism given that it cannot presuppose the sense in which the problem of the Other gains transcendental significance. This acknowledgment of what initially *appears* as solipsism is a place-holder for a question to come. As Husserl remarks, at the inception of

---

15 Hua VIII, 174; 68: "eine universale Kritik der Erfahrungen überhaupt muß *in gutem Sinne* solipsistische sein."

phenomenological reflection, "it is quite impossible to foresee how, for me in the attitude of reduction, other egos – not as merely worldly phenomena but as other transcendental egos – can become *positable as existing* and thus become equally legitimate themes of a phenomeno-logical egology" (Hua I 69 [30]). This confession of a transcendental naïveté with regard to the Other should not be taken as the affirmation of the ego's transcendental conceit, but rather as a confession of tran-scendental naïveté with regard to *myself*. A further exploration of the question that I become for myself in the face of the Other is required. As Husserl notes, "for children of philosophy, this 'I am' may be the dark corner haunted by the specters of solipsism and, perhaps, of psycholo-gism, of relativism. The true philosopher, instead of running away, will prefer to fill the dark corner with light" (Hua XVII, 237).

## A very puzzling possibility

Contrary to Ricoeur's proposal, the objection of solipsism does *not* mark an interruption in the course of the *Cartesian Meditations*. An attentive reading of the admittedly abrupt opening of the Fifth Medi-tation reveals a direct connection with the concluding discussion of transcendental idealism in the Fourth Meditation. In fact, Husserl already announces the concern that the phenomenological argument of transcendental idealism is "unstable, since we have not carried our methodic predelineations through to the point where the possibility of the being for me of others (as we all feel, a very puzzling possibility) and the more precise nature of their being for me are understandable since the complex problem of their being for me has not been explicated" (Hua I, 120 [87–88]). The "imaginary interlocutor" of the Fifth Meditation does not initiate, but rather specifies the objection of solipsism already set into motion from the inception of phenomen-ology, once again reiterated in the closing of the Fourth Meditation.[16] And yet, the opening of the Fifth Meditation nevertheless comes as a surprise. Our surprise is not that we suddenly stumble across the "grave objection" of solipsism for the first time, but rather the unexpected way in which the problem of transcendental solipsism – already known to us – becomes framed as nothing less than the phenomenological proof

16 This interlocutor is not "imaginary" but represents transcendental reflection becoming consequent to itself.

of transcendental idealism; the refutation of transcendental solipsism
delivers the final refutation of transcendental realism.[17]

17 Also surprising is this. In *The Critique of Pure Reason*, Kant defines his distinction between
transcendental realism and idealism in the following manner. "I understand by the
transcendental idealism of all appearances the doctrine that they are all together to be
regarded as mere representations and not as things in themselves, and accordingly that
time and space are only sensible forms of our intuition, but not determinations given for
themselves or conditions of objects as things in themselves. To this idealism is opposed
transcendental realism, which regards space and time as something given in themselves
(independent of our sensibility). The transcendental realist therefore interprets outer
appearances (if their reality is conceded) as things in themselves, which would exist
independently of us and of our sensibility and thus would also be outside us according
to pure concepts of the understanding" (A 369). Kant's understanding of transcendental
realism's conflation of appearances and things in themselves extends to all appearances,
including time (inner sense) (A 490–491/B518–519). Objects of cognition are "in us" in
the transcendental sense that objects of possible experience are necessarily conditioned by
a transcendental form of sensibility underpinning any cognition; yet, objects of
(perceptual) experience are outside of us in the empirical sense, and thus not "in us"
empirically. In failing to acknowledge that space and time are neither objects in themselves
nor objective determinations independent of the sensible forms of our intuition,
transcendental realism remains unable to distinguish between the transcendental and
empirical senses of "outside of us" and "inside of us" (A373). Indeed, once it is granted that
determinations of space and time are neither conceptual determinations nor aposteriori
synthetic judgments, the discovery of transcendental forms of intuition (transcendental
aesthetic) becomes central to the shape of transcendental idealism (along with, in Kant's
particular circumstance, the discovery of the dialectic of reason). Husserl follows Kant in
considering the rubric of transcendental realism as including a diverse range of theoretical
positions in contrast to transcendental idealism. For both philosophers, the "absurdity" of
transcendental realism – for which, as noted earlier, Husserl identifies an original
Cartesian responsibility – stems from a conflation between the empirical and
transcendental senses to the distinction "outside of us" and "in us." However, Husserl's
phenomenological appropriation of Kant's crucial distinction comes with a series of
qualifications, and most significantly: Husserl considers Kant's own critical enterprise as
failing to articulate a genuine form of transcendental idealism. Husserl rejects the
metaphysical distinction between appearance and thing in itself to which, on his view,
Kant still remains captive; Husserl rejects Kant's metaphysical derivation of space and time
as the basis for the transcendental aesthetic; Husserl rejects the "psychologism" of Kant's A-
deduction; Husserl faults Kant's critical project for its "mythical construction" due to its
lack of a genuine phenomenological method of reduction (Hua VI, 116). There is,
however, another significant difference between Kant and Husserl on the issue of
transcendental idealism. From Husserl's vantage-point, Kant fails to recognize the
transcendental problem of the Other, or inter-subjectivity. Indeed, it seems
incomprehensible from a Kantian point of view how the rejection of transcendental
realism would ultimately depend on the problem of (transcendental) solipsism, since it
would be impossible to understand how the problem of the Other would find a place
within the transcendental aesthetic. Yet, Husserl's contention is not only that the problem
of the Other extends the transcendental aesthetic, but that, in this manner, completes the
proof of transcendental idealism.

Husscrl's novel refashioning of transcendental idealism ascribes a double-significance to the transcendental problem of the Other: the Fifth Meditation opens with the impasse of transcendental phenomenology with regard to the Other, and, in so doing, casts this impasse as the final resolution to the question of transcendental idealism. The transcendental reduction discloses the egological field of transcendental experience in and through which objects of experience become constituted as actualities and potentialities of the universe of all being: "Transcendence in any form is a constituted sense of being with the ego. Every imaginable sense, every imaginable being, whether immanent or transcendent, falls within the domain of transcendental subjectivity that constitutes sense and being" (Hua I, 86 [84–85]). What of the transcendence of a life other than mine? Husserl readily acknowledges that the Other is "surely" not *merely* an intended object of my consciousness; if this were the case, the Other would be constituted "in me," and so reduced to a being that *only has sense for me* (Hua I, 117 [83–84]). This affirmation of the Other's irreducibility to a sense solely constituted in me stems from Husserl's insistence on the definition of the Other as an alter-ego who cannot be given to me in an original manner. The Other is not merely an ego that transcends me; the Other is also for herself given as a self-transcendence, as a life that constitutes itself and its world. And yet, if the Other is *for me* an ego that I myself am not, this sense of the Other's alterity must admit of phenomenological clarification even as the Other is "surely" not an object merely constituted in me.[18]

In this fashion, Husserl opens the Fifth Meditation on the horns of a transcendental dilemma – and precisely that dilemma for which commentators have often faulted Husserl's phenomenological approach. In the words of David Carr, "the alter ego is not susceptible to phenomenological treatment ... he cannot even be considered purely as meant; or, to the degree that he is, he is no longer an ego. The concept of the ego is in general incompatible with the concept of the given, at least if the alter ego is to be considered transcendental and not merely worldly."[19] In Husserl's own rhetorical formulation: "Have we not therefore done transcendental realism an injustice?" (Hua I, 121 [89]).

18 See David Carr's suggestive "The 'Fifth Meditation' and Husserl's Cartesianism," *Philosophy and Phenomenological Research,* 34, 1 (Sept. 1973), 14–35; 19: "The imaginary critic doubts not Husserl's ability to prove that others exist, which is not in question, but the ability to make phenomenological sense of alter-egos in the scheme *ego-cogito-cogitatum.*"

19 Carr, 21.

228 HUSSERL AND THE PROMISE OF TIME

Must we not concede to a form of transcendental realism in order to guarantee the alterity of the Other? And yet, as Husserl is quick to respond, the *entire* course of the *Cartesian Meditations* and its presentation of transcendental idealism argues for the "absurdity" of transcendental realism. On the one hand, we are convinced that transcendental realism lacks a foundation; on the other hand, we are convinced that transcendental realism appears "essentially right in the end" in the absence of a phenomenological account of the Other, which, it seems, cannot be given without forfeiting the sense of the Other as an alter-ego.

Faced with this dilemma, Husserl's attitude is characteristically pragmatic. Rather than appeal to "dialectical argumentations" or "metaphysical hypotheses," Husserl turns to the patient work of phenomenological description – small change, rather than large bills – and the "transcendental fact" of the Other's manner of givenness, as constituted "in and arising from my intentional life." As Husserl asks: "How else than by questioning [*my experience of the Other*] can I explicate *the sense* of existing others (den Sinn *seiender Anderer*) in all its aspects?" (Hua I, 90). Indeed, if we are committed to the ideal of phenomenological clarification as the refusal to take anything for granted, then any abandonment of questioning the Other would be tantamount to accepting the Other's givenness as *unquestionable*. The truth of solipsism, in its methodological *and* transcendental sense, is this radicalism of not taking for granted the unquestionable givenness of the Other insisted upon by the banality of mundane experience. The common-sense protest of the banality of the Other turns on the unquestioned – and apparently unquestionable – assumption that the life of the Other is neither probable nor demonstrable, that it remains inaccessible to me in an original manner, and thus obstinate to any questioning. But can we rest contented with the unquestionable givenness of the Other?

The Other is an ego who is not me. This sense of alter-ego should not, however, be accepted without question, that is, without an understanding of how such a possibility could at all be constituted for me. As Sartre remarks, idealism (for which the Other is nothing beyond my consciousness) and realism (for which the Other is beyond my consciousness) each implicitly conceive of the "constitutive negation" of the Other (as not-me) as an "exterior negation as if this nothing or negation was the element of separation *given* between me and the other."[20] Expressed in

---

20 Jean-Paul Sartre, *Being and Nothingness*, trans. H. Barnes (New York: Washington Square Press, 1992), 269.

Husserl's frame of thinking, idealism and realism are both forms of transcendental realism, each of which conflates a transcendental distance between myself and the Other with an empirical distance between myself and the Other, as when I perceive the Other standing over there, next to a tree. As Sartre astutely remarks, "the fact that the Other is manifest for us in the spatial world means that it is space – ideal or real – that separates us."[21] It is as if the distance separating, and thus distinguishing, myself and the Other was tacitly conceived as a third, independent element – a given – that introduced itself between us. The sense of "in-betweenness," however, cannot be taken as a given, empirical or otherwise; spatiality cannot be accepted as already constituted in framing an account of how the Other is given to me. The failure to distinguish between a transcendental and an empirical sense to the distinction between "outside of us" and "in us" condemns the problem of the Other in advance to incoherence: if the Other is constituted in us, we fail to secure the sense of the Other as not-me; if the Other is outside of us, the Other becomes a mystery.

Under the heading of the analysis of "foreign experience" or "empathy," Husserl sets his sights on the "puzzling possibility" of how my transcendental subjectivity can constitute something in me that nonetheless does not belong to me; in me and yet not in me; like me and yet unlike me. That this possible form of constitution presents a truly radical proposal is evidenced from Husserl's emendations in a typescript of the *Cartesian Meditations* ("Typescript C") that he gave to Dorian Cairns. Cairns reports two revealing corrections: "we must discover in what intentionalities, syntheses, motivations, the sense 'other ego' becomes fashioned in me ... These experiences and their works are facts belonging to my phenomenological sphere" (Hua I, 122 [90], my underline). In Husserl's Typescript C, both underlined references to myself ("in me," "my") are revealingly crossed-out; in a marginal notation to the crossed-out phrase "belonging to my," Husserl remarks, "The dangerous first person singular!"

This puzzling possibility of a transcendence that is constituted through me, but not in me, and that cannot be properly considered as solely my accomplishment, calls into question the phenomenological conception of immanence that resides at the basis of the transcendental idealism advocated in the Fourth Meditation. As Husserl

21 Sartre, 270.

rhetorically asks, "Is it not *self-understood* from the very beginning that my field of transcendental knowledge does not reach beyond my sphere of transcendental experience and what is synthetically comprised therein?" (Hua I, 122 [90]). This transcendental obviousness of immanence – that transcendence of any kind must be constituted in *my* immanence – becomes challenged. In the face of Other, I can no longer take myself for granted as a constituting immanence to which something could not be given that is not entirely constituted within me.

The decision to press forward and bring to constitutive clarity the sense of the Other requires the fashioning of a "primordial reduction" that further restricts, or "reduces," my egological subjectivity to the "sphere of my ownmost" (*Eigenheitssphere*) in view of the Other's transcendence.[22] As Ludwig Landgrebe reports Husserl to have remarked, "the discovery of what is mine is more basic than the discovery of the ego [*Die Entdeckung des mein geht der Entdeckung des Ich voran*]."[23] Rather than assume the transcendence of the Other as a given, Husserl's primordial reduction uncovers a more foundational stratum than hitherto described at the level of the egological consciousness. This reduction to my primordial sphere is a reduction to my primordial embodiment. If we recall Husserl's characterization of perceptual experience as the form of consciousness in which objects are given to me "in the flesh" (*leibhaft*), my own body, as it is experienced or "lived" by me, and as the "medium" for all perceptual experience,

22 See Jean-Toussaint Desanti, *Introduction à la phénoménologie* (Paris: Gallimard, 1994), 121: "Aucun *Ego* ne peut échapper à sa sphère d'appartenance. Elle consiste à apprendre à délimiter, dans la *sphère de ce qui m'appartient*, ce qui, dans son *essence*, s'offre comme n'étant *pas moi*." This is why, moreover, the primordial reduction is not an abstraction from the Other in the mundane sense – such abstraction from the Other, as when I extract myself from relations to Others, is in fact not radical enough since nothing regarding the sense of the world as there for everyone is altered. As Husserl notes, even if a "universal plague" destroyed the entire human population, the sense of the world as inter-subjective would still remain transcendentally in tact. This is the clearest statement that Husserl considers solipsism self-refuting within the natural attitude; it also means that there cannot be destruction of the world as he considered in the *Ideen*. A parallel can be drawn between the exclusion of time-consciousness from the *Ideen*. In a comparable vein, the introduction of the primordial reduction reveals to us that the egological reduction has yet to uncover the final constitutive foundation of my transcendental subjectivity. This is why the primordial reduction is not the same as double-reduction.

23 Cited in Lilian Alweiss, *The World Unclaimed: A Challenge to Heidegger's Critique of Husserl*, (Athens, OH: Ohio University Press, 2003), 157.

defines "my original ownmost" (*das ürsprünglichste Meine*). My lived-body is primordial in two senses, as that manner of being that most properly belongs to me and as the source of "primordiality," as that to which any perceptual object can at all be given "in the flesh."[24] An object can only be perceptually "in the flesh" to an embodied subjectivity of the flesh. In this regard, as Ricoeur notes, the primordial reduction recovers the transcendental meaning of incarnation within subjectivity.[25]

My lived-body defines what is most properly and always mine, in the sense of what I have of myself, and this having of myself is also a having of the world at the foundational level of "pre-givenness," in which the world is constituted through my engagements as an embodied subjectivity moving and acting in the world.[26] The reaching of my hand for a doorknob, the running after a bus, and the pulling of a chair are each ways in which the world becomes passively constituted in an intentional interplay of kinaesthetic and aesthetic (i.e., perceptual) experiences. The reduction to the primordiality of the lived-body allows us to distinguish between the constituted level of objective, spatial bodies that is the thematic subject of the natural sciences, as revealed in perceptual experience in the objects around us that are taken to be present at hand, and the foundational stratum (*Unterschicht*) of what Husserl in the Fifth Meditation calls "nature proper" (or "earth" in its later incarnation), as the "pre-objective" level of "pre-givenness."[27] As Husserl writes, "In any case, anything built by activity necessarily presupposes, as the lowest level, a passivity that gives

---

24 Hua IV, 56. Cf. Didier Franck, *Chair et Corps. Sur la phénoménologie de Husserl* (Paris: Les Éditions de Minuit, 1981), 43ff. As Franck points out, my lived-body, within the sphere of my ownmost, is my ownmost but also that which is closest to me: my lived-body is the origin of what is proper and close ("l'origine du propre et du proche") (95).

25 Cf. Hua XIV, 7. Ricouer aptly speaks, in this regard, of the primordial reduction as the "conquest of the meaning of incarnation" (243).

26 Indeed, we could argue that the primordial reduction leads to a primordial "engagement" with the world, not as the "theoretical" object of perception (looking at a hammer as an object), but as a "practical" object of concern (working with a hammer). See the excellent discussion in Jean-Luc Petit, *Solipsisme et intersubjectivité: quinze leçons sur Husserl et Wittgenstein* (Paris: Les Éditions du CERF, 1996), 93ff.

27 Depraz (117ff) suggests that the primordial reduction suspends the interior–exterior relation; my nature proper is, as a unitary milieu, at the same time interior and exterior, immanent and transcendent. The world is both in me and I am in the world, as Husserl places the terms "outside of me" ("Außer-mir") and "soul" ("in meiner 'Seele'"), as well as "drinnen," "Außenwelt," etc., in quotes to guard against any misreading of these terms. The primordial reduction discovers the "phenomenon of the world" and, in this sense, pushes farther than the double-reduction discussed previously.

something beforehand; and, when we trace anything built actively, we run into constitution by passive generation" (Hua I, 112 [78]).

As Husserl developed extensively in his writings, the spatiality of perceptual objects, as well as the fixed order of space as a system of positions and distances – an order of separation-in-simultaneity (*Außereinander-im-Zugleich*) – is constituted through the motility of my lived-body in its unique form of self-constitution, as both a constituted body that occupies a determinate spatial position vis-à-vis other spatial objects and a constituting center of spatial orientation, to which all possible spatial positions refer. As I walk around this table, the table is given to me through a continuous series of changing perspective; each perpective correlates to a determinate spatial orientation vis-à-vis where I stand: moving to my left, the table is revealed to me in a different perspective than if I moved to my right. As a spatial object, the table is necessarily given to me in an inadequate manner for I could never perceive all of its sides at once nor exhaust the given-ness of the table in my life-time of perceiving. This constitutional solidarity between a perceptual object's inadequate manner of given-ness and its spatiality (already remarked upon in chapter 1) is insep-arable from the spatiality of my own lived-body, from which I perceive the world and in which I am myself located in the world of other spatial objects. My lived-body functions as an absolute reference point for all possible spatial positions. A table can only meaningfully be to the left or the right *of me*. As Husserl argues, "The lived-body has, for its particular ego, the unique distinction of bearing in itself the zero-point of all these orientations. One of its spatial points, even if not an actually seen one, is always characterized in the mode of the ultimate central here; that is, a here which has no other here outside of itself, in relation to which it could be a 'there'" (Hua IV, 158). My lived-body is not merely a body (*Körper*) that occupies a definite position relative to other spatial objects. It is also an absolute here, a "Nullpoint" (*Nullpunkt*) or "Nullbody" (*Nullkörper*), to which any possible spatial position gains meaning as "there." This distinction between "here" and "there" is intrinsic to the double significance of my lived-body as both a constituted spatial object and a constituting "spatializing" center.

This intrinsic differentiation of my lived-body as both "here" and "there" provides the basic constitutional matrix for the constitution of spatiality. Indeed, this self-differentiation of the body as both here and there expresses an intrinsic bodily self-apperception, or self-consciousness, manifest in experiencing my own movement through kinaesthetic sensations. The self-propelled movement of my embodied subjectivity is a function of the ego as a center of possibilities, as the "I can" (*Ich kann*): I can move to the left, go to the door, run to that tree. In such movements, the spatiality of the world becomes constituted to the extent that any possible "there" can become a "here" of my own. In moving over there, the lived-body differentiates itself from itself as well as from the spatial objects of its surroundings much as, as argued earlier, retentional consciousness entails a dual differentiation; in the case of movement, my lived-body differentiates itself from itself in terms of being itself a "here" and a "there" while also differentiating itself, as a here, from the other spatial objects out there. When I move to occupy a position over there, the distance between "here" and "there" diminishes to a "zero-degree." Yet, once I have arrived at the end-station of my movement, the there that is now my here is still a there for me in so far as my own lived-body retains its character as "there" – as a spatial body that I perceive "in front of me" that occupies a spatial position relative to other spatial objects. Indeed, the absolute here of my lived body does not, strictly speaking, itself "move" since it is a here in an absolute sense. In a sense, my lived body is an unmoved mover of sorts that is always already there, as a here, even before I have in fact arrived there. The spatiality of the world is in this manner circumscribed by the horizon of my touch (*Tasthorizont*). Touching is the transformation of the there into a here without entirely suppressing the *difference* between the there that I touch and the here of my touching. Whereas touch sensations that delineate and reveal the surface, texture and temperature of an object are localized in the object as it is there, the touching itself – the sensing that is my hand and its kinaesthetic sensation of gliding over a surface – is localized in the absolute here of my lived-body (Hua XIV, 541; 543).

As an absolute axis of reference, my lived-body as here escapes perception in the sense that when I perceive my own body, the hand that I perceive is constituted as a spatial object, as there in front of me, and so given in an inadequate manner. I cannot see my own body as the vantage-point from which I see, for any part of my body that I can see is itself a "there" in relation to the absolute here of my own lived-body. My lived-body is paradoxically "inside" the world and "outside" the world, as an inside that is outside and an outside that is inside.[28] Indeed, this paradoxical condition defines the intimacy of my lived-body in two related senses as intimate to me (as "my ownmost") and as my vulnerability to an outside. As Husserl writes, "The same lived-body that serves me as means for all my perceptions obstructs me in the perception of itself and is a remarkably imperfectly constituted thing" (Hua IV, 159). And yet, when I move towards the tree, I sense my own body as moving itself. I am both the subject that moves as well as the object moved. This embodied self-manifestation, or embodied self-consciousness, is constituted in the form of self-affection in tactile sensations (Hua IV, 145).[29] As Husserl explains:

> Touching my left hand, I have touch-appearances, that is to say, I do not just sense, but I perceive and have appearances of a soft, smooth hand, with such a form. The indicational sensations of movement and the representational sensations of touch, which are objectified as features of the thing "left hand" belong in fact to my right hand. But when I touch the left hand I also find in it, too, a series of touch sensations, which are "localized" in it, though these are not constitutive of properties (such as roughness or smoothness of the hand, of this physical thing). (Hua IV, 144–145)[30]

Touching my left hand, I experience tactile sensations in my right hand that delineate the contour and surface of my left hand while simultaneously my left hand, as the hand constituted in being touched, is also a hand that senses itself being touched as well as itself sensing the contour and surface of my right hand. Likewise, my right hand senses itself touching my left hand while also sensing itself as being touched. The sense in which my left hand senses itself being touched is distinct from the sense in which my left is itself sensing, that is, touching my right hand. This self-affection blurs any sharp distinction between passivity and activity, between "sensed" and "sensing," within the constitution of *my own* hands as touched and touching.

---

28 Cf. James Dodd's excellent discussion of the paradox of transcendental embodiment in *Idealism and Corporeity: An Essay on the Problem of the Body in Husserl's Phenomenology* (Dordrecht: Kluwer, 1997).

29 For Husserl vision only gives us two dimensions; tactile sensations account for the third dimension of depth.

30 Cf. Hua I, 128: "Das wird dadurch möglich, daß ich jeweils 'mittels' der einen Hand die andre, mittels einer Hand ein Auge usw. Wahrnehmen 'kann', wobei fungierendes Organ zum Objekt und Objekt zum fungierenden Organ werden muß. Und ebenso für das allgemein mögliche ursprüngliche Behandeln der Natur und der Leiblichkeit selbst durch die Leiblichkeit, die also auch praktisch auf sich selbst bezogen ist."

Taken as a circuit of "double-touching," embodied consciousness differentiates itself in a two-fold manner: each hand, so to speak, differentiates itself from the hand that it touches but also differentiates itself from itself in sensing itself as sensing; since both hands belong to my own sphere of ownness, the lived experience of each hand must be constituted as an immanent temporality or "immanent time-object" in the guarded meaning argued in earlier chapters. Each intentional act of touching (along with its immanent sensible underpinning) is an immanent experience constituted in original time-consciousness. By virtue of belonging to one and same unity of absolute time-consciousness, the circuit of self-affection, in its dual differentiation, is constituted as my ownmost. When considered from the vantage-point of time-consciousness, the circuit of "double-touching" can be described as two intentionalities (i.e., the left and right hands) constituted within the same unity of absolute time-constituting consciousness.

## The intersection of the Other

I walk into a room and perceive another person – the Other – standing over there next to a table. The Other's perceptual givenness is akin to the givenness of other things in this room. Her body occupies a determinate spatial position relative to other spatial objects as well as to my own position. As I walk further into the room and circumscribe the Other's presence, her body is given to me through different perspectives much as this table is given to me as a system of inadequate, yet harmonious perspectives. In this sense, I see, and thus constitute, the Other's body much as I would see, and thus constitute, any spatial object as a function of the absolute here of my own lived-body. But now I spy some loose change on the ground, and as I reach to retrieve my fortunate find, a foreign hand suddenly arrests mine halfway. As I look up, I realize that that body over there, next to the table, is an animate body that moves in the world much as I do, and who, in perceiving the same world as "mine," enters into conflict (in this particular instance) with my own projects. I perceive that the Other is an alter-ego, a center of possibilities that is not mine, and towards which the world flows away from me, as if flowing into a drain, to invoke Sartre's striking metaphor.

The Other's hand is constituted as a spatial object through my touch: the texture and contours of the Other's hand are revealed to me in tactile sensations with the double significance outlined above: the tactile sensations touched by my hand are localized in the Other's hand (it is the clamminess of *her* hand that I feel) while the sensing of my own touch is localized in the absolute here of my lived-body. In touching the Other's hand, however, I do not touch an object like other objects in the room, for when I touch those objects, my hand is not in turn touched by them. As my hand touches her hand, however, I sense my own hand as being touched by hers. And yet, this sensing of being touched reveals to me a sensing that is not mine, which I do not immediately sense as a sensing; otherwise, I would consider the touching that touches me as my own touching, much as when I rub my hands together. In touching her hand that touches me, I am myself touched by a touch other than mine. The touching of the Other's hand is constituted neither in the manner in which I touch this table nor in the manner in which my hands touch each other. This touching of being touched, however, is ambiguous since although I feel her hand as touching mine, as itself a hand that senses, I do not touch her touching itself. We might therefore more properly speak of my hand as *caressing* her hand, and of her hand as caressing mine, rather than touching. As Lévinas remarks, "what is caressed is not touched, properly speaking."[31] In this light, I can never caress myself since when I rub my hands together I touch myself touching, as introduced above under the heading of "double-touching" of my self-affection. When I caress the Other, I touch nothing at all of the Other's own touching, and yet that sense of the Other as a touching that is not mine is nonetheless given to me, as other, as beyond the contact of my touching, in the form of a surrendering caress of the Other's inaccessibility.

The intentionality of the Other's touching *intersects* the intentionality of my own touching. When I touch the Other, I sense her hand as not only an object, but as a subject, as a hand that senses in touching me. The "subjectivity" of her hand, so to speak, is not constituted through my touching; her touching arrests my own activity of constitution since an inaccessible center of foreign constitution erupts in the midst of my own constituting activity, in the midst of the

---

31 We might define my touching of a foreign hand as a *caress*, as a manner or intentionality in which the Other who is in contact with me "goes beyond this contact" (*Time and the Other*, trans. R. Cohen [Pittsburgh: Dusquesne University Press, 1987]), 89.

immanent sphere of my ownmost. The Other as a constituting subjectivity does not become present and is not given as present within the immanent sphere of my primordial self-presence. The Other's distance from me remains the same: the Other remains inaccessible to me in an original manner.[32] The ambiguity of touching the Other can be characterized as an *intersection*: my touch is arrested or intersected by a foreign sensing that I myself do not genuinely sense. We do not have the ambiguity of self-affection, as when my left hand touches my right, but the ambiguity of an intersection, of a sensing that transcends my touch completely – and yet I must sense that the hand of the Other itself touches and senses me, without, however, making that foreign sense of touch present to myself; otherwise, it would be indistinguishable from the ambiguous self-affection of the double-touching of my left and right hands.

The constitutional difference between the circuit of self-affection in the "double-touching" of my own hands and the intersection of my own hand and the hand of the Other can be illustrated with the phenomenon of tickling. Much as I cannot caress myself, I am also unable to tickle myself. The sensation of being tickled is the sensation of being touched by a foreign hand that I sense as a sensing that is not mine. In being tickled, I sense myself through the sensing of the Other without thereby sensing myself as the Other or directly sensing the Other's sensing. It is this sensing of sense that is not me that takes me by surprise. Tickling is the manifestation of this surprise that I am being touched, and this touch that I sense, I sense in an intimate manner without thereby sensing it immediately. This sense of the Other as within me but not constituted in me provokes the spasmatic and uncontrollable reaction of my body. In this recognition of the Other as an alter-ego, the intimacy of being touched from the inside, so

32 Merleau-Ponty, *Signs*, 68: "My right hand was present at the advent of my left hand's active sense of touch. It is in no different fashion that the other's body comes to life before me when I shake another man's hand or when I just look at it." However, when I touch the Other's hand, the ambiguity of "sentir" and "sentant" is different from my own self-affection ambiguity; because in my case, the ambiguity of "sentir" and "sentant" is *at the same time* (Alweiss [154] stresses – rightly – "*Instantaneously* my left hand has certain sensations that are not qualities") – but what constitutes this temporality of "at the same time"? This "at the same time" is not simultaneity, but "zugleich" in the manner of self-differentiation, just like the constituting and constituted time-consciousness; however, if this is so, then the "zugleich" belongs to the same stream of inner time-consciousness. But this is not the case with the touching of the Other.

to speak, plays itself out across my skin and reverberates throughout my lived-body. The intimacy of being tickled is constituted through a disconnect between the anticipation of the Other's touch that I can never fulfill myself. This is the surprise of being touched: I anticipate the Other's touch yet cannot fulfill her sensing in a sensing or original presentation of my own. In this regard, I am unable to tickle myself because I anticipate myself as touching as well as fulfill that anticipation: my touch does not take me by surprise since it is constituted as a circuit of my self-affection. In the case of being tickled by the Other, I sense myself being touched by another sensing that I cannot anticipate; every tickling is a surprise and the eruption of a distance, or transcendence, of the Other's absence. In being tickled, I am thrown outside of myself: the center of gravity of my own body is thrown outside of itself in a movement of both contraction inwards and explosion outwards, as if my lived-body had temporarily lost its own orientation around its absolute here, and in flaying outwards and retreating inwards, I am trying to regain, within myself, my own center of balance.[33]

The experience of caressing and tickling exemplifies the ambiguity of the Other's givenness as "at the same time" (*zugleich*) an "objective subject existing in the world" *and* another subject "in the transcendental sense," that is, as another center of constitution.[34] The Other is a *Gegen-subjekt*, not a *Gegen-stand*, in the dual sense of a subjectivity given to me as a transcendence for me and as, for itself, a subjectivity that is given to itself as transcendence, as its own self-transcendence (Hua IV, 194; 318). The Other is another subject in the sense that it is another stream or horizon of possibility, another stream of temporality, that is other than mine. On the one hand, we have the perceptual given of the Other's material body as a constituted object in me; on the other, we have the Other as alter-ego, that is, as a constituting subjectivity, and thus, in this sense, as not-constituted in me: how to understand this "zugleich" and its constitution is the puzzling possibility of the Other's givenness.[35]

---

33  Of course, a feather can also be ticklish; but the feather is a prosthesis of the lived-body of the Other. Can I tickle myself with a feather?

34  This recognition of the Other as both "object" and "subject" must not be conflated with the distinction between body and mind, since the mind, as the psychological constitution of empirical consciousness, is as much an object in the world as its body.

35  On the basis of this initial delineation of the Other's subjectivity, Husserl develops a three-fold paradoxical form of Other, according to which the Fifth Meditation is organized. The first level: I recognize the Other as another transcendental subject,

As Husserl writes, "That my own proper-being can at all be contrasted with something other, or that I, who I am, can become aware of something other (who is not I but someone other than I), presupposes that not all modes of consciousness are modes of my self-consciousness" (Hua I, 135 [105]). That the ego can constitute the Other means there are modes of my own consciousness that are not modes of my self-consciousness, and thus, as Husserl argues, that there are intentionalities in me that radically exceed me – something more is paradoxically accomplished through me than what I can accomplish myself. As Husserl states this paradoxical problem: "And now the *problem* is how we are to understand the fact that the ego has, and can always go on forming, in himself such intentionalities of a different kind, intentionalities with an existence-sense [*Seinssinn*] whereby *he wholly transcends his own being*" (Hua I, 135 [105]). In other words, "how can something actually existent for me . . . be anything else than, so to speak, a point of *intersection* belonging to my constitutive synthesis?" (Hua I, 135 [105]).

This "new kind" of intentionality in and through which the sense of the Other is constituted as another sphere of ownness is a radical transcendence within my immanence. A clue towards understanding and describing this "new kind" intentionality is provided by the

---

another subject like me, but not me; second level: the world is not mine but inter-subjective, and this is the genuine sense of the objectivity of the world; third level: the cultural world of objects that refer to "foreign subjects" (*fremde Subjekte*); objects with "spiritual value." The primordial reduction reveals the different levels in the constitution of the objective world. As we discussed, the objective world is constituted on the basis of "my primordial world," with the first level that of the Other ego; in this fashion, the objective character of the world as for everyone is constituted. This "pure" Other – i.e., in the transcendental sense – has not yet acquired a "worldly" sense; in addition, it belongs, so Husserl claims, to the trajectory of this constitution that the Other does not remain an individual, but rather, that there is constituted a "community of egos" and a community of monads. This means that transcendental inter-subjectivity has, due to this "community" (*Vergemeinschaftung*), an inter-subjective "sphere of ownness" in which the objective world is constituted. Husserl then says that we recognize that the inter-subjective proper essence of the objective world is not a transcendence in the proper sense, but that it is an immanent transcendence (110); what he means is that the objective world as idea is the ideal correlate of an inter-subjective experience; each monad is constituted with a system of harmony with other monads (harmonizing of constitutive systems). This introduces the notion of monadic harmony; and so offers a view on the entire sequence of constitutional steps that provides the only possible solution to the transcendental problem and so the only way to realize the transcendental idealism of phenomenology.

ambiguity of the Other's perceptual givenness. Indeed, "the most general feature of the perception of the Other" is her ambiguity, since, as exemplified in our hands caressing, the Other is given and not-given to me. Husserl even speaks of the "strange noematic" form of the Other in order to stress the irreducibility of this ambiguity, that is, that the ambiguity of the Other's givenness belongs to the noematic sense of the Other.

This ambiguity of the Other's perceptual givenness is, in one sense, familiar to us since it shares in the form of an ambiguity, or pretension, that characterizes *any* perceptual experience. Every perceptual object is given in an inadequate manner as it is composed of aspects of an object that remain hidden from us; the presentation of an object is situated within a horizon of empty intentions, in which hidden sides of the object are given as absent. In this interplay of presentation and appresentation, the horizon of hidden sides traces, and thus, motivates lines or trajectories of possible intuitive fulfillment through a synthesis of identification that constitutes "confirmation" or "validation" (*Bewährung*), as for example, when I turn the object to discover its other sides.

The perceptual givenness of the Other also partakes in this interplay of presentation and appresentation in the formal sense that the sense of the Other transcends any actual perceptual experience. Yet the ambiguity of the perceptual givenness of the Other is unfamiliar to us, since it is radically *unlike* the perceptual transcendence of a perceptual object like a table. In the case of the Other's givenness, there is no possibility – not even an ideal – of ever having an originary intuition, or fulfillment, of the Other's absence as an alter-ego. That is: the sense in which the absence of the Other's sphere of ownness is perceptually given to me is *not* comparable to hidden sides of a table; in the latter case, the empty intentionality *motivates* and, indeed, when seen in its teleological dimension, *demands* fulfillment, for example, motivates me to turn the table and look at its other side. But this does not mean that the intentionality of the Other is mediated in the sense of being based on an image or representation.[36] The intentionality of the Other, as a perceptual experience, is not a species of "presentification" (*Gegenwärtigung*). And yet, the intentionality of the Other cannot take the form of a "re-presentification" (*Vergegenwärting*) in either of

36 Cf. Hua XIII, 187ff. where Husserl distinguishes between image-consciousness and consciousness of the Other.

its three possible species: image-consciousness, imagination and remembrance. It is not an image of the Other that I perceive, nor do I imagine the Other, nor is the Other not-now in the sense of having once been. The sense in which the not-now of the Other is given to me is radically unlike these other forms of absence. Perhaps the most important quality of this new kind of intentionality, and the sense of its absence, is that it is predicated on an absence or foreclosure of any possible verification and fulfillment of its intentional object. It is not simply that the Other is herself not given in an immediate and originary manner; its more radical sense as transcendence implies that the Other is the impossibility of an originary givenness, that is, as impossibility of presence.

As Husserl remarks, "properly speaking, neither the other ego himself, nor anything of his subjective processes … nor anything else belonging to his own essence, becomes given in our experience" (Hua I, 109 [75]). The intentionality of the Other does not render present another "now" but rather renders the Other as not-now, and, more strongly, as never able to be now. This kind of intentionality is *neither a making presence nor a making absent*, but an appresentation in the sense of "*making co-present*" or "*co-presentification*" (*Mitgegenwärtig-Machens*). As Husserl stresses, this mediated intentionality prohibits that the Other could ever be given as "Selbst-Da." Yet, the Other is therefore not rendered into a mystery to me since Husserl's claim is not that the Other is entirely inaccessible to me, but rather, that the Other is given to me as "Mit-Da" but not as "Selbst-Da." The sense of "being-with" strikes a course between the Charybdis and Scylla of "Selbst-Da" and "Nicht-Da."

The absence of the Other is not comparable to the absence of a hidden side of an object, in which case, there is a motivation or expectation to encounter the what is not actually given, and, indeed, a possibility that I can turn the object and discover the other side. But also, the intentionality of the Other is also not a re-presentification, in which case, we would have an imaginary Other. The intentionality of the Other, as appresentation, is neither a *Gegenwärtigung* (the relationship between what is perceptually given to me of the Other and what is withheld from me of the Other is not comparable to the relationship between visible and invisible sides of a thing) nor a *Vergegenwärtigung*. This positioning in-between these two fundamental forms of intuitive consciousness raises two questions, only one of which will be pursued in the Fifth Meditation. What kind of consciousness is "rendering co-present" (*Mitgegenwärtigung-Machen*)? In the course of working out this structure, Husserl will continually move back and forth, making contrasts with both perceptual experience and re-presentification (for example, remembrance). But as we discussed in our chapter on the imagination, the structure of *Gegenwärtigung* and *Vergegenwärtigung*, as well as their difference, is ultimately rooted in absolute time-consciousness.

## In-between us

The givenness of the Other as an alter-ego wihin the consciousness of "co-presentification" has the form of a double-constitution with a three-fold structure of intentionality. In this regard, the constitution of the Other exhibits a *formal* similarity with various acts of *Vergegenwärtigung*, since, as explored in chapter 4, remembrance, the imagination and image-consciousness each represented a particular kind of double-consciousness. And although Husserl emphatically rejects any comparison between the givenness of the Other and an image, it is nonetheless striking that both forms of consciousness are structured in terms of a stratified intentionality folded three ways. A significant difference, however, exists between "co-presentification" (*Mitgegenwärtigung*) and "re-presentification" (*Vergegenwärtigung*): the latter type of intentional acts are structured by the interplay of empty and fulfilled intentions. In the case of "co-presentification," however, the intentionality of the Other is uniquely predicated on the *structural impossibility* of any intuitive fulfillment of its intended "object." The sense in which the Other is intended forfeits the possibility of fulfillment from within the immanence of its own constitutional departure. The Other is given to me as the impossibility of being given as she is given to herself, that is, as itself self-given. This impossibility, as we shall discuss, is neither a restriction imposed on my constitutional aim from the outside nor a failure in the meaning or performance of constitution from within. The Other remains a promise of givenness – of presence – for which I refrain from any determinate or definite expectations.

The stratified three-fold apprehension of the Other in perceptual experience has a footing in an underlying perceptual apprehension of the Other's body as a constituted material object. As with any spatial object in this room – this chair, this table, etc. – the body of the Other, when described from within the primordial reduction, belongs to the primordial sphere of ownness for constitutional reasons already adduced above: the constitution of spatiality is the accomplishment of *my lived-body*. All spatial objects, including the bodies of Others, are immanent in a transcendent sense. In this manner, the perceptual apprehension of the Other is grounded in a perceptual experience (*Gegenwärtigung*).

On the basis of this perceptual presentation, a second act of consciousness introduces a "re-presentification" (*Vergegenwärtigung*) of how my own body would appear from the perspective of the Other's location, were I to inhabit her perspective on me, and thus see myself from where she stands – from an absolute here that is not mine. Such an act of "re-presentification" is partly of the imaginary as I must imagine myself as being seen from over there, as if I actually stood in the Other's shoes. And yet, this apprehension of myself as seen from another perspective, as a form of self-objectification through a perspective other than mine, is not entirely of the imaginary; otherwise, I would ultimately only be imagining the Other as Other as opposed to perceiving that there is actually another life of consciousness, other than mine, in this room. This act of "re-presentification" is thus modified *through my own lived-body* in the form of what Husserl calls a "pairing association" between my body and the body of the Other. That body over there moves and acts much as my lived-body here; this recognition of that body over there as a like-bodied body, so to speak, "awakens" an associative intentionality whereby a "re-presentification" of how I look from over there becomes generated. Much as with the self-induced modification of my own consciousness through the imagination, I project myself over there, and imagine my own body as seen from elsewhere. An incompatibility arises between my actually standing here and my imagined perspective over there. Husserl, however, does not argue the phenomenologically implausible position that I actually perceive the Other as perceiving, as if I were actually looking through the eyes of the Other. Instead, his point is that I perceive that someone else other than me can perceive in the manner that I perceive, and thus, that my own perceptions are one of many *possible* perceptions, and not possible perceptions pointing back to myself, but as pointing away to the lives of Others. The awakening of a consciousness of subjectivity is the awakening of my lived-body: for example, when I unreflectively mimic the physical habits of others.

On the basis of such a "pairing association," a third, distinct act of re-presentification (*Vergegenwärtigung*) enters into the play of

constitution in the form of what Husserl calls an "apperceptive transfer." The pairing association between my body and the body of the
Other motivates the transfer of my embodied (self) apperception to
that other body over there. Of course, in one sense, it is not that my
consciousness transfers *my apperception* to the Other; if I gave myself to
the Other, the Other would become myself, and I would have only
succeeded in duplicating myself in the Other. I would see that body
over there as belonging to me, and as my lived-body. Instead, Husserl
contends (though, to sure, Husserl would remain unsatisfied with this
proposal well into his numerous manuscripts on empathy written
during the 1920s) that I "re-presentify" (*vergegenwärtige*) to myself the
Other as a consciousness (or ego) who "holds sway" (*walten*) in that
lived-body over there. It is less that I give or transfer myself to the
Other as that I dissociate the sense of being-for-myself from myself,
and transfer this sense of consciousness to a pole or sphere other
than mine. Moreover, this apperceptive transfer is fused with the
underlying perceptual apprehension of the Other's body, otherwise
I would not recognize that the Other has itself in that body over
there, as opposed to my body that is here, in which *I* hold sway.
Indeed, if my apperceptive transfer was not fused with the perception
of the Other's body, and instead, fused with the perception of my own
body, I would either "perceive" an alter-ego bereft of an *essential*
embodiment, and thus perceive a "ghost in the machine," or else
confusingly perceive the Other's ego as holding sway in the same
lived-body in which I hold sway.

Throughout this challenging analysis of "the experience of the
foreign" (*Fremderfahrung*), Husserl's central insight turns on recognizing that the appresentation of the Other's ego as "holding sway" in
*her own lived-body* is given to me as itself a constituting ego, in other
words, as itself a constituted and constituting lived-body much as the
self-constituting form of my own lived body, as manifest in the self-
affection of "double-touching." And yet, the lived-body of the Other
is also constituted for me as *the same* lived-body that the Other constitutes for herself. As Husserl remarks, "the body [*of the Other*] is the
same, given *to me* as the body there, and *to him* as the body here, the
central body" (Hua I, 153). Husserl extracts from this dual sense of
the Other's body as constituted by both of us (much as mine is also
constituted by both of us) the thought that the primordial sphere of
ownness of Other is "re-presentified" (*vergegenwärtigt*) to me without,

however, being given to me as it is given to itself. The body of
the Other becomes constituted, or "co-constituted," through the
intersection of two primordial spheres of ownness: mine and hers.
These two primordial spheres of ownness are discontinuous as they
are separated into two "absolute" streams of time-consciousness that
cannot coincide with each other. A phenomenologically intriguing
situation arises in which there obtains a shared noematic unity of the
Other's lived body, in so far as it is constituted by both of us, with a
*noetic* discontinuity between both of us, in so far as each reserves for
herself the prerogative to constitution to the degree that each recognizes that prerogative as also belonging to the Other.

In this manner, Husserl argues that the Other is constituted as a
"primordially unfulfillable experience – an experience that does
not give something itself originally but that consistently verifies something indicated" (Hua I, 115). In order to further develop how the
Other is both "primordially unfulfillable" in an original manner and
yet "verified" *as indicated*, that is, verified in its emptiness, Husserl
suggests an "instructive comparison" with remembrance. The osten-
sible point of this comparison is to illuminate how the Other (as alter-
ego) is given to me through an intentional modification – in the same
way in which the remembered past is given to me as past, as an
absence, on the basis of the modification of the present in which
I remember. As Husserl writes: "Somewhat as my memorial past, as a
modification of my living present, 'transcends' my present, the appresented other being 'transcends' my own being (in the pure and most
fundamental sense: what is included in my primordial ownness)"
(Hua I, 115).

This comparison between the appresentation of the Other as a
transcendence within my own immanence and the transcendence of
*my own* memorial past, as given again in reproductive consciousness,
within my immanence is, however, of little service in understanding
the peculiar sense in which the Other is constituted through me, but
not in me, as an alter-ego. The limitations of this comparison should
be apparent from the double-intentionality of remembrance as constituted in inner time-consciousness. When I remember something
from my past, I implicitly remember *myself* as having once constituted
that experience in its original temporal form as present. The reproductive modification of my own consciousness, in which my past
consciousness is given again, but as transcendence within the

immanence, or living present, in which I remember is predicated on an underlying unity of my consciousness. It is only because my original act of perception as well as its intentional object were both constituted temporally in and through my absolute time-consciousness that I can, in the future, recall myself as having perceived this object. But this is *precisely* what is lacking in the appresentation of the Other: the appresented Other transcends my own primordial ownness in a more radical manner since the primordial ownness that belongs to and defines the Other as Other was never, and could never, have been originally constituted in my absolute stream of time-consciousness. Whereas a memorial object is given as past in the temporal index of having once been present, the appresented Other is given as *absent* in the temporal index of an impossibility to ever be present for me. The absence of the Other is, so to speak, irreducible to any presence within my stream of consciousness; it is an absence that is more ancient that any memory and beyond any expectation that emerges from my self-constituting stream of temporality.

Despite the severe limitations on Husserl's proposed "instructive comparison" between the remembrance of time past and the appresentation of the Other (a comparison made once again in *Crisis of the European Sciences* [Hua VI, 189]), another possible comparison nevertheless suggests itself in the wake of this comparison's demise, and in light of Husserl's elaborations of original time-consciousness in the Bernau Manuscripts. Husserl would be better served by fashioning a comparison between the appresentation of the Other and *retentional consciousness*, albeit in a modified form, and specifically in terms of *far retention*, as we shall presently suggest and briefly explore.

Let us first recall the structural difference between remembrance and retention. Remembrance is a reproductive modification of consciousness in which a memorial object is given again as past along with an implicit reproduction of my past consciousness; in remembering something from the past, I implicitly remember myself. In contrast, retentional consciousness also possesses a double intentionality since, in retaining the now-phase as just-past, consciousness retains itself as having just heard. However, the constitutional function of retentional consciousness consists in the "de-presentification" or "emptying" of the fulfilled presence of a protention in an original presentation. As argued earlier, retentional modification "reverses" or withholds the givenness of the present from within; rather than constitute givenness in terms of presence, retention constitutes an original givenness of absence, in the form of the just-past or "no longer." In this regard, retention is not the re-production or giving again of an object that was already constituted as present for me. Rather, as an original form of constitution, the now-phase is constituted as "no longer" or "just-past," and, in this sense, as an absence that is irreducible to the original presentation of which it is nonetheless its modification. Retention is the modification of original presentation in the sense of constituting a transcendence from within; in throwing its center of gravity outside of itself. This reversal of original presentation is inscribed within a tension or original difference between retention and protention, which, in the reflections of the Bernau Manuscripts, passes *through* the original presentation

Yet, even with this established difference between remembrance and retention, a straightforward comparison between retention and appresentation of the Other would not work since, in the fold of retentional modification, absolute time-consciousness retains itself and is itself retained. Moreover, retentional consciousness is a modification of an original presentation within my primordial immanence; thus, any comparison with the appresented Other would falter on the same stumbling block as Husserl's proposed comparison with remembrance: in both cases, remembrance and retention are modifications of *my presence*. However, a guiding thought does emerge from this brief recapitulation of the distinctive temporalization of retentional consciousness. An initial comparison is possible between the "de-presentification" (*Entgegenwärtigung*) or "emptying" (*Entleerung*) of retentional consciousness and the appresented "co-presentification" (*Mitgegenwärtigung*) of the Other since in both instances, we are dealing with an empty consciousness in which something is given as not-now, as absent. What is needed to fashion a more illuminating comparison and, indeed, enter into the temporalization implicit in the constitutive accomplishment of appresentation, would be a form of retentional consciousness that was "headless," so to speak, that is, bereft of an original presentation within my sphere of ownmost.

Towards this end, let us recall the distinction formulated in chapter 5 between near and far retentions, and our example of forgetting my wife's name. To a significant degree, the givenness of the Other as an alter-ego is manifest to me in a comparable manner as my forgotten wife's name. Indeed, we could argue that even when I remember my wife's name or see her "in flesh and blood" standing before me, her alterity as a life other than mine is always given to me in an empty consciousness, as if the Other were always on the tip of my tongue. As discussed earlier, it is through a far retention that an empty consciousness gives an object as both an opacity of motivation as well as in its palpable or tangible absence, as what I have without having or possessing. And yet, in the case of my wife's forgotten name, the empty consciousness of having her name on the tip of my tongue strives for fulfillment and, indeed, demands fulfillment given that the object of far retention was once constituted for me as a living present. In the case of the Other, however, I intend the Other in an empty consciousness, as transcending my own primordial ownness, without a striving for fulfillment; indeed, if a far retention were unmoored, so to speak, from a genealogical origin in a living present within my sphere of ownmost, it would give me an object in its absence without any demand for fulfillment. By the same token, far retention would motivate a far protention, but not with any determinate content, but in an entirely, and radical, empty manner.

If we think back to our discussion of the imagination, let us recall the unique headless temporality of the imaginary. As we argued, Husserl's argument is implicitly committed to a temporalization without an original presentation; retentional and protentional dimensions of temporalization are entwined without passing through any fulfillment in an original presentation. In the case of the imaginary, the original presentation of the imaginary consciousness was "neutralized" and "set at a distance." Consciousness produced an incompatibility within itself in inducing a semblance of its consciousness. Of course, in the case of the appresentation of the Other, I do not imagine that the Other is given to me as an alter-ego. But this is precisely the insight that we have been searching for: what distinguishes the headless temporalization of the appresentation of the Other and my imagining the Other is that in the latter instance, the neutralization of the original presentation – its decapitation – was self-produced in and through my own primordial sphere. In the case of the appresentation of the Other, a headless temporalization is produced as an *intersection between myself and the Other*. Moreover, the temporalization of appresentation has the form of far retentions and protentions without a stabilizing center in the arc of my living present, without, that is, near retentions and protentions. With this headless form of temporalization, the appresentation of the Other is an empty intentionality that neither motivates nor demands the presence of the Other. The proper name of the Other, as a sphere of ownness other than mine, is given to me as a word on the tip of my tongue, but which I could never remember (since far retention never emerged from a near retention within the arc of my living present) nor which I could ever expect to speak (since the far protention is not motivated towards fulfillment). Rather than speak of a reversal of constitution in such a far retention of the Other, we can speak of an inversion of time-consciousness: far retention inverts its relationship with the living present. The living present is on the side of the Other even though the far retentions and protentions are on the side of my own sphere of ownness; in this manner, far retention throws my center of gravity outside myself; by the same token, far protention always misses the presence of the Other.

In light of this proposal, let us return to the appresentation of the Other as it is portrayed in the Fifth Meditation. Husserl reconfigures the dynamic of intention and fulfillment in the face of the Other such that it is not the case that the Other resists fulfillment or evidence, but that, from the beginning, the movement towards the Other in the very intending of the Other as an alter-ego refrains from demanding or making a claim on the Other to show itself or become present as itself. No demand is made on the Other to be given as itself and this restraint is a novel form of transcendence and constitution. The primary intentionality of the Other is a "letting the Other" be other than me – it is an intersection of my intentionality, not from the outside, but an intersection that meets my own refusal to demand of the Other that she show herself as herself. It is an intersection of the Other that meets my own self-interruption half-way, and this meeting half-way is both *in-between us* and *between us*. As Husserl remarks, "my primordial ego constitutes the ego who is other for him by an appre-sentative apperception, which, according to its intrinsic nature, *never demands and never is open to fulfillment by presentation*" (Hua I, 119, emphasis mine). It is in this sense that the proper name who is the Other remains perceptually on the tip of my tongue.

# 7

# THE LIFE OF CONSCIOUSNESS

*An authentic analysis of consciousness is, so to speak, a hermeneutics of the life of consciousness.*

— Husserl

## Time and genesis

The transcendental constitution of the Other, as discussed in chapter 6, allows us to motivate the question of how to negotiate the relationship between the primordial transcendence of the Other and the primordial transcendence of original time-consciousness, on the basis of which transcendence as such – the transcendence of the world, including the lives of Others, and my own self-transcendence – is grounded. In both instances, time-consciousness is opened in a radical form of transcendence – the original transcendence within immanence within the retentional and protentional dimensions of my self-temporalization and the transcendence within the immanence of the Other, which, as suggested earlier, finds a promising phenomeno-logical comparison with the intersection of *far* retention and *far* protention, bereft, however, of a living present. The Other's proper name, or, in other words, her sphere of ownmost, forever remains on the tip of my tongue, beyond the arc of my living present, as given in an empty consciousness of far retention predicated on the impossi-bility of fulfillment in an equally distant far protention. As we further explored in chapter 6, the phenomenological argument of transcen-dental idealism, cast in the form of a refutation of transcendental solipsism, critically depends on Husserl's analysis of time-consciousness and the constitution of the Other as an alter-ego. Indeed, each

problem – time and the Other – mirrors the question of how to construe
the constitution of absence without undermining the phenomenological principle of original evidence as the foundation for any and all
constitution: in both instances, the challenge consists in how the present
is opened to the irreducible givenness of the not-now. The problems
of time-consciousness and the Other each provoke a fundamental
questioning of the "manner of being" (*Seinsweise*) of transcendental
subjectivity from different angles that converge on its transcendental
prerogative as the origin of the world. And yet, how can we reconcile the
claim that "the intrinsically first being, the being that precedes and bears
every worldly Objectivity, is transcendental intersubjectivity" with the
claim that my consciousness constitutes itself in an absolute, and thus
primary, sense as original time-consciousness – which, indeed, is the
form in which consciousness as such, my own or the Other's, constitutes
itself as the openness of a life? (Hua I, 182 [15]).

Of course, the point is not to frame this question – central to
assessing what is philosophically distinctive of transcendental subjectivity for Husserl's brand of transcendental idealism – in terms of
either/or, since Husserl argues that the phenomenological reduction
leads to "two entwined, and, in this sense, founded universal structures of life," namely, "my transcendental life" and "foreign subjectivity" (Hua VII, 188). This entwinement of myself and the Other, we
suggested, can be seen in terms of an entwinement of a headless
temporalization of far retentions and protentions, as if the stabilizing
wellspring of an original presentation was inverted in its constitutive
function. This "inverted" temporality of the Other constitutes the
Other as the impossibility of fulfillment; the Other is given to me in
a consciousness of emptiness that intends the Other as this impossibility, and without the demand of fulfillment. The reduction leads to
a "first" and a "second" transcendental life: in each case, however,
what constitutes subjectivity as a life is that "only subjectivity can in an
authentic and absolute sense be for itself [*für-sich-sein*]," by which
Husserl understands that "being-for-itself is appearing for itself
[*für-sich-sein ist sich-selbst-erscheinen*]" (Hua VII, 188). It is this constitution of subjectivity as "being-for-itself" that we want here, in our
concluding chapter, to better understand through the resources of
genetic phenomenology. This question regarding the intersection of
the primordial transcendence of time-consciousness and the primordial transcendence of the Other can be reformulated into a question
regarding the relation between time-consciousness and the problem

of genesis, or "genetic phenomenology," to the extent that the phenomenological constitution of the Other depends on the accomplishment of a complex form of association, apperceptive transfer, and double-constitution (detailed in chapter 6). We are also able to frame the confrontation between the claim of time-consciousness and the claim of the Other on transcendental phenomenology since the problem of genetic phenomenology, as we shall explore in this chapter, addresses the genesis, or becoming, of subjectivity as a life. This is yet another way to raise the question of in what sense is time-consciousness the ultimate transcendental foundation or origin. In order to answer this question, we must turn to Husserl's genetic phenomenology and investigate the relationship between time-consciousness and genesis: what is the relationship between the problem of genesis and the problem of time-consciousness?

In one regard, it would seem that the answer to this question is straightforward: does Husserl not clearly argue that association, passive synthesis and genesis are based on time-consciousness, in other words, that time-consciousness is the universal form of all synthesis, and, thus, equally the foundation for the constitution of the Other? In another regard, however, the sense in which time-consciousness is a foundation is not without ambiguity; as discussed in earlier chapters, Husserl's analysis inconclusively rested on the "impossible puzzle" of absolute time-constituting consciousness.

## An impossible puzzle revisited

A central issue that emerges in our discussion of Husserl's phenomenological analysis of time-consciousness is the "impossible puzzle" of how absolute time-consciousness is both constituted and (self-) constituting. Much of this difficulty stems from the unique form of "transcendence within immanence" that Husserl first discovers with the retentional modification of an original presentation, but which is also implicit in the double intentionality of protentional consciousness that Husserl develops in the Bernau Manuscripts. If we follow Husserl in placing an emphasis on the retentional dimension, absolute time-consciousness, as a continuous self-differentiation or flow, differentiates itself in a two-fold manner along the lines of the double intentionality of retentional consciousness: as a differentiation *from* itself in terms of the transcendence of constituted time-objects ("cross-intentionality") vis-à-vis constituting immanent consciousness ("length-intentionality"); as a differentiation *of* itself in terms of the transcendence of absolute time-constituting consciousness vis-à-vis constituted immanent consciousness. Absolute time-consciousness retains itself and is retained, to adopt Rudolf Bernet's felicitous insight, and, in this manner, differentiates itself from itself. The non-coincidence of the retentional consciousness of the immediate past with the consciousness of the now structures the non-coincidence of consciousness with itself, in so far as the experienced, and constituted, immanent content of consciousness does not coincide with the constituting and experiencing consciousness. As Husserl argues, absolute time-constituting consciousness does not itself appear as constituted; there is thus a distance, or difference, between the constituted immanent temporality of consciousness and the constituting consciousness, and this self-transcendence within consciousness itself is grounded in the double-structure of retentional consciousness (crosswise and lengthwise intentionalities). However, this self-transcendence does not succumb to the vicissitudes of Husserl's earlier "reflection theory" since the sense in which consciousness has differentiated, and thus distanced, itself from itself does not imply a distance between an "object" and a "subject" (cf. chapter 4). A direct consequence of this phenomenological state of argumentation is that absolute time-consciousness, in its constituting expression, is itself not "in time" in the manner in which its immanent noetic acts and sensual contents are temporally constituted (Hua X, 112 [116]). Absolute time-consciousness possesses another form of temporality that, to be sure, remains unexplored in the span of writings with which we are concerned in this study, but which we can nonetheless mark with the concept of *life* – as opposed to consciousness. Expressed differently, the tension between the poles of retention and orignal presentation in Husserl's early analysis of time-consciousness constituted an original difference, or temporal spacing. This original self-differentiation of time-consciousness becomes shifted in the Bernau Manuscripts to a tension between retention and protention such that the original presentation is caught in-between – as the edge – the colliding streams of retention and protention. This tension between the "de-presentification" of retentional modification and the fulfillment of protention can be taken as the tension between *life* and *consciousness*. The life of consciousness, as the event of its self-temporalization, hides itself within its own accomplishment as consciousness. The concept of life, or better, the stress on the *life* of consciousness increases in prominence in Husserl's later writings, especially in the C-manuscripts on time and temporality.[1] Absolute time-consciousness does not coincide with itself, and is thus not simultaneous with itself; and yet the constituting consciousness and the constituted consciousness are "at the same time" (*zugleich*) precisely as non-coinciding, as an "Entgegebleben," much like the original difference between retentional modification and original presentation.

If we are looking for a clearer sense of what this might mean, Husserl's insight can be characterized as the phenomenological discovery that consciousness eludes and escapes itself as an event of original constitution. As Husserl remarks: "the *cogitationes*, which we regard as simple givens and in no way mysterious, hide all sorts of transcendencies" (Hua II, 11 [8]). Indeed, as we have been tracking throughout our discussion, absolute time-consciousness, as an impossible puzzle, hides itself within its own self-temporalization. Contrary to Husserl's initial conception of the immanence of consciousness as defined by the coincidence of "being" and "perception," an analysis of time-consciousness reveals the sense in which consciousness constitutes itself as an opacity; its being is prior to its own perception. But this consequence must be parsed in its transcendental significance. The sense of "prior" is emphatically here not one of succession as if the event of consciousness – even of living in its lived experience – was "before" or "earlier." The event of consciousness as lived is prior to consciousness in the sense of an opacity. There is at least two senses in which this constitution of opacity can be understood: as the covering over of its own constitutional accomplishment as an origin; and as the unanticipated renewal of its own self-constitution, despite its own self-anticipation.

But how can we reconcile this claim that absolute time-consciousness remains invisible for itself in its constituting activity – a medial "activity" that disrupts the traditional distinction between activity and passivity in the eruption of an in-betweenness of a pure (self-) differentiation – with Husserl's insistence that, along the lengthwise intentionality of retentional consciousness, absolute time-consciousness is itself manifest? On the one hand, Husserl needs to argue for the self-manifestation of the stream of time-consciousness in order to avoid succumbing to the infinite regress of inner perception that haunted

---

1 For Husserl's C-manuscripts on time-consciousness, see Klaus Held, *Lebendige Gegenwart* (The Hague: Martinus Nijhoff, 1966); Gerd Brand, *Welt, Ich und Zeit* (The Hague: Martinus Nijhoff, 1955).

Brentano or the untenable notion of an unconscious self-awareness –
a paradoxical thought that Husserl explicitly rejects (Hua X, 119
[123]). Husserl thus contends that absolute self-constituting consciousness constitutes itself as an immanent stream of consciousness,
and yet does not appear as constituted; it is that for which "all names
are lacking." As Husserl argues, "There is one, unique flow of consciousness in which both the unity of the tone in immanent time and
the unity of the flow of consciousness itself become constituted at
once [zugleich]. As shocking (when not initially even absurd) as it may
seem to say that the flow of consciousness constitutes its own unity, it is
nonetheless the case that it does" (Hua X, 80 [84]). Husserl concedes
this conclusion to be "shocking" or even "absurd" because it implies
that consciousness, as the source of all givenness, is also the source of
its own givenness; even though this consciousness that gives itself to
itself is itself not *given* as an immanent object. Husserl thus discovers
that absolute time-consciousness cannot entirely become the object of
reflection; indeed, as discussed in chapter 4, consciousness "swims"
in a horizon of its buried constitution and always remains haunted
by an original self-forgetting which, on the one hand, conditions the
possibility, and indeed, the sense, of remembrance and, on the other
hand, prevents consciousness from a complete self-understanding
(even if Husserl posited total remembrance as an ideal possibility).
As Husserl writes, "The constituting and the constituted coincide, and
yet naturally they cannot coincide in every respect. The phases of the
flow of consciousness in which phases of the same flow of consciousness become constituted phenomenally cannot be identical with these
constituted phases, nor are they" (Hua X, 83 [88]). Consciousness
cannot apprehend itself as a unity; only aspects or phases of its life can
be apprehended in reflection, but never life as a unity.[2] Husserl's
impossible puzzle, then, is that absolute time-consciousness is self-
constituting – and this secures its transcendental prerogative as
absolute – yet this absolute time-consciousness, as an origin, escapes
itself, as its own beginning: it escapes itself in the sense of being *too late*

---

2 Husserl thus considers that the unity of consciousness can only be apprehended as an
"idea" in the Kantian, regulative sense (*Ideen*, § 83). However, Husserl cannot therefore
mean that the consciousness apprehends itself as an idea in the same manner in which
it would apprehend the thing as a unity in terms of a Kantian idea, since this would
conflate the unity of the thing, as object, with the unity of the stream of consciousness.
This thought will be reaffirmed and further developed in the passive synthesis lectures,
namely, of how consciousness constitutes for itself as an "An-Sich."

for itself. The double-structure of retentionality performs a dual service since retentionality provides both a solution to the specific issue of the apprehension of time-objects and reveals what is fundamental to consciousness – the sense in which consciousness constitutes itself as an inheritance of itself since it retains itself as past. It is along the lengthwise intentionality of "self-retention," in the form of a "de-presentification" of my own consciousness, that Husserl identifies the "self-appearance" of consciousness. The lengthwise retentional consciousness is consciousness of myself in so far as I have already escaped myself while at the same time being buried or covered over with myself. Paradoxically, it is as if consciousness buries itself alive in its past while surviving into a renewed future.[3] In this regard, retentional consciousness, in its essential form of double-retentionality, accounts for the sense in which I become other than myself, and so transcend myself, in my own past. But it also accounts for how constituting consciousness is itself not given as a constituted immanent object, and thus, in this sense, is invisible, or hidden ("de-presentified") in its own self-constitution. This is the sense, as explored in chapter 5, in which retentional consciousness, when taken in its dual aspect as far and near retention, represents the constitution of an original forgetting and sedimentation.

Given the protentional dimension developed in the Bernau Manuscripts, absolute time-consciousness is also *too early* for itself in so far as it anticipates itself, and, in this manner, takes itself to be what it is not yet. Along with this protentional dimension, the characterization of an original presentation as the eruption of the new allows us to attain a clearer understanding of the phenomenological significance of its impossible puzzle, not by disarming it of its puzzling character, but, on the contrary, by drawing out the significance of this puzzle by revealing what is essential to subjectivity itself. In the reflections from the Bernau Manuscripts in which Husserl patiently describes the "entwinement" (*Ineinander*) of protentions and retentions in his effort to describe more fully the constitution of time-consciousness as the continual interplay of "emptying" and "fulfillment," time-consciousness is centered on the unceasing renewal of the now, characterized as the new. Yet, this characterization of each now (more specifically, each now-phase *qua* original presentation) is further strengthened through

---

3 My thanks to Claudio Majolino for our numerous discussions in which this thought first took shape for me.

the representation of the now as a "cut" or "crease" that both divides and unites, like a stitch in time, the continuum of time-consciousness. To think of each original presentation as an "edge consciousness" (*Kantenbewußtsein*) can mean that the new is the eruption of a pure difference, and that our consciousness of the now, taken in abstraction from its necessary temporal density in terms of its retentional modification, is nothing but *this ever renewed sense of a difference between past and future* – a difference that is lived anew and without end. As proposed in chapter 5, consciousness of the now is the indecidability of whether consciousness is ever on the mark – in the now – or on the edge of the now. This indecidability is the opacity of absolute time-consciousness in the self-differentiation of meeting itself halfway or what Husserl calls "Entgegenleben."

The arrival of each original presentation as the irruption of the new is here the event of a renewed intuitiveness of experience itself in conjunction with the renewal of consciousness. On the one hand, the arrival of the new is the fulfillment of a protention that emerged from a retentional modification of the previous original presentation. As we discussed in chapter 5, Husserl tightens, so to speak, his phenomenological grasp on the de-presentification of retentional consciousness in the Bernau Manuscripts through a consistent and illuminating description of the retentional accomplishment as the "emptying" of the fulfillment of the now. Retentional consciousness is, in this sense, the reverse of the constitution of presence in fulfillment of the protention. On this account, the novelty of each original presentation is the fulfillment of a protention, and thus, the now is always framed by its necessary expectation just as much as each original presentation becomes retentionally modified. On the other hand, if Husserl is to make good on the robust and *absolute* sense in which each original presentation is the eruption of the new, the arrival of each now cannot merely be the fulfillment of its protentional expectation. Indeed, Husserl implicitly operates in the Bernau Manuscripts with a distinction between *two kinds of protentions*: the protention of a determinate empty intentionality towards the next now; a far protentional horizon as an openness towards the future *as such*, and which is not *motivated by a retentional past*. Consciousness repeatedly expects another now, yet the arrival of each now always surpasses or saturates any determinate expectation of the now on the basis of the past; in this fashion, each now catches us from behind; the future never arrives from the direction in which we expect it.

This refined Husserlian conception of the now as the eruption of novelty has significant ramifications for the "impossible puzzle" of self-constituting time-consciousness. The unexpected eruption of each now, as the eternal recurrence of temporal self-differentiation, is the eruption of consciousness itself: the original presentation is the "source-point" for intuitiveness as such, on the basis of which the possibility of any givenness is constituted, but it is also the source of consciousness as itself given in the visibility of time-consciousness. Another way in which we can characterize the sense in which each original presentation is the eruption of an unexpected novelty is to think of each original presentation as the arrival of a renewed now that cannot be entirely captured by time, that is, that cannot be captured within time-consciousness. This is not to deny the necessary retentional modification of every original presentation nor the necessary expectation of each now in the form of protentional consciousness. Yet, we can paradoxically argue that the feel of each new now is itself perpetually renewed, and, in this sense, always new; we never lose the freshness of what it is to sense the now, yet this freshness or vivacity of the now is the opacity of its slender difference between a now that once was and a now that is yet to be. Since it is consciousness itself that is erupting in the now, consciousness constitutes itself as the opacity of its own self-presence; in a significant sense, consciousness fails to anticipate *itself* in the character of its renewed awakening, but consciousness also fails to be entirely transparent to itself, that is, to fulfill itself entirely in any phase of its own temporalization.

With the three-fold declension of time-consciousness, each original presentation is absolute in the sense that it is a new beginning, and a genuine beginning in the sense of the unexpected, or, in other words, in the sense of *facticity* or *contingency*. Yet, each original presentation is necessary in the sense that another now must come; there is no final now, there is only the next to last, that now that we always necessarily *hope for*, even in the final hour, or now, of consciousness. In this regard, each now is relative to the series of past nows that have been ordered and sedimented through the iterative process of retentional modifications. If the beginning is the now, and the now is the new, and the new always arrives as surpassing our expectation, I would suggest that the transcendence of the new is the transcendence of consciousness as absolute, as origin: it is due to the newness of each beginning, of each now, that consciousness must miss itself and not coincide with itself. Otherwise, consciousness would always just

be itself, and be condemned to being what it already has been. There
is here a double consequence. Consciousness must always obscure its
own origins in "jumping over itself" and obscuring itself in its own
constitutional accomplishment. An origin can only be reactivated and
retrieved, yet it never returns fully constituted since it never was fully
constituted from the beginning *as* a beginning. Origin is the iteration
of distance, as is evident by the necessary retentional modification
of an original presentation. In addition, this self-obscuration of
consciousness as an origin equally constitutes itself as the *possibility
of retrieval and reactivation.* Retentional modification as the de-
presentification of the given is not only the withholding of full givenness; it is also the constitution of giving again what was never, and can
never, be fully given. If we connect these consequences to the form of
a beginning as erupting in the now, as the novelty of an original
presentation, this withholding of consciousness from itself in order
to find itself again constitutes the opacity of consciousness as the
movement of life itself, not the *failure* of consciousness to coincide
with itself but rather the *success* of missing itself in such a way that
consciousness remains open to itself and to the world. The impossible
puzzle of absolute time-consciousness resolves into the impossibility of
*not beginning again*; every beginning is the promise of beginning again,
yet not from where we just started.[4]

## The deference of absence

The interpretation proposed above offers a new way to understand
the sense in which the self-constitution of time-consciousness breaks
its immanence from within. The transcendence of subjectivity with
regard to itself in relation to the transcendence of the world, itself
constituted in a temporal form in and through the primordial
temporality of subjectivity, opens consciousness from its beginning:
consciousness is always and already transcendence. There are two
important ramifications of this proposed interpretation: the question
of the constitution of the primordial transcendence of the Other; and
the challenge to an established interpretative tradition, stemming
from Heidegger, and amplified and sharpened by Derrida, that sees

---

4 Indeed, we could furthermore say that consciousness never escapes its own origin, by
which I mean that consciousness is a relation to origins; we are indeed creatures of
distance – the distance of an origin.

in the impossible puzzle of absolute time-consciousness an inherent limitation and "destruction" from within of Husserl's conception of transcendental subjectivity. According to this critique, delineated here in general terms, Husserl's commitment to the idea of rigorous science requires that transcendental subjectivity explicate itself as to its constitutive function in the form of self-presence; yet, the most immediate consequence of the impossible puzzle points to the *impossibility* of absolute self-givenness, or self-presence, since although absolute consciousness appears to itself, it can never apprehend itself fully as itself in its own self-constituting performance. Self-presence is perpetually fissured, or fractured, to such an extent from within as to prohibit the fulfillment of its own promised givenness.

The interpretation just proposed responds to this line of critique in two ways (and I shall take Derrida's influential critique as setting its standard): first, by taking issue with a critical misunderstanding of the function of retentional modification; and, second, by investigating the failure to take into consideration the emphasis on the absolute novelty of the now as the constitution of opacity *as* the origin of constitution itself.[5]

Derrida's interpretation – mainly its articulation in *Speech and Phenomenon* where the critique of Husserl's analysis of time-consciousness is presented – is heavily indebted to Heidegger's critique of Husserl and the Heideggerian conception of metaphysics of presence. Even though Husserl insists on the suspension and neutralization of metaphysics, Derrida contends that phenomenology nonetheless adheres to a metaphysical presupposition that in fact constitutes the possibility of phenomenology, namely, the principle of all principles according to which original evidence is defined as presence of meaning in fulfilled and originary intuition. Derrida further contends (we hear echoes of Heidegger) that the idea of a theory of knowledge is in itself metaphysical. My comments here shall not engage directly with the other substantial component of Derrida's reading of Husserl, namely, his critique of ideality, since Derrida sees the presupposition of

---

5 For comparable reservations against Derrida's construal of retentional consciouness, see Rudolf Bernet, "La voix de son maître," in *La vie du sujet*, 267–296; Lilian Alweiss, *The World Unclaimed: A Challenge to Heidegger's Critique of Husserl*, 41ff. On the general rapport between Derrida and Husserl: Leonard Lawlor, *Derrida and Husserl: The Basic Problem of Phenomenology* (Bloomington: Indiana, 2002).

self-presence as the ultimate form of ideality itself. Derrida's critique of the form of ideality as the indefinite repetition of the same is grounded in his critique of time-consciousness; in so far as Derrida considers the constitution of temporal presence, through the repetition of the original impression, as the transcendental basis for Husserl's conception of meaning, he aims to show it cannot be (truly) meant. It is this principle of the repetition of identity that secures the non-*reell* character of objectivity. In this sense, for Derrida, presence is the form of meaning in terms of which we have a diversity of meanings. This is why Derrida thinks that the "inaugural" opposition of metaphysics between form and matter "finds in the concrete ideality of the living present its ultimate and radical justification" (SP, 5 [6]). As Derrida remarks, the unity of life, "which diffracts its light in all the fundamental concepts of phenomenology, escapes the transcendental reduction," yet Husserl fails, and, indeed, is unable to question the concept of life because it eludes transcendental reflection and self-presence (SP, 9 [10]). Derrida argues that the reduction of the sign to the act of meaning depends on the constituted identity of the living present, which he will pursue in Husserl's analysis of time-consciousness and ultimately "deconstruct" by way of arguing that the temporalization of the living-present presupposes an original givenness of the past that cannot itself be given as present in an originary manner. Derrida thinks that the ideality of meaning is grounded in the ideality of self-presence that in turn must be grounded in a temporal self-presence in the form of the instant or a pointless now. Indeed, this premise underpins Derrida's curious belief that Husserl's thinking is "menaced from within" should the now not be that simple in its constitution; if the concept of presence as the instant can only be understood as myth or metaphor, or even if this presence has a synthetic form, the principle of all principles becomes "menaced" in principle and in its principle.

Derrida acknowledges that the concept of the punctual now (or instant) cannot, as Husserl recognizes, be constituted in isolation from the interlacing of other, different nows in the stream of time-consciousness. Nonetheless, the now, *qua* original impression, is the source-point or absolute beginning for all constitution. Derrida wrongly takes Husserl to mean that the actual now is punctual in character and, as the source-point of presence, that it is this punctual now that is originary, or, in other words, that the origin must have the presence of a punctual now as the "form" for ever renewing "matter."[6] Interestingly, Derrida argues, after remarking that the self-identity of the actual now is the basis for Husserl's principle of all principles and his notion of meaning, that there is no possible objection to this privilege within philosophy because this privilege "defines the very element of philosophical thought, it is *evidence* itself, conscious thought itself, it governs every possible concept of truth and sense" (SP, 69 [62]).

For Derrida, the dominance of the actual now is connected to the founding metaphysical opposition between form and matter in the sense of the opposition between act and potentiality. That is why Husserl's lectures "confirm" the dominance of presence by rejecting the "becoming-aware" of an unconscious content as an "after-effect." As Husserl argues, consciousness is consciousness through and through in each of its phases. Derrida rightly notes that the retentional phase is aware of the preceding phase in a pre-objective manner, and that the original impression is also consciousness, without being an object: we have self-awareness or consciousness in each phase, that is, original consciousness (SP, 71 [63]).

Derrida, therefore, sets up a tension: we have just seen that, for Derrida, the actual punctual now is the original form of the presence of consciousness, yet the description of temporalization goes against the idea of a simple identity of the present. On the one hand, Husserl's lectures confirm the irreducibility of re-presentification to perception. As Derrida points out, Husserl argues that the presence of the perceptual present can only appear to the extent that it is continually composed of a non-presence and a non-perception, that is, a retention and a protention. These non-perceptions do not add themselves in hindsight, so to speak, but rather participate essentially in the possibility of the present now. For Husserl, retention is a perception of a past content as the modification of a present: we "see" the past in "primary memory." Retention thus delivers and "enables us to see a non-present, a past and unreal present" (SP, 72 [64]). As Derrida notes, Husserl speaks of retention as perception because he wants to hold onto the "discontinuity between retention and reproduction, between perception and imagination, etc. and not between perception and retention. This is the *nervus demonstrandi* of Husserl's critique

6  As Husserl writes in the *Ideen* (§ 81), the actual now is the form that remains for a matter that is always new. As we discussed in chapter 3, Husserl operates with a dual concept of the now.

of Brentano" (SP, 72 [64]). Derrida worries about this position because if we accept this "continuity of the now and the not-now, perception and non-perception, in the zone of primordiality common to primordial impression and primordial retention, we admit the Other in the identity of the *Augenblick*; nonpresence and nonevidence are admitted within the *blink of an instant*" (SP, 73 [65]). For Derrida, this "alterity is in fact the condition for presence" (SP, 73 [65]). The "radical distinction that Husserl wanted between perception and non-perception" is not "the difference between retention and reproduction"; instead, it is the "difference between two modifications of non-perception" (SP, 73 [65]). For Derrida, retention and reproduction are two different relations to a non-present that is irreducible to another now. This non-presence, on the one hand, does not dissolve the actual now, but rather allows its event and its always renewed "virginity." Yet Derrida thinks that this non-presence "destroys radically any possibility of identity in its simplicity" (SP, 74 [66]). On the one hand, Husserl, in Derrida's view, in being faithful to the things themselves, shows that the living now is constituted with retentional modification, that is, by including a non-perception. On the other hand, if the originary living present is the source of certainty, we must include retention in the sphere of originary givenness, and thus displace the boundary between originarity and non-originarity such that this boundary no longer depends on the distinction between actuality and inactuality but on "two forms of the re-turn or restitution of the present – re-tention and re-presentation" (SP, 75 [67]).

This last statement by Derrida reveals what is symptomatic of his basic misunderstanding. Derrida characterizes retention and reproduction both as "re-turn" or "restitution"; retention, however, is not a "re-turn" of presence, but, a "de-presentification" and reversal of original presentation in mid-flight, so to speak. Whereas Derrida thinks that retention and reproduction are two different relations to non-presence that, in a sense, share the same "form," it is, we have at length discussed, more accurate to recognize a radical difference between the constitution of non-presence, or the not-now, in remembrance and as retention. The distinction between retention and remembrance cannot be considered a distinction between two types of "restitution." In this way, Derrida fails to see that reproduction, as a re-presentification, is based on original time-consciousness and the "emptying" (*Entleerung*) or forgetting of retentional consciousness (which we also designated as "de-presentification" [*Entgegenwärtigung*]); indeed, it is only in terms of these diverse, yet unified constitutive functions that retentional consciousness is the condition for reproductive conscious-ness of remembrance. Derrida is therefore blind to the possibility that retention is an intentionality and a transcendence within immanence, and moreover, that the relation between original impression and reten-tion (as well as protention) is that of fulfillment – the claim that retention destroys "simple" identity is nonsensical, since there is no "simple" identity of the now. Derrida admits that there are a host of questions regarding the relation between retention and reproduction, and that this relation is nothing less than that of the history of life and the becoming conscious of life; yet Derrida claims that the shared and apriori common root of retention and reproduction is the possibility of repetition in its most general form – the trace in the most universal sense – and that this is a possibility that no longer inhabits the actual now, but constitutes it through the very movement that it introduces. This trace, according to Derrida, is more originary than phenomeno-logical originarity; the ideality of the form of presence implies that it can be repeated infinitely, and that its return is infinitely necessary and inscribed in presence itself; the re-turn is the return of a present that is retained in the finite movement of retention. Without this non-identity of presence that is said to be originary, how can we explain the possibi-lity of reflection and reproduction? Derrida argues that, for Husserl, the presence of what is present is conceived in terms of the fold of return, of the movement of repetition, and not the other way around.

The consequence is that the primordial immanence of time-consciousness is fissured; the concept of solitude is "infected" in its own proper origin, in the condition of its own self-presence. This is the sense of presence and non-presence in the same moment (*im selben Augenblick*), and this, as Derrida mentions, is also the sense of Husserl's analogy in the *Cartesian Meditations* between "alter-ego as constituted in the interiority of the monad" and remembrance "as constituted in the actuality of the absolute living present" (SP, 77 [68–69]). When Derrida speaks of *Augenblick*, we should read more correctly *zugleich*, as discussed in chapter 3. Derrida will argue that consciousness is the voice and that the voice is auto-affection: this pure auto-affection is not spatial but temporal, and this means that this auto-affection is the "possibility for what is called *subjectivity* or the *for-itself*, but, without it, no world *as such* would appear" (SP, 89 [79]). Derrida explains that the concept of auto-affection imposes itself on this reading of Husserl and references Heidegger's use of the concept

in his *Kantbuch* with regard to time. On Derrida's reading, the original
source point is already pure auto-affection, and it is pure because
temporality is never "the predicate of a being . . . it is a receiving that
receives nothing. The absolute novelty of each now is therefore engendered by nothing . . . The new now is not a being, it is not a produced
object; and every language fails to describe this pure movement other
than by metaphor" (SP, 93 [83–84]). Derrida reads the famous
Husserlian passage – "for all this names are lacking" – as meaning
that Husserl cannot pose the question of the being of absolute
subjectivity (SP, 94 [84]). Derrida writes that the living present is
produced through spontaneous generation yet, to be a now, it must
be retained "in another now," and thus affect itself, of a new actual
now in which it becomes "a non-now, a past now – this process is
indeed a pure auto-affection in which the same is the same only in
being affected by the other, only by becoming the other of the same.
This auto-affection must be pure since the original impression is here
affected by nothing other than itself, by the absolute 'novelty' of
another primordial impression which is another now" (SP, 95 [85]).
Derrida continues by arguing that this "pure difference" – the differentiation of auto-affection – reintroduces the impurity of non-
presence into originarity; the trace of retention cannot be considered
in terms of the simplicity of the living present; "the self of the living
present is originally a trace" (SP, 95 [85]).

As our interpretation has shown, however, retention accompanies
the original presentation, but does not follow it, and thus retention
is not an additional moment attached to the now from the outside.
In this regard, one must also question the identification of retention
and original presentation as a form of *auto-affection*. As argued in
chapters 3 and 4, the auto (or self-)affection of original presentation
and retentional modification is a "throwing outside of oneself"
in which time-consciousness becomes other than itself. As a pure
differentiation of consciousness, the original presentation is always
(necessity of retentional modification) and already (retentional modification is not an after-thought following the original presentation,
but its "zugleich" modification) folded into retentional horizons.
Contrary to Derrida's reading, retention and non-presence do not
"infect" the primordial presence; instead (near) retention constitutes
within the three-fold temporal declension the living present of
"mental presence-time" (to hark back to Stern's terminology). Derrida's
cherished metaphor of "infection" betrays the degree to which

retentional modification is misconstrued as the entry of a foreign agent from the outside rather than as the *necessary* self-transcendence of the immanence of original presentation *from within*. Rather than speak of an infection or menace, retentional modification represents the *fruition* of presence necessarily punctuated with absence. This explains why retention is not *negation* but *iteration* or *iterated modifications* (Hua X, 116 [120]).[7] Strictly speaking, if retention did function like a negation, we would have to speak of retentional *modalization* and not modification; indeed, the distinctive character of a *modification of sensible content* (and not a determining attribute) as a "de-presentification" is already recognized in Brentano's original association. The original presentation is the point of tension or "edge consciousness" in between retention and protention; it is neither "outside" nor "inside" consciousness, neither matter impinging from the outside nor form from the inside. In the Bernau materials, this insight is developed more fully: the now is a "cut" or "fold," the point of tension between two planes (retention and protention), and the original presentation is characterized as the "new" and the "unexpected," as a discontinuity or "rupture" in the sense of beginning anew. But this rupture, like stitches in time, perforates time in order to constitute its continuity. The now is always folded into the folds of retention and protention; it is folding – a "pli selon pli." The renewed emergence of the original presentation underpins the constancy of the form of presence even while it cannot retain that form; the continuity of one and the same stream of consciousness is determined by the periodicity of original presentation. The abiding form of flow resides in the abiding change or ceaseless renewal of its primordiality, or originality, in the living present.

Derrida thinks phenomenology is "tormented" (another metaphor of a threatening exteriority!) within itself by its own analysis of temporalization and inter-subjectivity because what links these two themes together is an irreducible non-presence that is given a constitutive value; thus, we discover a non-presence or a "non-vie" within the living present, that is, a "non-originarity" at the basis of "original" presence. In both cases, the modification of presentation is added as a supplement to presentation, but instead conditions it by fissuring or

---

7 Derrida's Hegelian dialectical reading of Husserl's time-consciousness (i.e. retention as negation) might have been inspired by Yvonne Picard "Le temps chez Husserl et chez Heidegger," *Deucalion*, 1, 95–124, see Alweiss, 45.

fracturing presence (SP, 5 [7]). As we have seen, however, in the discussion of Husserl's analysis of the constitution of the Other in chapter 6, even within the framework of the Fifth Cartesian Meditation, which, when compared to subsequent revisions and reworkings of the constitution of the Other in later manuscripts, can be seen as the *weakest* Husserlian formulation of the transcendence of the Other, the absence of the Other does not "torment" my primordial living-presence from within. On the contrary, if, as I suggest, we realign Husserl's analysis of the Other away from Husserl's own proposed analogy with remembrance towards the original impression as the absolute eruption of the new, one discovers within oneself *a deference towards the absence of the Other*. As argued, the peculiarity of that intentionality responsible for the robust transcendence of the Other from within my primordial sphere is the sense in which one refrains or withholds the demand for fulfillment. In other words, the intentionality of the Other should be seen in an analogical relation with the entwinement of *far and near retentions* within a headless temporality bereft of an original presentation within my midst. The conceptual advantage of formulating an analogy between the transcendence of the Other and the far transcendence of a headless temporality allows us to articulate more sharply the sense in which the demand for the Other to show herself is *never made*. The Other is constituted as the impossibility of presence. The Other is constituted through an empty intention, or through an emptying of intentional fulfillment, such that the Other is a presence that one awaits without an expectation of arrival. I wait for the Other who will never arrive. The Other is thus the yet-to-come, but not in the form of an expectation that the Other could ever be made present within my primordial sphere. The constitution of the Other within my primordial sphere is the *converse* of the impossible puzzle of my own self-constitution – and this is the sense in which we can understand Husserl's metaphor of mirroring in his analysis of empathy, especially in his manuscripts collected in the "Intersubjectivity" volumes (Hua XIV and XV). The formal comparison turns on the sense in which I escape myself as constituting, and the Other, as herself a constituting subjectivity, eludes, and cannot be made present within, my primordial sphere even though the Other is "co-constituted" by me. The intersection of intentionalities in the constitution of empathy (the structure of double intentionality) mirrors the double intentionality of retentional consciousness and my own self-differentiation; the opacity of my own consciousness for itself, as the condition of its appearance, mirrors the opacity of the Other for me. In both instances, in terms of the "first" and "second" subjectivity and their intertwining, life transcends its own grasp or presence. We thus recognize two different senses in which the immanence of consciousness is broken from within, that is, two different senses of "transcendence in immanence" (recall that Husserl also uses this characterization to speak of the transcendence of the Other within my primordial sphere). We therefore learn how to relate more concretely the relationship between the transcendence of the Other and the self-transcendence of absolute consciousness, the sense in which one is unknown to oneself, and constitutes oneself as this opacity, to the extent that one cannot fully recuperate oneself entirely within oneself as an origin. Yet, it is important that this comparison does not mean that the *sense* in which one is a question to oneself – in so far as one constitutes oneself as the opacity of one's self-transcendence – is the *same* as the sense in which one awaits the Other. What Derrida fails to recognize is that though the mystery of the Other is the mystery of myself, I am not a mystery for myself in the same sense in which the Other is a mystery for me. Indeed, it is this *difference* between two senses of self-transcendence, of myself and of the Other, that marks the site of my own becoming. I am in-between myself and the Other. In order to grasp this point, we need to engage with the Husserlian problem of genesis as the becoming of subjectivity as a life.

## The becoming of subjectivity

Husserl delineates two parallel lines of phenomenological research that crystallize around the structure of intentionality – *ego-cogito-cogitatum* – as the structure of transcendence, which, as we have seen, is grounded in the temporal transcendence of absolute time-consciousness. These two correlating sides of intentionality (noetic and noematic) encompass the duality of subjectivity as both "immanence" and "transcendence." In our general presentation of phenomenology in chapter 1, we followed Husserl's procedure in the *Ideen* of omitting an investigation into the "subjective" direction of phenomenological reflection that, over the course of his thinking from the *Ideen* to the early 1920s, would progressively lead to the realm of genetic phenomenology and the self-constitution of the ego. The guiding thought behind the problem of genesis is the recognition that "the ego is himself *existent for himself* in continuous evidence; thus, in himself, he is *continuously constituting himself as existing*" (Hua I, 100 [66]). The ego *constitutes itself* in the sense discovered as the dual self-differentiation of absolute time-consciousness. As Husserl argues, "the ego grasps himself not only as flowing life but also as I" both "in relation to objects as a pole" and "the ego grasps itself also as I" (Hua I, 100 [66]). As already indicated in chapter 1, the sense of the constitution of an object for consciousness is different, though related, to the constitution of the consciousness for itself as an ego. Whereas the ego's transcendence towards the noematic object-pole operates through the synthesis of identification (the synthesis of the manifold of appearances within the system of the object as a unity of sense), consciousness is constituted for itself through a "second polarization" with regard to its own immanent manifold, as unified and belonging to an identical ego. In one sense, the self-constitution of consciousness was already developed within the analysis of time-consciousness, without any reference, however, to the constitution of the ego, which, in the context of genetic phenomenology, presupposes the pre-reflexive "operative intentionality" of original time-consciousness. The phenomenological concept of the ego comprises three basic structures of constitution: the ego as the identical pole of acts of consciousness; the ego as the substrate of habitualities; and the ego as the full concretion of consciousness as a monad (a conceptual term that Husserl borrows from Leibniz in order to designate the stream of consciousness as a whole, as both the unity and uniqueness of a life).[8] Genetic phenomenology thus articulates – as a transcendental problem – the question of self-constitution in two related senses. First: in terms of the self-givenness of the ego for itself. As Husserl writes, "I am for myself and am continually given to myself, by experiential evidence as '*I myself*'." Second: in terms of the monadic ego as containing "the total, actual, and potential life of consciousness." It is due to this latter consideration that "the problem of self-constitution embraces all constitutive problems" (Hua I, 102 [68]).

But, why does the problem of self-constitution entail or contain (*befaßt*) *all* problems of constitution? This is not the claim that all problems of constitution are *reduced* to the self-constitution of the ego, but rather, that the self-constitution of the ego "entails" or "contains" all problems, that is, it situates all problems of constitution within the

---

8 For a more thorough study of the ego in Husserl's thinking, see Eduard Marbach, *Das Problem des Ich in der Phenomenologie Husserls* (Den Haag: Martinus Nijhoff, 1974).

life of subjectivity as the movement of constitution. We have seen that Husserl's basic claim is that consciousness has the structure of temporality, as the stream of noetic acts and their sensual content in an intentional correlation to objects of experience. The ego is a pole in the sense that these experiences – the acts of consciousness in the temporal flow – belong to an ego. Yet this ego-pole is not empty; as a pole, it serves as the center for the concretion or "enduring ego" (*verharrendes Ich*) that becomes constituted, as discussed earlier, on the basis of the sedimentation of retentional consciousness as legacy of self-constitution. This self-constitution of the ego as the substrate of habitualities leads to a higher form of self-constitution as a person, which Husserl addresses in his ethical writings as well as in *Ideen II.*

The ego "grasps" itself as a temporal stream such that consciousness is "being for itself," yet this being for itself is itself a "movement" or becoming: transcendental subjectivity is not a being, but a becoming, in the specific manner of continually *constituting*. Consciousness is the becoming of constitution. In the discussion of time-consciousness, however, we recognize the "self-appearance" of time-consciousness, yet since the analysis of time-consciousness remained formal, that is, without any content or concretion, the self-appearance of consciousness only retained an abstract significance. As noted in chapter 5, Husserl rules out the possibility that absolute time-consciousness has the form of a "something," or a "what," that appears. The problem of self-constitution and genesis is thus the posing of the question, "who is transcendental subjectivity?" Genesis enters into the field of phenomenological research by way of a reflection on the self-givenness of subjectivity; we cannot take for granted *that* subjectivity is, or who subjectivity is, and thus we must give an account of the *becoming* of subjectivity such that the problem of genesis is the problem of the conditions of possibility for subjectivity.

Let us now turn to our first question: how does the problem of self-constitution "contain" all problems of constitution, and, indeed, what does "contain" mean? The problem of genesis is not simply a one-sided account of the becoming of subjectivity because subjectivity is also self-transcendence towards the world. Taken from this angle, the problem of genesis formulates the issue of "pre-givenness." The constitution of objects rests "on the basis of pre-given objects." As Husserl formulates the distinction between the "static" analysis of the constitution of objects for consciousness and an analysis of the "genesis" of that constitution,

If the theme of constitutive analyses is to make understandable how perception brings about its sense-giving and how the object is constituted through all empty intending as always only exhibiting optimal appearance-sense in a relative manner, and to make this understandable from perception's unique intentional constitution according to intimately inherent components of lived-experience itself, according to the intentional noema and sense, then it is the theme of genetic analyses to make understandable how, in the development proper to the structure of every stream of consciousness, which is at the same time the development of the ego – how those intricate intentional systems develop, through which finally an external world can appear to consciousness and to the ego. (Hua XI, 24 [62])

In this regard, the theme of genetic phenomenology is the problem of pre-givenness and, in this sense, embraces all problems of constitution since it functions as the presupposition for all other forms of constitution, prior to the thematic activity of the ego.

There is, however, another sense in which self-constitution "contains" all problems of constitution, and this sense has something to do with the idea of the concrete monad. Already in the *Ideen*, we noted that Husserl identifies two distinct, yet related themes for the phenomenological analysis of consciousness: the constitution of individual experience and the constitution of the unity of consciousness as such. In the discussion of genesis, Husserl introduces the idea of the concrete monad, or "life," by which he means, the unity of my life as whole and as concrete, that is, in its facticity. The monadic ego entails the totality of the actual and potential life of consciousness that leads Husserl to introduce the idea of compossibility. As Husserl writes, "The *universe of lived experiences*, which are the 'really inherent' consciousness-constituents of the transcendental ego, is a universe of compossibilities only in the universal *unity-form of the flux*, in which all particulars have their respective places as processes that flow within it" (Hua I, 109 [75]). The "universe" of experience, as the unity of life as such, is the totality of what I can possibly experience. The world is the emergence or genesis of possibilities as the concreteness of the unfolding life of consciousness. The question of how possibility enters into the world is equivalent to how consciousness becomes its own past and future.

Though ideas of compossibility, harmony and monad are drawn from Leibniz, Husserl stresses that he seeks to translate these concepts into a phenomenological context of significance by jettisoning their metaphysical meaning. The conditions for the possibility of experience are, however, also an issue of what can be meant, or intended, or experienced, along with other things within the horizon of the world as such. Husserl has established that meaning is embedded in a horizon of unfolding possible experience. This foundation establishes that experience must "harmonize" or cohere with the form of the world as such, as the world of possible experience. In other words, the perceptual object is a *system* of coherence unfolding in time, and this is the sense in which time is the form of objectivity. Within the system of a perceptual object, the object is constituted as the harmony of experiences (different sides, colors, etc.), yet each system of appearance must in turn cohere with other systems of possible experience. The basic idea is that objects can only be constituted that are compossible with each other. The idea of compossibility thus defines the idea of one world: my world and myself must cohere. The problem of compossibility within the unity of my own life, and as the genesis of my own ego, is parallel to the compossibility among monads in the constitution of one inter-subjective world of possible experience.

The question of how possibility enters the world is equivalent to the question of how consciousness has a past and future. Of course, we cannot take for granted the categories of actuality and possibility, that is, we need to tell a phenomenological story of their origin and motivation: this is one of the features of an analysis of time-consciousness. A real possibility is the possibility of something empirical and contingent: real possibility is equivalent to empirical possibility. This view of possibility requires an idea of event or content: a something that may or may not happen. In Husserl's scheme, we only have this contingent event with passive synthesis, since time-consciousness as such is abstract and without content. The concept of the actual and the possible, under the rubric of real possibility, only enters with genesis; yet genesis is based on time-consciousness since time is the universal form of all genesis. Husserl tells us that time-consciousness remains abstract, and it is the theme of genesis that provides "content" or "matter." Time-consciousness "enables" actuality and possibility. In other words, possibility and actuality cannot be presupposed in a phenomenological manner – their phenomenological origin must be clarified through the problem of genesis.

## Time-consciousness as the promise of subjectivity

Husserl argues that time-consciousness is "obviously" (*selbstverstän-dlich*) the basis for association and genesis, that is, the universal form for egological genesis, and thus the foundation for all constitution. When he writes that time-consciousness is "selbstverständlich" the basis, we must understand and read this coded term as meaning that time-consciousness is the "presupposition" or "foundation," not in the sense of being taken for granted, but rather, as we have learned in previous chapters, in the sense in which time-consciousness is the wellspring and origin of consciousness in its essential constitution as "other-directed" and "self-directed." The short answer to this question is that time-consciousness *is* consciousness; since association is itself a form of consciousness, then in so far as consciousness is fundamentally time-consciousness, time-consciousness is the basis for association as a form of consciousness. Time-consciousness is the "primordial crucible" (*Urstätte*) for all forms of constitution and thus the general form of constituting, or establishing, consciousness (*allgemeine Form einer herstellende Bewußtsein*). Not only is time-consciousness the general form of any constituting consciousness, but the "being" of consciousness is time-consciousness; in other words, time-consciousness is both the "self-temporalization" of consciousness and the consciousness of temporal givenness. The basic form of givenness as such is temporal givenness, and this is also the form for the "self-givenness" of consciousness. Time-consciousness is the "event" of consciousness itself.

Time-consciousness, therefore, is the promise of subjectivity in the becoming of its constitution. There are four essential ways in which time-consciousness contributes to the transcendental constitution of subjectivity and the world.

First: time-consciousness is the form of "identity-unity" or "objectivity as such." Any object of experience is the synthetic and intentional unity of a manifold; time is the form of objectivity, for objects of all types, in so far as each region of being is a form of temporality. Second: time-consciousness is not only the form of objectivity as such, in so far as any object is the unity of a manifold, but it is also the "connecting form" for objects within the common-time form of world time. Time is thus the form of individual objects as well as the form of their connection to other possible objects of experience. Third: time-consciousness is also the form of the self-givenness of consciousness itself in two senses: the form of any individual experience (the constitution of immanent content in the stream of consciousness) and the form or the unity of the stream as such. In this regard, time-consciousness is responsible for the "encompassing synthesis" of the living present (*Lebensgegenwart*) of the concrete monad; in other words, it is the universal form of egological genesis. Thus, we realize that time-consciousness is necessary for both the genesis of subjectivity as well as the constitution of objectivity *in the* world. Fourth: time-consciousness also structures, as discussed in chapter 4, different kinds of acts of consciousness – remembrance, perception, imagination etc. – in constituting the primordial distinction between the different characters of consciousness "presentification" (*Gegenwärtigung*) and "re-presentification" (*Vergegenwärtigung*).

Given these central and various accomplishments, Husserl nonetheless, and perhaps surprisingly, claims that time-consciousness remains "abstract" and "meaningless" (*bedeutungslos*) (Hua XXI, 124 [169]). In what sense, then, is time-consciousness "meaningless"? Let us first recall that time-consciousness is the form of individual objects, and yet time is itself not a predicate in the sense of other possible determinations of an object. Brentano's distinction between "idea-association" and "original association" already signaled this separation in that temporal determinations were recognized as *modifying attributes* of an object's givenness and not as determining attributes of the "content" of the object. This Brentanian insight, carried over into Husserl's own analysis, further established that time, under the heading of temporality (*Zeitlichkeit*), is not a sensation or part of a sensation, even though it is intuitive in a sensible manner, as discussed in chapter 2 on the expansion of sensibility beyond sensation. Husserl further develops this key distinction within the context of his phenomenological reflections. On the one hand, Husserl argues that when I hear a melody, my intuitive apprehension of the temporality of the melody – its temporal duration – cannot be parsed out in terms of sensual content as, for example, the sensations of tone quality. As with Brentano, there is no sensation of time as a determination of the object's "whatness." On the other hand, Husserl does insist on a "seeing" of time itself, and thus, on a phenomenological description of our lived experience of time. This requires an "intuition" of time, a grasp of time in its intuitive givenness. As presented in chapter 3, Husserl fashions a distinction between the object and its manner of givenness – the identity of the note as middle C does not change over time. This distinction allows Husserl to introduce the idea of temporality as the determination of the object's manner of givenness (*Objekt im Wie*).

It is important to note that this "object in the how" (*Objekt im Wie*) is not a second object, but the time-object seen or taken in its temporal manner of givenness (which Husserl calls "Zeitlichkeit" or duration as opposed to "Zeit"). The "object in the how" is disclosed through a reduction and is an object of phenomenological reflection; this is another way to understand the claim that the phenomenological reduction, as the disclosure of the immanent stream of consciousness, is the disclosure of temporality – it is only with the reduction that we have the basis for understanding and describing temporality. Husserl thus refuses the Kantian idea of an intuition of time as such, as an empty transcendental form of intuition, yet accepts that there is an intuition of temporal givenness within phenomenological reflection. In mundane experience, we do not "see" time, but only things in time; through phenomenological reduction and reflection, we do "see" time – and that is why we can provide phenomenological descriptions. But we also see time in the sense of eidetic intuition, since the status of Husserl's analyses is that they are descriptions of eidetic forms. Returning to the distinction between temporality as the manner of givenness of the object and the object considered as what it is (with its appropriate noematic sense and characteristics) means that the content in the register of the description of time-consciousness is not the same as the content of the object *qua* blue, green, red; Husserl argues that sensation itself is a time-object. The fields of sensations are opened through time-consciousness since sensations must first become constituted as time-objects; in other words, the concreteness or specificity of a sensation presupposes its temporalization, yet it is only at the concrete level of sensations, as correlated to particular systems of sensations (visual perception, etc.), that we can meaningfully speak of something happening. In this regard, time-consciousness is the event of a possibility in which nothing yet happens.

The transition from the abstract character of time-consciousness to the dimensions of pre-givenness in genetic analysis is a story of how the accomplishment and origin of meaning is covered over and obscured in the process and progression of the accomplishment itself. The "explanatory" gap between time-consciousness and genesis means we cannot "explain" the formation of facticity merely on the basis of time-consciousness; yet the structures of time-consciousness must nonetheless contain the basic structures that are also discernible – at a more concrete level – in genesis. Ultimately, these structures are also reiterated and reappropriated at the level of judgment and active synthesis, which means that at the basic level of time-consciousness we already have an "image" or anticipation of thinking itself.

If, as we noted, time-consciousness is abstract and meaningless, this means, in so far as time-consciousness is presupposed by concrete subjectivity, that the subjectivity of time-consciousness is also meaningless and abstract: it is meaningless subjectivity – but what gives subjectivity its meaning as subjectivity? The answer is its movement of self-constitution. Subjectivity becomes; yet time-consciousness is not yet the meaningful becoming of subjectivity and thus there is not yet any meaningful sense to becoming. That is why Husserl struggles with the sense in which we can speak of the flux: not as the becoming of anything, but as the becoming of nothing, or the becoming that is in-between being and nothingness. Subjectivity is the movement of becoming a life.

### Genesis of subjectivity

Our aim here is neither to develop an exhaustive presentation, or interpretation, of Husserl's genetic phenomenology – as primarily articulated in his lectures on passive synthesis – nor to explore the full scope, or range, of the genetic problematic. Rather, my aim is to present and interpret the accomplishment of genesis, that is, the significance of the theme of genetic phenomenology in such a manner that connections to the theme of time-consciousness are established and explored. Our main interest, in this respect, is to sketch how the problem of genesis articulates the genesis of subjectivity as well as to understand how the problem of genesis challenges, or at least modifies, the *sense* in which time-consciousness is at the foundation of all constitution.

Let us first highlight the many senses in which passivity functions within the scope of genetic phenomenology. Passivity, in each of its different meanings, contains both noetic and noematic components, given the noetic–noematic structure of intentionality. We can speak of passivity in the sense of the passive modalization of experience; but we can also, in this regard, speak of passivity within primordial time-consciousness. We can speak of passivity in contrast to the activity of the ego, that is, of passive intentionality in the form of association, which Husserl contrasts to the synthesis of identification that serves as the basis for constitution within static phenomenology. However, this conception of passivity is not to be equated with the traditional notion of receptivity and thus does not set into relief the distinction between receptivity and activity, since this phenomenological understanding of passivity undercuts the distinction between receptivity and spontaneity. As stressed in our earlier discussion, the retentional modification of original presentation is neither active nor passive, but "medial" – in other words, the undercutting of the traditional distinction between receptivity and spontaneity is already prepared from within time-consciousness, and the accomplishment of genesis is to pursue the consequences of this discovery by exploring this newly opened dimension of passivity. To return to the different meanings of passivity, we can further state that passivity also functions as the pre-predicative and pre-reflective sedimentation of meaning and delineates the dimension of pre-givenness in contrast to constituted objectivities; finally, passivity is the origin or genesis of reason and logic, as Husserl developed in *Experience and Judgment.*

Much is covered in the passive synthesis lectures, and to cover this ground entirely would expand beyond the scope of this chapter. Husserl begins, however, with the description, familiar to us from chapter 1, of the inadequate character of external perception. The basic character of perceptual experience is its pretension, in the sense that more of the object is given to us than what actually appears in an originary manner. As discussed in chapter 1, perceptual experience is a synthesis of authentic and inauthentic aspects of an object's givenness. This accounts for the "strange duality" (*Zwiespalt*) that, on the one hand, perceptual experience is an "original consciousness," by which Husserl means that the object itself is given to us in an original manner, yet, on the other hand, this original givenness is situated in a context of various perspectives of the object that are not given in an original manner, but which nonetheless shape consciousness in an intentional manner. These unseen sides of the objects are given to me, not in an originary manner (as, for example, are the sides that face me); but neither are these unseen sides "known" to me through inferential knowledge or through the imagination. The first option, for Husserl, represents a higher-order act of judgment; the second option denies the *intuitive* character of empty intentions and its horizonal structure. These empty intentions allow me to see without seeing; I see the *absence* of those sides that remain hidden from view. Expressed from the noematic point of view – and, again, this basic insight is familiar to us from chapter 1 – the perceptual object as such is a system of references (*ein System von Verweisen*) centered on a noematic sense and "object X."

In the passive synthesis lectures, Husserl adds an element to his description that was missing from his static analysis, as for example in

*Ideen.* Husserl speaks – from the noematic perspective – of the hidden sides of an object "calling us" to turn the object around, "telling us" that there is more to be seen. This description of *affection* can be seen as an insight into the movement of perception as the movement of *curiosity.* Perception is curiosity; this is another way to state the basic thrust of Husserl's genetic expansion of his description of perceptual experience: the increased attention to perception as a movement or process, and in the specific sense, as introduced above, of the "actualization" (or realization) of potentiality. In this regard, perceptual experience intends the "objective sense" (*gegenständliche Sinn*) or "identity" of the object X, as the substrate of its appearing moments; those aspects of the perceptual object that authentically appear are surrounded by intentional empty horizons that have the form of a "determinate indeterminacy" (*bestimmbare Unbestimmtheit*). Two significant lessons can be drawn from this description: the object is a system for something always new; and it is the system for the disclosure of the object's manifold ways of appearing. From this observation stems the idea that perception, as movement, is the movement towards increased clarification and determination of the object. As Husserl writes, "Perception is a process of acquaintance [*Kenntnisnahme*]" (Hua XI, 8 [44]). Moreover, perception, as movement from generality to particularity, is a temporal flow or stream of experience, and this flow is understood as the "interplay" (*Gefüge*) of fulfillment and emptying (*Entleerung*) (Hua XI, 8 [44–45]).

An object is always perceived, or given as itself, in a system of references that variably prescribe additional and continuous ways in which that same object can be perceived time and time again. An object is thus embedded in a system of references to possible future perceptions of the same object. This lamp on the table, for example: not only does it have a reverse side that cannot be perceived, but there is more to be known of the object than what can be taken in at a single glance. Those aspects of the object that are not directly given to me are nonetheless present to me in the form of possible lines of discovery, anticipation and contestation. An object is always situated in a nexus of horizons. These horizons of possible perceptual experience are implicit, and, in this sense, passive, or potential, forms of consciousness. These implicit horizons are nonetheless forms of intentionality: an object is intended as something; moreover, these implicit horizons can be explicated or made explicit – these lines delineate the shape of discoveries to come. These horizons are also passive in the sense that such consciousness is not constituted through an active or spontaneous act of the ego.

As Husserl notes, these empty references that permeate and texture the course of perceptual experience are pointers into an emptiness: "They are pointers into an emptiness since the non-actualized appearance are neither consciously intended as actual nor presentified" (Hua XI, 6 [42]). Yet the empty horizons that are thereby brought into relief are not entirely empty or nothing. We are not facing the complete or total appearance of nothing, but of a nothing into which we can step, into which we are drawn, and into which we are invited. The invitation here is that the possibility that attends or surrounds the actuality of the object calls us towards a further actualization of the object's appearance. In these lectures, especially when Husserl looks more closely at the phenomenon of affection and association, he uses expressions such as the "call." An affection calls on the ego to take note of it and to turn towards it. There is a kind of "Alice in Wonderland" quality to Husserl's descriptions: "'There is still more to see here, turn me so you can see all my sides'" (Hua XI, 5 [41]). Husserl does not mean to imply that there is something "magical" about these experiences. Quite the contrary, to speak of things as calling or beckoning us towards them is to adopt a phrasing of experience that inflects the passivity of giving oneself over to the allure of the world.

A *Leerhorizon* is the appearance of emptiness in the form of a horizon and an implicit, passive consciousness of possibility that extends perceptual experience beyond the here and now of the given, opening consciousness in a future. *Leerhorizon* is the appearance of a "Nichts, die nicht Nichts ist," of an emptiness that gives itself as something to be pursued and filled in. In this sense, the horizon of emptiness is "eine bestimmte Unbestimmtheit" (Hua XI, 6 [42]). These empty horizons belong to the basic structure of intentional consciousness. We may speak here of a distinction between the directedness of intentional consciousness towards an object given in perceptual fullness and the orientation of intentional directedness with regard to what circumscribes the actual perception of objects. Moreover, it is by virtue of an attending empty horizon that a rule is prescribed that regulates the transition from actualized perception to the next. Everything that has already happened and been perceived is in the grips of anticipating intention, rooted in near and far protentions, which provides a frame for the encounter with the new, for closer detailing or filling out of what has already been seen and understood. Thus the horizon-structure of perception, the reach of perception that runs ahead of itself, is guided and shaped by a previous seeing, by a history of seeing, that has become the knowledge of what it knows to see again. The complementary aspect to such projected empty horizons of possibility and continued perception is the possessing and habit of what has been already seen. What I see stays with me, becomes part of me, and accrues into a basic kind of understanding. I become what I have seen – I am a history of having seen. Perception becomes me. Perception is thus not only the movement of curiosity but also the becoming of visibility, and in this sense a knowledge that becomes (*werdende Kenntnis*). Perception becomes what it perceives; in this becoming, perception comes into pre-possession of what has yet to be seen while retaining what it can no longer see.

Perception is therefore movement: the actualization of potentiality, and the movement towards a *telos*. Two moments in Husserl's analysis reveal an underlying insight into the accomplishment of passivity. The lines of possible perceptual experience are characterized as leading into the future; these possibilities prescribe rules that determine the type of continued perceptions that are available, or, in the case of perceptual disappointment, that break and interrupt – realign – the course of perceptual experience. As Husserl expresses the matter, an empty horizon or anticipatory presentation is "a determinate indeterminacy" (*eine bestimmbare Unbestimmtheit*). It is indeterminate because nothing determinate has been presented, and, in this sense, it is empty. On the other hand, it is a determinable form of indetermination; but in what sense is the future here determinable in its indeterminancy? It is on account of the past, for it is in terms of the continuity of the past and its determinate shape, or course, that the various possible futures or horizons are projected. It is thus in view of the past that the future is determinable. The possible lines of determination that delineate the constitution of the future issue from the past; it issues from the past as what is thrown-ahead. It is on account of the dynamic interplay of filled and empty perceptual presentations that experience can be said to be in flux. When Husserl speaks of the "flow" or "stream" of consciousness he does not mean the mere succession of states, that is, as suggested by the metaphor of time as a flow or river; instead, flow means the *interplay of fulfillment and emptying* intentions that constitutes the synthesis of fulfillment and "harmony" (*Einstimmigkeit*).

Given this description of perception as movement, as the actualization of potentiality, the temporality of perception is recognized as constitutive of perceptual experience. This movement of perceptual experience is constituted through temporalization or, in other words, through time-consciousness, in two related senses: (a) time-consciousness is the origin of the distinction between authentic and inauthentic, and, in this regard, the basis of perceptual experience as the weave of fulfillment and emptying; (b) time-consciousness is at the basis of potentiality and actuality; or, in other words, actuality and possibility are constituted within time-consciousness. In our discussion of time-consciousness in earlier chapters, we established the first point (a), since this point can be articulated in the context of static phenomenology. It is, however, the expanded perspective of genetic phenomenology that allows Husserl to articulate the second point (b), though, to be sure, this point is already implied in static analysis. Retentional consciousness constitutes potentionality in two senses: retentionality as "emptying" (*Entleerung*), as we discussed in the chapter on the Bernau Manuscripts, is a de-presentification of the intuitiveness of the original presentation (which itself is the source of intuitiveness or authentic givenness); yet this de-presentification is now understood as the modification of actuality into potentiality, that is, potentiality becomes constituted in terms of the becoming of the past. And secondly, retention is the constituted sedimentation of possibility, that is, "latency," in the sense of potentiality. In other words, the dual-action or significance of retentional consciousness is that it "empties" and thus liberates possibility as a possibility (*Vermöglichung der Möglichkeit*). But it does so in so far as it retains the possible as such, in such a manner that what I have perceived becomes a permanent "property" (*bleibendes Eigentum*) for me; the past becomes "freely available" to me through remembrance, and, in this sense, retention can become reactivated and "re-filled" through remembrance. Moreover, to the extent that retentionality not only retains the object of perception (i.e., the tone) but also my own act of perception, since I retain myself as having perceived, retentionality is also the manner in which I become available to myself and become my own "possession" (*bleibendes Eigentum*). But I retain myself as my own history always with regard to what I have experienced or perceived; in other words, there is a de-presentification of my own consciousness. I myself become a potentiality, as the basis for the structure of habits, which Husserl, as noted earlier, identifies as essential to the constitution of the ego.

In this regard, Husserl's discussion of modalization is important for two reasons. First, it is at the conclusion of this section that

Husserl formulates his all-important distinction between passivity and receptivity; and second, it is in this section that Husserl discusses possibility, and its two forms as motivated and open. Of course, this discussion of modalization is important for Husserl's proposed genealogy of logic, since it accounts for the pre-predicative basis for modalization at the level of judgments. This attention to modalization stems from Husserl's focus on the "passive *doxa*" that is an essential moment in perceptual experience; every perceptual act contains a moment of belief that can undergo possible modalization – and the point of focusing on the modalization of belief in perceptual experience is to establish a distinction between the passive accomplishment of experience and the spontaneous accomplishment of thinking. We cannot enter the full the scope of Husserl's genetic analysis of disappointment with regard to anticipating intentions or doubt – though it is in the context of his discussion of doubt that Husserl remarks that, "the destiny of consciousness" is its "history" (Hua XI, 38). What is important is Husserl's discussion of possibility (*Möglichkeit*). This discussion centers on a distinction between "open possibility" (*offene Möglichkeit*) and "motivating" possibility (*anmutliche Möglichkeit*) (Hua XI, 41). The modalization of certainty, or belief, in the form of possibility, is not a modalization in the sense of negation and doubt. Both types of possibility are modalizations of certainty; but whereas, when faced with an open possibility, there is no particular "weight" in the case of motivating possibility, the particularity of one possibility over another "attracts me" or pulls me in its direction. This allows Husserl to introduce the idea of an "affective force" (*affektive Kraft*) that motivates the ego. As Husserl remarks, an affective force is a "tendency that exercises a pull on the ego, to which the ego responds with an opposite force" (Hua XI, 50 [99–91]).

This discussion leads to the discussion of passive and active modalization. Passive modalization is a passive intentionality in which there is the modal transformation of passive *doxa*; in contrast, the ego must decide for or against this affective force, and in so doing, adopts a responding posture (*Stellungnahme*); this is the sense in which the ego is active, and as Husserl stresses, this movement or transition from passive modalization to active modalization does not have the form of rendering explicit what is implicit. This is because the act of decision is an act of "Zueignung," but also, a self-realization of the ego itself; the ego constitutes itself as the history of its decisions. These considerations allow Husserl to argue that "perception has its own intentionality that does not contain anything from the active attitude of the ego and its constitutive accomplishment, which, indeed, presupposes passive intentionality such that the ego can at all have something for and against which it can decide" (Hua XI, 54 [94]). Moreover, this leads to the crowning distinction between passivity, receptivity and activity.

The distinction between passivity and receptivity is drawn by Husserl in the following manner: "We can include receptivity in this first level [of passivity] namely as that primordial function of the active ego that merely consists in making patent, regarding and attentively grasping what is constituted in passivity itself as formations of its own intentionality" (Hua XI, 64 [105]). Husserl thus distinguishes two forms of passivity: the passivity of receptivity and an underlying passivity on the basis of which receptivity is itself possible. In this more original sense, passivity is not to be confused with receptivity, by which Husserl means Kantian receptivity. In Husserl's formulation, receptivity is characterized by the function of "rendering patent" (*patent machen*) and, in this regard, designates the way in which content of experience is already given, or pre-given, within the scope of the ego's attentiveness. The function of passivity as receptivity is to render visible and open structures of experience to the ego, and thus to render consciousness open for itself and its experiences. But if the sense of passivity as receptivity is to render open or make manifest on the basis of a prior form of "pre-constitution," this prior form is a passivity that "renders latent" (*latent machen*). To "render latent" is to open the ground not simply for the possibility of being made patent, but to open the ground of possibility itself. Original passivity would thus account for the creation and entrance of possibility into the life of consciousness.

As we have already discussed, perceptual experience is a synthesis, or movement, of fulfillment that strives for a unity of harmony among the object's manifold ways of appearance. This movement of perception can undergo breaks and modalization, as when, for example, an anticipating intention is disappointed, or not fulfilled, in experience. The fulfillment of perception, however, requires "empty horizons," for it is these empty horizons that pre-delineate the possible continued manners of appearance, or, in other words, that project anticipated ways of appearance that may or may not be confirmed, or validated, over the course of experience. As Husserl writes, "where there are no horizons, and no empty intentions, there is also no fulfillment"

(Hua XI, 67 [108]). The synthetic interplay, or movement, between empty and fulfilled intentionalities, as well as the basis for the distinction between intuitive presentation (original or authentic presentations) and empty presentation, is to be found in time-consciousness. The genesis of such empty intentions is rooted in time-consciousness and, specifically, in the different kinds of empty-presentations that are constitutive of time-consciousness, namely, retentions and protentions. Whereas protentions are empty intentions that are directed towards the anticipated arrival, so to speak, of an object, as prescribed by past experiences, retentions do not intrinsically possess a "direction of sight" (*Blick-Richtung*) (Hua XI, 76 [118]). Retentions are indeed a form of intentionality, yet consciousness is not *directed towards a possible fulfillment* of the intended object, and it is precisely for this reason that the retentional modification of an original presentation is an "emptying" of intuitiveness. Fulfillment is oriented towards the future.

And yet, Husserl contends that retentions can nonetheless take on, or be the bearers of a directed-intention ("Blick-Richtung") through another form of passive intention called association or associative synthesis. For Husserl association is "a class of empty presentation that is intended, and which is specifically directed towards its object" (Hua XI, 76 [118]). In other words, the basic function of retentionality is not to objectify its object, or content, but, on the contrary, to "de-objectify" and retain the intentional object in a non-thematic and pre-objective form of apprehension. In this regard, as Husserl expresses, "nothing actually happens" in an empty presentation ("in der Leervorstellung spielt sich eigentlich nichts ab"), that is, an objective sense is not constituted in retention (Hua XI, 72 [114]); in a significant sense, an objective sense is *not authentically given*, that is, not actually given, in retentional consciousness, and yet, the objective sense is nonetheless retained – but in what sense? The objects of past experience are not retained in the proverbial metaphor of memory as a storehouse in which objects are stored and kept. Within the meaning of this metaphor, an object is fully constituted as actual, and its storage in memory is analogous to the storing of an object in a closet, hidden from my immediate and actual view, and yet nonetheless actually constituted. Retentional consciousness, it should be recalled, is not, however, a form of remembrance; instead, it is the basis for the possibility of remembrance. The sense in which the past is retained is not that actual objects are kept hidden from view, but rather, that objects are kept in an "empty" manner, that is, as possibility. It is thus through retentional consciousness that possibility or potentiality itself emerges into the world, and it emerges as the life or becoming of subjectivity itself. As Husserl writes, "the empty [*das Leere*] is the potentiality of what can be actualized in a corresponding intuition and syntheses of disclosure [*Synthesen der Enthüllung*]" (Hua XI, 94 [138]).

Thus, although retentional consciousness is empty in the senses introduced above, it is an emptiness, or possibility, that can be fulfilled in an intuitive manner in remembrance. On Husserl's account of remembrance, an empty presentation of retentional consciousness can "awaken" a "direction-synthesis" (*Richtungssynthese*); the theory of association is meant to capture this kind of passive intention as awakening. In light of Husserl's argument that association, as an intentionality of awakening, gives retentional consciousness a direction to an object "in hindsight," so to speak (Husserl speaks of "hinterher nachkommende Assoziation"), we can revisit Husserl's polemic against Brentano's theory of original association. The passive synthesis of retentional modification and original impression is *not* an association since an association is an intentionality and synthesis that bridges a distance already constituted as distance; association thus presupposes the constitution of possibility and actuality. In contrast to the lack of an objective intentionality, that is, intending in an objective manner, that characterizes retentional consciousness, protentions are objectively directed presentations. The protention is directed, and in this sense, strives for, or moves towards, its intentional object; for Husserl, perceptual experience – intentionality – is a striving for the object itself. Intentionality is movement in the sense of a striving (*Streben*) towards increased determination with its intentional object.

As explored within genetic phenomenology, Husserl increasingly adopts the language of striving and tendency to describe the movement of perceptual experience and intentionality. As suggested earlier, perception is the movement of curiosity: we are led into the presence of things; we strive towards the presence of things themselves, and, in this sense, subjectivity is the movement of constitution within the given. But this tendency of intentionality towards the givenness of its intentional object, this striving towards self-givenness, and, in this regard, towards "finality," become connected to the reverse movement of those intentional objects "affecting" consciousness; the ability of the ego to direct itself towards an object is a *response to a prior solicitation of the world*. In other words, the ego is affected by a "proto-object" in its passive form of pre-givenness; these objects solicit the directedness of consciousness, and although not every consciousness is necessarily a striving, every consciousness can become motivated to strive towards an intentional object. If this is so, as Husserl concludes, "Every intention is an anticipation" (Hua XI, 86 [130]).

These empty intentions, in the forms of retentional consciousness from which potentiality emerges and forms of protentional consciousness, project such potentionality as the future; in this way consciousness is surrounded by a general "milieu" of "always newly awoken empty intentions," and this is what Husserl calls "the world." But if the world is constituted through the projection of possibilities, and if the world is the correlate of synthesis of fulfillment and empty intentions, is it possible that the world as such might come to an end? In other words, "could it be that an external experience is the last, while consciousness would nonetheless continue?" (Hua XI, 107 [152]).

Husserl identifies the theme of "confirmation" or "validation" (*Bewahrheitung*) as a significant problem within the domain of passivity. As we have seen, perceptual experience is a synthesis of intuitive and non-intuitive presentations, a synthesis of identification in the form of fulfillment (or disappointment) of empty intentions. The synthesis of empty and fulfilled intentions conforms to the "lawfulness" of time-consciousness. Within this framework, Husserl describes intentional consciousness as a striving or tendency towards increased clarification, or self-givenness, of its intentional object. Intentionality is movement and is grounded in time-consciousness; we might emphasize this point more vigorously and state that intentionality is a form of temporalization: an intention is directed towards its object in the form of anticipation; in other words, intention is an "anticipation of the self-realization of the future" (*vorgreifend auf Selbstverwirklichung des Künftigens*) (Hua XI, 86 [130]).

Against this backdrop, Husserl raises the issue of the "universal synthesis" of "the world" in itself (*Ansich*). The thought here is that there can be local breaks within the experience of the world, yet the world as such is a "continually enduring world (*fortdauernde Welt*). Husserl speaks in this regard of the "fact of the world" (*Faktum der Welt*) and recognizes that this cannot be grounded with recourse to the passive basis of confirmation (Hua XI, 108 [153]). The question that Husserl raises, without, however, fully resolving it, is how to ground the world in itself (*die Welt an sich*) as reaching out beyond "its actually having been experienced" (*aktuelle Erfahrenheit*), that is, how to ground in advance the determinateness of the world as such, that "alles Seiende in der Welt" is "an sich bestimmt" (Hua XI, 108 [153])? This problem recalls the problem of inter-subjectivity; yet even if we recognize, as we did in chapter 6, the primary transcendence of foreign subjectivity, it is not immediately evident that inter-subjectivity itself answers completely the question raised here. This is because the problem emerges from a basic decision in Husserl's conception of time-consciousness; to the extent that the monadic immanent temporal streams of each consciousness are structured in the same manner, the harmonious collectivity of the "monadic community" remains bound by the basic conception of time-consciousness. The problem is that one must account for the openness of the future as a whole without recourse to the already constituted unity of the world as having been experienced. This recognition that the "Faktum" of the world cannot be grounded, that is, constituted, with recourse to passive confirmation signals that the "Faktum" of the world must already be pre-constituted, but not through the becoming of subjectivity and its history of constitution. As we have seen, the figure of absence – of empty intention and possibility – possesses the fundamental temporal form of the past; absence erupts from within time-consciousness through retentional modification; potentiality emerges from actuality through retentional consciousness; the constitution of objectivities has its source in remembrance; subjectivity itself becomes itself in becoming its own past. As we have seen, this conception of the past, or becoming past, as the *origin of absence itself*, as the givenness of absence, has its roots in Husserl's own past, in his point of departure from Brentano's theory of original association.

It is only with the theme of association that the full significance of Husserl's genetic phenomenology comes into view. Much of its conceptual force must be seen in the context opened up by the earlier distinction between passivity and receptivity. Husserl's theory of genesis is a theory of the genesis of pure subjectivity; its aim is to describe the "form and lawfulness of immanent genesis" (Hua XI, 118 [163]) within the transcendental field of the stream of consciousness. The problem of genesis is thus situated at the foundational level of passivity that had been broached by the analysis of time-consciousness. As Husserl remarks, "the phenomenology of association is so to speak a higher continuation of the theory of original time-constitution" (Hua XI, 118 [163]). Under the heading of association, Husserl understands a necessary form of lawfulness that makes subjectivity possible; this lawfulness structures the becoming of subjectivity, but since the theory of association treats both noetic and noematic elements, the becoming of subjectivity is also the becoming of the world as such. Two primary forms of the accomplishment of association are identified: the genesis of reproductive consciousness (remembrance) and, on this basis, the genesis of expectation and apperception. As Husserl writes, "What belongs to the possibility of subjectivity is having its essential sense as subjectivity, and without which subjectivity could not be subjectivity, that is, in the sense of a being that is for itself constituting [*den Sinn einer seiend für sich selbst seiend konstituierenden*]" (Hua XI, 124 [169]).

As noted, the formal temporal stream would be "meaningless" for the ego without "awakening," since the retentional consciousness is empty of intentional content and sinks into the indifference of the past. As we established in an earlier chapter, the intentional character of remembrance is a self-transcendence. Within the immanence of the living present, acts of remembrance are immanent, yet their objects are transcendent – my past experience is not given as present within the present in which I remember, but rather is given to me as not-present in the specific form of having once been present. This means that my entire past, which could, in principle, as we saw in an earlier chapter, be given to me – and recall that remembrance is a form of self-givenness because I can give myself "back" to myself – is transcendent, given to me as transcendent, in my living present. I am transcendent with regard to myself in an original manner, and this self-transcendence in remembrance would itself not be possible without the original self-transcendence of retentional modification – as we discussed at length in chapter 4. But as Husserl concedes, this state of affairs introduces a "curious paradox" (Hua XI, 204 [256]).

The primordial or original transcendence is the stream of consciousness and immanent temporality. Consciousness transcends itself in its temporal self-constitution in such a manner that consciousness becomes available to itself; since it is consciousness itself that constitutes its own temporal givenness, this transcendent "self" – transcendent because it does not coincide with its own immanence – constitutes itself as temporal in an original manner so as to become available for itself through remembrance. As detailed earlier, when I remember an event from the past, I implicitly remember myself as having once perceived that event as present. Remembrance is thus a self-objectification, a giving of oneself again. The stream of consciousness "lives in its streaming" (*das Bewußtseinsstrom lebt mit Strömen*), yet not in a self-enclosed manner that would prevent self-objectification. Husserl considers that "for the ego there corresponds the idea of a true self, the true past of consciousness, as the idea of complete self-givenness" (Hua XI, 204 [256]). This idea of complete self-disclosure is a consequence of Husserl's argument that my entire past is contained or retained within myself, as myself, through the sedimentation of retentional modifications, investigated in detail in chapters 3 and 5. The idea of my own being as self-given, however, is regulative. It defines the ideal horizon of self-clarification as well as the scope of expansion, since from any given memory, it is in principle possible to extend to another memory. Remembrance occurs when a retention is brought into relief and given again through an associative "awakening" (*Weckung*); this relief or saliency is either from the retention or given to it in hindsight, so to speak. This awakening of the past takes the form of an awakened self-givenness, which constitutes again the stretch of original time-consciousness. Retention is a "process" of emptying out the intuitiveness of the original impression and, in this sense, every retention is an empty intention or presentation. The recalled episode from my past is given to me with a relative degree of clarity within the scope of the idea of absolute fulfillment or self-givenness. This theory of remembrance has significance for Husserl's conception of consciousness as such. Husserl argues that the living stream of consciousness "contains" (*birgt*) in itself an increasingly expanding and enriched "empire of true being," an "empire of objectivity in itself," that is present for the ego and its active identification (Hua XI, 207 [259]).

At this point Husserl notes that we have to be clear about this "wonderful singularity of this state of affairs" (Hua XI, 208 [258]) Consciousness is not only a streaming original present, in which remembrances occur, nor is it only the continual ordering or connecting of experience with experience in a fixed temporal form, which constitutes the unity of consciousness. As Husserl writes, "The stream of consciousness up until the now is a true being [*ein wahres Sein*] and it is for the ego, whether or not the ego insists on itself. Every past experience is past [*gewesen*]: in itself [*an sich*]." But this "in itself" of past experience is present as a true being (*ein wahrhaft seiendes*) that is intelligible. If this were not the case, we could not speak of a stream of consciousness at all; moreover, it is only because the essence of consciousness is to constitute itself as an *"Ansich"* in the sense of being its own being (*ein wahres Sein seiner selbst zu tragen*), and according to the conditions of passivity, that active knowledge is possible. This statement is crucial since Husserl contends that it is only on the basis of having constituted myself as an *Ansich*, that is, as a past that I am and carry within myself, that I can "ascribe" (*zuweisen*) a stream of consciousness to another ego within the objective world.

For Husserl, the main theme of transcendental philosophy is consciousness as a constitutive accomplishment, in which different kinds of objectivities are constituted, and ever new forms of self-givenness are developed. As Husserl argues, "nothing can become conscious in a stream of consciousness or for an ego without this consciousness structured by essential laws, and without corresponding intentional genesis, and thus in terms of which we have as consequence [*Ausschlag*] consciousness of objectivity" (Hua XI, 218 [270]). But this becoming of consciousness, its streaming, is not the succession of experiences, that is, a flow, in the manner in which one thinks of an objective flow. Nor is consciousness *metaphorically* an objective flow. What Husserl discovers is that consciousness is a ceaseless becoming as the ceaseless constitution of objectivities; consciousness is "an uninterrupted history" (*eine nie abbrechende Geschichte*) (Hua XI, 219 [270]), that is, the history of its own constitutional accomplishment of self-temporalization in a two-fold manner along the lines of the double intentionality of retentional and protentional consciousness: as a differentiation *from* itself in terms of the transcendence of constituted time-objects ("cross-intentionality") vis-à-vis constituting immanent consciousness ("length-intentionality"); and as a differentiation *of* itself in terms of the transcendence of absolute time-constituting consciousness vis-à-vis constituted immanent consciousness. Within this self-differentiation of consciousness, consciousness escapes itself only to find itself anew, but not from where it just started. With this self-temporalization, consciousness is caught in-between the retentional sedimentation of itself as past – burying itself within itself – and the protentional unearthing of itself as a future without end. In this fashion, as Husserl writes, "an authentic analysis of consciousness is, so to speak, a hermeneutics of the life of consciousness" (Hua XXVII, 177). Indeed, this perpetual striving to become myself, which is possible only through the unceasing renewal of the now, is the true hope, or promise, of time.

# APPENDIX: NOTE ON
# TEXTUAL SOURCES

Husserl never originally produced a completed manuscript for the 1905 lectures "Phenomenology of Inner Time-Consciousness." An entire lecture manuscript is not extant and many indications suggest that Husserl composed these lectures in an ad hoc fashion, patching together materials from earlier lectures and notes along with material newly written. The ad hoc nature of these lectures is furthermore apparent from the fact that Husserl may have even read directly to his students from his own notes from Brentano's "unforgettable" Vienna lectures of 1885/86. Over a period of years after 1905, Husserl subsequently incorporated newly produced manuscripts into the folder entitled "Zeitbewußtsein" containing the original bundle of lecture manuscripts. Husserl was in the habit of removing individual pages from his original lecture manuscript and replacing them with revised material.[1] In 1917, Husserl entrusted this folder of manuscripts to his assistant Edith Stein with the expressed wish that she prepare the materials for publication – a request that granted Stein a degree of latitude to compose the envisioned work, but also placed on her the substantial burden of working out the details of what were often fragmentary and ever-changing analyses. Husserl's vision of his phenomenological movement contributed to this habit of parceling out individual domains of research to his assistants and charging them with the task of pursuing the intricacies of phenomenological investigations. Undoubtedly inspired by the example set by Brentano's scientific conception of psychology and his relationship to his students,

---

1 For example, §§ 8–10 (24–29) of the 1928 text consists of five pages written during November 10–13, 1911, which Husserl expressly arranged or introduced into the lecture manuscript, cf. Hua X, 389.

Husserl conceived of phenomenology as a rigorous science, not only in the methodological sense, but as importantly in the institutional sense of a collective program of philosophical research.

The manuscripts entrusted to Stein were in a chaotic state. As Stein wrote to Roman Ingarden: "the external state is pretty sad" (*der äußere Zustand ist ziemlich traurig*).[2] Stein's choice of wording is deliberate, for in noting explicitly that the "external" or textual condition of Husserl's folder was in a less than desirable state, she implies that the conceptual state (or "inner state") of Husserl's analysis was not as incoherent as their literary expression. In fact, in a subsequent letter to Ingarden, Stein characterizes Husserl's manuscripts as "beautiful things, but not entirely developed" (*schöne Sachen, aber noch nicht ganz ausgereift*).[3] The folder entrusted to Stein presented a complicated situation as it contained a diverse range of manuscripts, from as early as 1898 (Stein was only aware of manuscripts dating from as early as 1903) with other manuscripts as late as 1911 (with a few from 1917, presumably making their way into the folder while under her management). Towards the end of 1917, and not entirely in isolation from Husserl's involvement, Stein stitched together a draft text, which she considered an "Ausarbeitung" of Husserl's manuscripts. The division of the text into sections as well as the organization and headings of individual paragraphs were fashioned by Stein, but many indications suggest that she tried to retain the discernible framework of the "Gedankengang" of Husserl's original lectures in 1905 while also accommodating Husserl's constant revisions. Stein was very much aware of the significant progress in Husserl's thinking since the original lectures of 1905, and she accordingly attempted to bring the level of the analysis in her draft text in line with the condition of Husserl's thinking around the date of 1909.

That draft remained untouched until 1926 when, at the behest of Husserl, Heidegger agreed to publish Stein's composition in Husserl's *Jahrbuch*, albeit only after first seeing to the publication of his own *Sein und Zeit* in the *Jahrbuch*. Aside from adding a brief editorial note, Heidegger published Stein's "Ausarbeiting" in 1928 with minimal

---

2 Letter to Ingarden of July 6, 1917. Roman Ingarden, "Edith Stein on her Activity as an Assistant of Edmund Husserl (Extracts from the Letters of Edith Stein with a Commentary and Introductory Remarks)," *Philosophy and Phenomenological Research*, 23 (1962), 155–175; 171.

3 Letter of August 7, 1917: "Ich habe im letzten Monat Husserls Zeitnotizen ausgearbeitet, schöne Sachen, aber noch nicht ganz ausgereift," Ingarden, 173.

changes, but included a set of "Addenda and Supplements" for which, however, the original manuscripts have not been located.[4] To what extent Edith Stein's version is or is not a faithful expression of Husserl's envisioned publication is a difficult matter to discern given their close working relationship and the absence of an "Urtext" for the 1905 lectures with which to compare the 1928 edition.[5] Unambiguous, however, is that the 1928 publication does not make good on its claim to represent the original lecture course of 1905. As Rudolf Boehm established, the 1928 edition is only narrowly based on lecture manuscripts from 1905.[6] Stein's text is mainly composed of manuscripts written during 1907–1911 (with some as late as 1917). Rather than a faithful reconstruction of Husserl's 1905 lectures, the 1928 edition represents instead an amalgamation of Husserl's thinking spanning the years 1905–1917, during which the analysis of time-consciousness underwent considerable changes.

Evidently, the patchwork condition of the 1928 edition raises a number of issues for understanding both the substance and the historical evolution of Husserl's thinking. Because the 1928 edition fails to represent accurately the original lectures of 1905, it effectively "masks the development" of Husserl's thinking during the critical period 1905–1911 by compressing together texts from different stages in the evolution of Husserl's thinking.[7] On account of its peculiar pedigree, the 1928 text contains significant variations in terminology (for example: primary memory and retentional consciousness), shifting accents of analysis and unequal treatment of the principal parts of its subject matter – a condition that is comparable to Aristotle's treatise of time in *Physics*, IV, 10–14. The 1928 edition on its own, without the materials included in the appendices to *Husserliana X*, is consequently an insufficient source for understanding Husserl's analysis since substantial portions of that analysis are contained in

---

4 *Jahrbuch für Philosophie und phänomenologische Forschung*, IX, 367–498.
5 For the debate surrounding Stein's editorial function: Roman Ingarden, "Edith Stein on her Activity as an Assistant of Edmund Husserl (Extracts from the Letters of Edith Stein with a Commentary and Introductory Remarks)," *Philosophy and Phenomenological Research*, XXII (1962): 155–175; Henri Dussort's editorial comments in: Edmund Husserl, *Leçons pour une phenomenology de la conscience intime du temps*, (Paris: PUF, 1964).
6 §§ 1–6, § 7 (partially), §§ 16–17, § 19, § 23 (partially), § 30, § 31 (partially), § 32, § 33 (partially) and § 41. A notable omission from the 1928 edition of 1905 lectures are sections or pages containing Husserl's critique of Meinong.
7 In addition to editorial comments in Hua X, see the excellent introduction and discussion of these issues by John Brough in his translator's introduction.

manuscripts that were either excluded or only partially included in the 1928 edition. In providing a critical apparatus along with the majority of manuscripts from the "time-consciousness folder," *Husserliana X* provides for a clearer understanding of the development of Husserl's thinking from 1903 to 1917. In our own study, we shall not concern ourselves with commenting on the historical evolution of Husserl's analysis of time-consciousness directly, as it has already been well established and adequately studied. Rudolf Boehm's chrono-logical ordering of these manuscripts in *Husserliana X,* supplemented by Rudolf Bernet's minor corrections, offers a definitive portrait of the developmental stages in Husserl's analysis during these formative years.[8]

*Husserliana X* is not the only collection of texts specifically dedi-cated to the phenomenology of time-consciousness. A second collec-tion of manuscripts written in 1917/18 – the Bernau Manuscripts – has been published as *Husserliana XXXIII* (2002). The textual condi-tion of these manuscripts is more complicated than the *Husserliana X* texts. The Bernau Manuscripts offer less by way of a framework in which to situate Husserl's detailed analyses; they contain a wealth of what we may call "microscopic" analyses of key issues in the phenom-enology of time-consciousness, yet they make no claim to offering a systematic view of the time-consciousness problem. A third group of texts commonly known as the "C-manuscripts"[9] are working manu-scripts like the two other collections, but with the important differ-ence that in these manuscripts, Husserl expands considerably the range of his phenomenology of time-consciousness in ways that reflect the expansion of his phenomenological thinking during the late 1920s and early 1930s. In addition to these three concentrations of manuscripts, numerous other fragmentary sources and self-contained discussions in lectures exist in which time-consciousness is the subject matter of reflection.

Given the disparate condition of Husserl's writings on the subject of time-consciousness, a systematic interpretation of Husserl's phenom-enology of time-consciousness should not be motivated by the goal of providing a framework in which every available text can be reconciled

8 *Texte zur Phänomenologie des inneren Zeitbewußtseins (1893–1917),* ed. Rudolf Bernet (cf. his illuminating/instructive introduction): Bernet proposes new dates for the following texts: Nr. 18, 36, 37, 38, 39, 49, and 50.
9 Edmund Husserl, *Späte Texte Über Zeitkonstitution (1929–1934). Die C-Manuskripte,* Husserliana Materialien, III (Dordrecht: Springer, 2006).

with the proposed interpretation. In fact, it may be both an illusory and unproductive task to reconcile completely the range of Husserl's analyses with any one assessment of its meaning as a whole. When considering solely the three substantial collections of Husserl's writings on time-consciousness, we face a complex and diverse set of writings that span the entire period of Husserl's intellectual development.

In order to keep this study to a manageable size, material from the C-manuscripts, as well as other writings from the 1930s, will not be included in the scope of the proposed interpretation. There are also substantive reasons for this restriction. In the C-manuscripts, Husserl expands the scope of his earlier work on time-consciousness in a dramatic, if not unproblematic, fashion. We find in these writings a wide consideration of time-consciousness that either continues on previous lines of reflection (for example, the "impossible puzzle" of absolute consciousness) or renders them more explicit (for example, the analysis of sleep). In addition to a consideration of familiar themes, Husserl also undertakes a set of new analysis, introducing such themes as "the immortality" of the transcendental ego and the "primordial-ego" (*Ur-Ich*). The C-manuscripts thus need to be read with other texts from the 1930s as well as with earlier texts where certain themes (for example, the notion of the monad) had been developed. A further problem with the C-manuscripts is the uneven quality of their written thoughts. Husserl's power of concentration frequently wavers in these writings; themes often appear to disappear in the form of after-thoughts, tangents and "hunches." More significantly, what differentiates Husserl's thinking in the C-manuscripts from his previous work on time-consciousness is Heidegger's shadow. Another set of interpretative issues would therefore have to be addressed since even though Husserl may not be directly responding to the rising star of his precocious student, it is nonetheless apparent that Husserl's thinking is indirectly responding to the Heideggerian ontological critique of transcendental subjectivity, which I hope to investigate in a subsequent work.

# BIBLIOGRAPHY

Alweiss, L. *The Wold Unclaimed: A Challenge to Heidegger's Critique of Husserl* (Athens, OH: Ohio University Press, 2003).

Augustine. *Confessions*. Trans. R. S. Pine-Coffin (London: Penguin, 1961).

Benoist, J. *Représentations sans object: aux origins de la phénoménologie et de la philosophie analytique* (Paris: Presses Universitaires de France, 2001).

Berger, G. *The Cogito in Husserl's Philosophy*. Trans. K. McLaughlin. Northwestern University Studies in Phenomenology and Existential Philosophy (Evanston, IL: Northwestern University Press, 1972).

Bergson, H. *Oeuvres*. Ed. A. Robinet (Paris: Presses Universitaires de France, 1959).

Bernet, R. *Conscience et existence: Perspectives phénoménologiques*. Épiméthée: Essais Philosophiques (Paris: Presses Universitaires de France, 2004).

*"Einleitung,"* in: *Edmund Husserl, Texte zur Phänomenologie des inneren Zeitbewußtseins (1893–1917)*, ed. Rudolf Bernet, (Meiner Hamburg, 1985).

*La vie du sujet: Recherches sur l'interprétation de Husserl dans la phénoménologie* (Paris: Presses Universitaires de France, 1994).

Bernet, R., I. Kern and E. Marbach. *An Introduction to Husserlian Phenomenology*. Northwestern University Studies in Phenomenology and Existential Philosophy (Evanston, IL: Northwestern University Press, 1993).

Berti, E. "Brentano and Aristotle's Metaphysics," in R. W. Sharples (ed.), *Whose Aristotle? Whose Aristotelianism?* (Aldershot: Ashgate, 2001), 135–149.

Brand, G. *Welt, Ich und Zeit* (The Hague: Martinus Nijhoff, 1955).

Braun, E. *What, Then, Is Time?* (New York: Rowman & Littlefield, 1999).

Brentano, F. *Deskriptive Psychologie*. Ed. R. Chisholm and W. Baumgartner (Hamburg: Felix Meiner, 1982).

*Descriptive Psychology*, Trans. B. Miller (London: Routledge, 1995).

"Die Vier Phasen der Philosophie und ihr augenblicklichen Stand," in O. Kraus (ed.), *Die Vier Phasen der Philosophie* (Leipzig: Felix Meiner, 1926).

*Geschichte der griechischen Philosophie*. Ed. F. Mayer-Hillebrand (Hamburg: Felix Meiner, 1977).

*Grundzüge der Ästhetik*. Ed. F. Mayer-Hillebrand (Hamburg: Felix Meiner, 1959).

*Kategorienlehre* (Leipzig: Meiner, 1933).

*Meine Letzen Wünschen für Österreich* (Stuttgart: Cotta, 1985).

*Philosophische Untersuchungen zu Raum, Zeit und Kontinuum.* Ed. R. Chisholm and S. Körner (Hamburg: Felix Meiner, 1976).

*Psychologie vom empirischen Standpunkt.* Ed. O. Kraus (Hamburg: Felix Meiner, 1973).

*Über Aristoteles.* Ed. R. George (Hamburg: Felix Meiner, 1986).

*Über die Zukunft der Philosophie.* Ed. O. Kraus (Hamburg: Felix Meiner, 1968).

*Über Ernst Machs "Erkenntnis und Irrtum."* Ed. R. Chisholm and J. Marek (Amsterdam: Rodopi, 1988).

*Wahrheit und Evidenz.* Ed. O. Kraus (Hamburg: Felix Meiner, 1958).

Brough, J. "Husserl on Memory." *The Monist*, **59** (1975), 40–62.

"The Emergence of an Absolute Consciousness in Husserl's Early Writings on Time-Consciousness," in F. Elliston and P. McCormick (eds.), *Husserl: Expositions and Appraisals* (Notre Dame: Notre Dame University Press, 1977).

Brück, M. *Über das Verhältnis* E. Husserls zu F. Brentano (Würzburg: K. Triltsche 1933).

Capek, M. *The New Aspects of Time: Its Continuity and Novelties* (Dordrecht: Kluwer, 1991).

Carr, D. "The Fifth Mediatation and Husserl's Cartesianism." *Philosophy and Phenomenological Research*, **34**, 1 (September), 14–35.

*The Paradox of Subjectivity: The Self in the Transcendental Tradition* (Oxford: Oxford University Press, 1999).

Casey, E. *Imagining: A Phenomenological Study* (Bloomington, IN: Indiana University Press, 1979).

Chisholm, R. "Brentano's Analysis of the Consciousness of Time." *Midwest Studies in Philosophy*, **6** (1981), 3–16.

Chrudzimski, A. "Die Theorie des Zeitbewußtseins Franz Brentano im Licht der unpublizierten Manuskripte." *Brentano Studien*, **7** (2000), 149–161.

Dahlstrom, D. (ed.) *Husserl's* Logical Investigations (Dordrecht: Kluwer, 2003).

Dastur, F. *Telling Time: Sketch of a Phenomenological Chrono-logy.* Trans. E. Bullard (London: Athlone Press, 2000).

de Warren, N. "The Significance of Stern's *Präsenzzeit* for Husserl's Phenomenology of Time Consciousness." *The New Yearbook for Phenomenology and Phenomenological Philosophy*, vol. V. Ed. B. Hopkins and S Crowell (Seattle: Noesis Press, 2005), 81–122.

"Tempo e memoria in Agostino e Husserl." Trans. N. Scapparone, in A. Ferrarin (ed.), *La realità del pensiero.* (Pisa: ETS, 2007), 93–142.

Depraz, N. *Transcendance et Incarnation: le statut de l'intersubjectivité comme altérité à soi chez Husserl* (Paris: Vrin, 1995).

Derrida, J. "Hospitality, Justice and Responsibility: A Dialogue with Jacques Derrida," in R. Kearney and M. Dooley (eds.), *Questioning Ethics* (London: Routledge, 1999).

*L'écriture et la différence* (Paris: Seuil, 1967).

*La voix et le phénomène* (Paris: Presses Universitaire de France, 1967).

Desanti, J. T. *Introduction à la phénoménologie* (Paris: Gallimard, 1994).

*Réflexions sur le temps (Variations philosophiques 1)* (Paris: Grasset, 1992).

Descartes, R. *Meditations on First Philosophy.* Trans R. Ariew and D. Cress (Indianapolis: Hackett, 2006).

Dodd, J. *Crisis and Reflection: An Essay on Husserl's* Crisis of the European Sciences. Phaenomenologica 174 (Dordrecht: Kluwer, 2004).

"Expression, Ideality and the Ego: Some Remarks on the 1913 Revisions of Husserl's Logical Investigations." in D. Dahlstrom (ed.), *Husserl's* Logical Investigations. (Dordrecht: Kluwer, 2003) 167–187.

*Idealism and Corporeity: An Essay on the Problem of the Body in Husserl's Phenomenology* (Dordrecht: Kluwer).

Drummond, J. *Husserlian Intentionality and Non-foundational Realism* (Dordrecht: Kluwer, 1990).

Dubosson, S. *L'imagination légitimée: La conscience imaginative dans la phénoménologie proto-transcendentale de Husserl* (Paris: l'Harmattan, 2004).

Elliott, B. *Phenomenology and Imagination in Husserl and Heidegger* (Oxford: Routledge, 2005).

Ferrarin, A. (ed.) *Passive Synthesis and Life-World: Sintesi passiva e mondo della vita.* Filosofia 91 (Pisa: Edizioni ETS, 2006).

Fink, E. *Studien zur Phänomenologie* (The Hague: Martinus Nijhoff, 1966).

Franck, D. *Chair et Corps. Sur la phénoménologie de Husserl* (Paris: Les Éditions de Minuit, 1981).

Frank, M. *Zeitbewußtsien* (Tübingen: Neske Pfullingen, 1990).

Gadamer, H.-G. *Philosophical Hermeneutics.* Trans. D. Linge (Berkeley, CA: University of California Press, 1976).

*Truth and Method.* Trans. J. Weinsheimer and D. Marshall (London: Continuum, 2004).

Gallagher, S. *The Inordinance of Time.* Northwestern University Studies in Phenomenology and Existential Philosophy (Evanston, IL: Northwestern University Press, 1998).

Gilson, É. "Franz Brentano's Interpretation of Medieval Philosophy." *Medieval Studies,* 1 (1939), 1–10.

Gilson, L. *Méthode et métaphysique selon Franz Brentano* (Paris: Vrin, 1955).

*La Psychologie descriptive selon Franz Brentano* (Paris: Vrin, 1955).

Giovannangeli, D. *Le Retard de la conscience: Husserl, Sartre, Derrida* (Brussels: Ousia, 2001).

Götschl, J. "Brentanos Analyse des Zeitbegriffes," in R. Chisholm and R. Haller (eds.), *Die Philosophie Franz Brentanos: Beiträge zur Brentano-Konferenz Graz, 4–8 September 1977.* (Amsterdam: Editions Rodopi, 1978), 225–248.

Granel, G. *Le Sens du temps et de la perception chez E.* Husserl (Paris: Gallimard, 1968).

Gurwitsch, A. *Studies in Phenomenology and Psychology.* Northwestern University Studies in Phenomenology and Existential Philosophy (Evanston, IL: Northwestern University Press, 1966).

Held, K. *Lebendige Gegenwart* (The Hague: Martinus Nijhoff, 1966).

Hermberg, K. *Husserl's Phenomenology: Knowledge, Objectivity, and Others.* Continuum Studies in Continental Philosophy (London: Continuum, 2006).

Holenstein, E. *Phänomenologie der Assoziation* (The Hague: Martinus Nijhoff, 1971).

Housset, E. *Husserl et l'énigma du monde* (Paris: Seuil, 2000).

Hume, D. *Treatise of Human Nature* (Oxford: Clarendon Press, 1965).

Husserl, E. *Briefe an Roman Ingarden. Mit Erläuterungen und Erinnerungen an Husserl* (The Hague, 1968).

*The Crisis of the European Sciences and Transcendental Phenomenology.* Trans. D. Carr (Evanston: Northwestern, 1970).

*Experience and Judgement.* Trans. J. Churchill and K. Ameriks (Evanston: Northwestern, 1973).

*The Idea of Phenomenology.* Trans. L. Hardy (Dordrecht: Kluwer, 1999).

*Ideas Pertaining to a Pure Phenomenology and to a Phenomenological Philosophy.* Trans. F. Kersten (The Hague: Martinus Nijhoff, 1982).

*Leçons pour une phénoménologie de la conscience intime du temps.* Trans. H. Dussort (Paris: Presses Universitaire de France, 1964).

*La terre ne se meut pas.* Trans. D. Franck, D. Pradelle and J.-F. Lavigne (Paris: Éditions de Minuit, 1989).

*Philosophie als strenge Wissenschaft,* (Klostermann: Frankfurt, 1965). English Translation: *Philosophy as Rigorous Science,* trans. P. McCormick, in: Husserl: Shorter Works, (University of Notre Dame: Notre Dame, 1981).

*Texte zur Phänomenologie des inneren Zeitbewußtseins (1893–1917).* Ed. R. Bernet (Hamburg: Felix Meiner Verlag, 1985).

Husserliana *Cartesianische Meditationen und Pariser Vorträge.* Ed. S. Strasser. Husserliana I (The Hague: Martinus Nijhoff, 1973).

*Die Idee der Phänomenologie. Fünf Vorlesungen.* Ed. W. Biemel. Husserliana II (The Hague: Martinus Nijhoff, 1973).

*Ideen zu einer reinen Phänomenlogie und phänomenlogischen Philosophie. Erstes Buch: Allgemeine Einführung in die reine Phänomenologie.* Ed. W. Biemel. Husserliana III (The Hague: Martinus Nijhoff Publishers, 1950).

*Ideen zu einer reinen Phänomenologie und phänomenologischen Philosophie. Erstes Buch: Allgemeine Einführungin die reine Phänomenologie 1. Halbband: Text der 1–3. Auflage – Nachdruck.* Husserliana III.1 Ed. K. Schuhmann (The Hague: Martinus Nijhoff, 1977).

*Ideen zu einer reinen Phänomenologie und phänomenologischen Philosophie. Erstes Buch: Allgemeine Einfuhrung in die reine Phänomenologie, 2. Halbband: Ergänzende Texte (1912–1929).* Ed. K. Schuhmann. Husserliana III.2 (The Hague: Martinus Nijhoff, 1988).

*Ideen zur einer reinen Phänomenologie und phänomenologischen Philosophie. Zweites Buch: Phänomenologische Untersuchungen zur Konstitution.* Ed. M. Biemel. Husserliana IV (The Hague: Martinus Nijhoff, 1952).

*Ideen zur einer reinen Phänomenologie und phänomenologischen Philosophie. Drittes Buch: Die Phänomenologie und die Fundamente der Wissenschaften.* Ed. M. Biemel. Husserliana V (The Hague: Martinus Nijhoff, 1971).

*Die Krisis der europäischen Wissenschaften und die transzendentale Phänomenologie. Eine Einleitung in die phänomenologische Philosophie.* Ed. W. Biemel. Husserliana VI (The Hague: Martinus Nijhoff, 1976).

*Erste Philosophie (1923/4). Erste Teil: Kritische Ideengeschichte.* Ed. R. Boehm. Husserliana VII (The Hague: Martinus Nijhoff, 1956).

*Erste Philosophie (1923/4). Zweiter Teil: Theorie der phänomenologischen Reduktion.* Ed. R. Boehm. Husserliana VIII (The Hague: Martinus Nijhoff, 1959).

*Phänomenologische Psychologie. Vorlesungen Sommersemester. 1925.* Ed. W. Biemel. Husserliana IX (The Hague: Martinus Nijhoff, 1968).

*Zur Phänomenologie des inneren Zeitbewusstesens (1893–1917).* Ed. R. Boehm. Husserliana X (The Hague: Martinus Nijhoff, 1969).

*Analysen zur passiven Synthesis. Aus Vorlesungs- und Forschungsmanuskripten, 1918–1926.* Ed. M. Fleischer. Husserliana XI (The Hague: Martinus Nijhoff, 1966).

*Philosophie der Arithmetik. Mit ergänzenden Texten (1890–1901).* Ed. L. Eley. Husserliana XII (The Hague: Martinus Nijhoff, 1970).

*Zur Phänomenologie der Intersubjektivität. Texte aus dem Nachlass. Erster Teil. 1905–1920.* Ed. I. Kern. Husserliana XIII (The Hague: Martinus Nijhoff, 1973).

*Zur Phänomenologie der Intersubjektivität. Texte aus dem Nachlass. Zweiter Teil. 1921–28.* Ed. I. Kern. Husserliana XIV (The Hague: Martinus Nijhoff, 1973).

*Zur Phänomenologie der Intersubjektivität. Texte aus dem Nachlass. Dritter Teil. 1929–35.* Ed. I. Kern. Husserliana XV (The Hague: Martinus Nijhoff, 1973).

*Ding und Raum. Vorlesungen 1907.* Ed. U. Claesges. Husserliana XVI (The Hague: Martinus Nijhoff, 1973).

*Formale und transzendentale Logik. Versuch einer Kritik der logischen Vernunft.* Ed. P. Janssen. Husserliana XVII (The Hague: Martinus Nijhoff, 1974).

*Logische Untersuchungen. Erster Teil. Prolegomena zur reinen Logik. Text der 1. und der 2. Auflage.* Ed. E. Holenstein. Husserliana XVIII (The Hague: Martinus Nijhoff, 1975).

*Logische Untersuchungen. Zweiter Teil. Untersuchungen zur Phänomenologie und Theorie der Erkenntnis. In zwei Bänden.* Ed. U. Panzer. Husserliana XIX (The Hague: Martinus Nijhoff, 1984).

*Logische Untersuchungen. Ergänzungsband. Erster Teil. Entwürfe zur Umarbeitung der VI. Untersuchung und zur Vorrede für die Neuauflage der Logischen Untersuchungen (Sommer 1913).* Ed. U. Melle. Husserliana XX.1 (The Hague: Kluwer Academic Publishers, 2002).

*Aufsätze und Rezensionen (1890–1910).* Ed. B. Rang. Husserliana XXII (The Hague: Martinus Nijhoff, 1979).

*Phantasie, Bildbewusstsein, Erinnerung. Zur Phänomenologie der anschaulichen Vergegenwartigungen. Texte aus dem Nachlass (1898–1925).* Ed. E. Marbach. Husserliana XXIII (The Hague: Martinus Nijhoff, 1980).

*Einleitung in die Logik und Erkenntnistheorie. Vorlesungen 1906/07.* Ed. U. Melle. Husserliana XXIV (The Hague: Martinus Nijhoff, 1985).

*Aufsätze und Vorträge. 1911–1921. Mit ergänzenden Texten.* Ed. T. Nenon and H. R. Sepp. Husserliana XXV (The Hague: Martinus Nijhoff, 1986).

*Vorlesungen über Bedeutungslehre. Sommersemester 1908.* Ed. U. Panzer. Husserliana XXVI (The Hague: Martinus Nijhoff, 1987).

*Aufsätze und Vorträge. 1922–1937.* Ed. T. Nenon and H. R. Sepp. Husserliana XXVII (The Hague: Kluwer Academic Publishers, 1988).

*Die Krisis der europäischen Wissenschaften und die transzendentale Phänomenologie. Ergänzungsband. Texte aus dem Nachlass 1934–1937.* Ed. R. Smid. Husserliana XXIX (The Hague: Kluwer Academic Publishers, 1992).

*Aktive Synthesen: Aus der Vorlesung "Transzendentale Logik" 1920/21 Ergänzungsband zu Analysen zur passiven Synthesis.* Ed. R. Breeur. Husserliana XXXI (The Hague: Kluwer Academic Publishers, 2000).

*Die "Bernauer Manuskripte" über das Zeitbewußtsein (1917/18).* Ed. R. Bernet and D. Lohmar. Husserliana XXXIII (Dordrecht: Kluwer Academic Publishers, 2001).

*Zur phänomenologischen Reduktion. Texte aus dem Nachlass (1926–1935).* Ed. S. Luft. Husserliana XXXIV (Dordrecht: Kluwer Academic Publishers, 2002).

*Transzendentaler Idealismus. Texte aus dem Nachlass (1908–1921).* Ed. R. Rollinger in cooperation with R. Sowa. Husserliana XXXVI (Dordrecht: Kluwer Academic Publishers, 2003).

*Wahrnehmung und Aufmerksamkeit. Texte aus dem Nachlass (1893–1912).* Ed. T. Vongehr and R. Giuliani. Husserliana XXXVIII (New York: Springer, 2005).

Ingarden, R. *Briefe an Roman Ingarden. Mit Erläuterungen und Erinnerungen an Husserl* (The Hague, 1968).

"Edith Stein on her Activity as an Assistant of Edmund Husserl (Extracts from the Letters of Edith Stein with a Commentary and Introductory Remarks)." *Philosophy and Phenomenological Research*, 23 (1962), 155–175.

James, W. *Principles of Psychology* (Cambridge, MA: Harvard University Press, 1981).

Kelly, M. "On the Mind's 'Pronouncement' of Time: Aristotle, Augustine and Husserl on Time Consciousness." *Proceedings of the American Philosophical Association*, **78** (2005), 249–262.

Kern, I. "The Three Ways to the Transcendental Reduction in the Philosophy of Edmund Husserl," in P. McCormick and F Elliston (eds.), *Husserl: Expositions and Appraisals* (Notre Dame: Notre Dame University Press, 1977).

Kraus, O. *Franz Brentano* (Munich: Oskar Beck, 1919).

*Franz Brentano: Zur Kenntnis seines Lebens und seiner Lehre.* Ed. O. Kraus (Munich: Oskar Beck, 1919).

"Zur Phänomenognosie des Zeitbewußtseins." *Archiv für die gesamte Psychologie*, **75** (1930), 1–22.

Landgrebe, L. *Der Weg der Phänomenologie: Das Problem einer ursprünglichen Erfahrung* (Gütersloher Verlagshaus Gerd Mohn, 1963).

Lawlor, L. *Derrida and Husserl: The Basic Problem of Phenomenology* (Bloomington, IN: Indiana University Press, 2002).

Lévinas, E. *Basic Philosophical Writings.* Ed. A. Peperzak, S. Critchley and R. Bernasconi. Studies in Continental Thought (Bloomington, IN: Indiana University Press, 1996).

*Le temps et L'autre* (Paris: Presses Universitaites de France, 1983).

*Time and the Other and Additional Essays.* Trans. R. Cohen (Pittsburgh, PA: Duquesne University Press, 1987).

Locke, J. *An Essay Concerning Human Understanding* (Oxford: Clarendon Press, 1975).

Lohmar, D. "What does Protention Protend? Remarks on Husserl's Analysis of Protention in the *Bernau Manuscripts* on Time-Consciousness." *Philosophy Today*, **46**, 5 (SPEP Supplement), 154–167.

MacDonald, P. *Descartes and Husserl: The Philosophical Project of Radical Beginnings* (Albany, NY: State University of New York Press, 2000).

Mach, E. *Beiträgen zur Analyse der Empfindungen* (Jena: G. Fischer, 1886).

Marbach, E. *Das Problem des Ich in der Phenomenologie Husserls* (The Hague: Martinus Nijhoff, 1974).

Marrati, P. *Genesis and Trace: Derrida Reading Husserl and Heidegger* (Stanford, CA: Stanford University Press, 2005).

Marty, A. *Die Frage nach der geschichtlichen Entwicklung des Farbensinnes* (Vienna: Gerold, 1879).

Mensch, J. *Intersubjectivity and Transcendental Idealism*. SUNY Series in Contemporary Continental Thought (Albany NY: State University of New York Press, 1988).

Merleau-Ponty, M. *Phenomenology of Perception*. Trans. C. Smith (London: Routledge, 1962).

*In Praise of Philosophy and Other Essays*. Trans. J. Wild and J. Edie. Northwestern University Studies in Phenomenology and Existential Philosophy (Evanston, IL: Northwestern University Press, 1963).

*The Primacy of Perception*. Trans. J. Edie. Northwestern University Studies in Phenomenology and Existential Philosophy (Evanston, IL: Northwestern University Press, 1964).

*Signs*. Trans. R. C. McCleary. Northwestern University Studies in Phenomenology and Existential Philosophy (Evanston, IL: Northwestern University Press, 1964).

*The Visible and the Invisible*. Ed. C. Lefort. Trans. A. Lingis. Northwestern University Studies in Phenomenology and Existential Philosophy (Evanston, IL: Northwestern University Press, 1968).

Mezei, B. "Brentano and Husserl on the History of Philosophy." *Brentano Studien*, 8 (1989), 81–94.

Mohanty, J.-N. *The Possibility of Transcendental Philosophy*. Phaenomenologica 98 (Dordrecht: Martinus Nijhoff, 1985).

"Time: Linear or Cyclical, and Husserl's Phenomenology of Inner Time-Consciousness." *Philosophia naturalis*, 25 (1988), 123–130.

Möhler, J. A. *Kirchengeschichte von J. A. Möhler*, 3 vols. (Regensberg, 1867–68).

Montavont, A. *De la passivité dans la phénoménologie de Husserl* (Paris: Presses Universitaires de France, 1999).

Moran, D. *Edmund Husserl: Founder of Phenomenology* (Cambridge: Polity Press, 2005).

Mulligan, K. "Brentano on the Mind," in D. Jacquette (ed.), *The Cambridge Companion to Brentano* (Cambridge: Cambridge University Press, 2004), 66–97.

Mulligan, K. and Smith, B. "Franz Brentano on the Ontology of Mind." *Philosophy and Phenomenological Research*, 45 (1985), 627–644.

Natorp, P. *Einleitung in die Psychologie nach kritischer Methode* (Freiburg, 1888).

Patocka, J. *An Introduction to Husserl's Phenomenology*. Trans. E. Kohák. Ed. J. Dodd (Chicago, IL: Open Court, 1996).

Petit, J. *Solipsisme et intersubjectivité: quinze leçons sur Husserl et Wittgenstein* (Paris: CERF, 1996).

Picard, Y. "Le temps chez Husserl et chez Heidegger." *Deucalion*, **1**, 95–124.

Piedade, J. I. *Der bewegte Leib: Kinästhesen bei Husserl im Spannungsfeld von Intention und Erfüllung* (Vienna: Passagen Verlag, 2002).

Richir, M. *Phénoménologieen esquisses: Nouvelles fondations* (Brussels: Jérôme Millon, 2000).

Ricoeur, P. *Husserl: An Analysis of His Phenomenology*. Trans. E. Ballard and L. Embree. Northwestern University Studies in Phenomenology and Existential Philosophy (Evanston, IL: Northwestern University Press, 1967).

*La mémoire, l'histoire, l'oubli* (Paris: Éditions du Seuil, 2000).

*Temps et récit*. 3 vols. (Paris: Seuil, 1985).

Rodemeyer, L. *Intersubjective Temporality: It's About Time* (Dordrecht: Springer Verlag, 2006).

Sartre, J.-P. *Being and Nothingness*. Trans. H. Barnes (New York: Washington Square Press, 1992).

*L'imagination* (Paris: Presses Universitaire de France, 1936).

Scheler, M. *Selected Philosophical Essays*. Trans. D. Lachterman. Northwestern University Studies in Phenomenology and Existential Philosophy (Evanston, IL: Northwestern University Press, 1973).

Scruton, R. *The Aesthetics of Music* (Oxford: Clarendon Press, 1997).

Sokolowski, R. *Husserlian Meditations*. Northwestern University Studies in Phenomenology and Existential Philosophy (Evanston, IL: Northwestern University Press, 1974).

Sommer, M. *Evidenz im Augenblick: Eine Phänomenologie der reinen Empfindung*. Suhrkamp-Taschenbuch Wissenschaft 1291 (Frankfurt: Suhrkamp, 1996).

*Lebenswelt und Zeitbewußtsein*. Suhrkamp-Taschenbuch Wissenschaft 851 (Frankfurt: Suhrkamp, 1990).

Sorabji, R. *Aristotle on Memory*. 2nd edn (London: Duckworth, 2004).

Spiegelberg, H. *The Phenomenological Movement* (The Hague: Martinus Nijhoff, 1960).

Staiti, A. "Il luogo della verità: La presenza di Agostino nella fenomenologia di Husserl." *Quaestio*, **6** (2006), 373–402.

Stein, E. *On the Problem of Empathy*. Trans. W. Stein. *The Collected Works of Edith Stein*, vol. III (Washington, DC: ICS Publications, 1989).

Stern, W. "Psychische Präsenzzeit." *Zeitschrift für Psychologie und Physiologie der Sinnesorgane*, **13** (1897), 325–349. "Mental Presence-Time." Trans. N. de Warren. *The New Yearbook for Phenomenology and Phenomenological Philosophy* V (2005). Ed. B. Hopkins and S. Crowell (Seattle: Noesis Press, 2005), 325–351.

Stumpf, C. *Tonpsychologie*. 2 vols. (Leipzig: Hirzel, 1883–1890).

Tháo, T. D. *Phénoménologie et Matérialisme Dialectique* (Paris: Gordon & Breach, 1971). Trans. D. Herman and D. Morano, as *Phenomenology and Dialectical Materialism* (Dordrecht: Riedel, 1951).

Thévenaz, P. *What is Phenomenology? and Other Essays*. Ed. J. Edie (Chicago: Quadrangle Books, 1962).

Welton, D. (ed.) *The New Husserl: A Critical Reader.* Studies in Continental
   Thought (Bloomington, IN: Indiana University Press, 2003).
Zahavi, D. *Husserl's Phenomenlogy* (Stanford: Stanford University Press, 2003).
   *Husserl and Transcendental Intersubjectivity.* Trans. E. Behnke (Athens, OH:
   Ohio University Press, 2001).
   *Self-Awareness and Alterity: A Phenomenological Investigation.* Northwestern
   University Studies in Phenomenology and Existential Philosophy
   (Evanston, IL: Northwestern University Press, 1999).

# INDEX

CPSIA information can be obtained
at www.ICGtesting.com
Printed in the USA
LVHW042237210323
742238LV00005B/202